Designing with Natural Materials

Designing with Natural Materials

Edited by
Graham A. Ormondroyd and Angela F. Morris

CRC Press is an imprint of the
Taylor & Francis Group, an **informa** business

CRC Press
Taylor & Francis Group
6000 Broken Sound Parkway NW, Suite 300
Boca Raton, FL 33487-2742

© 2019 by Taylor & Francis Group, LLC
CRC Press is an imprint of Taylor & Francis Group, an Informa business

No claim to original U.S. Government works

Printed on acid-free paper

International Standard Book Number-13: 978-1-4987-8270-8 (Hardback)

This book contains information obtained from authentic and highly regarded sources. Reasonable efforts have been made to publish reliable data and information, but the author and publisher cannot assume responsibility for the validity of all materials or the consequences of their use. The authors and publishers have attempted to trace the copyright holders of all material reproduced in this publication and apologize to copyright holders if permission to publish in this form has not been obtained. If any copyright material has not been acknowledged please write and let us know so we may rectify in any future reprint.

Except as permitted under U.S. Copyright Law, no part of this book may be reprinted, reproduced, transmitted, or utilized in any form by any electronic, mechanical, or other means, now known or hereafter invented, including photocopying, microfilming, and recording, or in any information storage or retrieval system, without written permission from the publishers.

For permission to photocopy or use material electronically from this work, please access www.copyright.com (http://www.copyright.com/) or contact the Copyright Clearance Center, Inc. (CCC), 222 Rosewood Drive, Danvers, MA 01923, 978-750-8400. CCC is a not-for-profit organization that provides licenses and registration for a variety of users. For organizations that have been granted a photocopy license by the CCC, a separate system of payment has been arranged.

Trademark Notice: Product or corporate names may be trademarks or registered trademarks, and are used only for identification and explanation without intent to infringe.

Visit the Taylor & Francis Web site at
http://www.taylorandfrancis.com

and the CRC Press Web site at
http://www.crcpress.com

Contents

Preface vii

Editors ix

Contributors xi

1. An Introduction to the Aesthetics of Design 1
Bruce Wood

2. Selection of Natural Materials Using CES EduPack 15
Martin P. Ansell

3. Natural Materials – Composition and Combinations 29
Morwenna Spear

4. Designing with the Life Cycle in Mind 111
C. Skinner

5. Restoring Credibility to Natural Materials 133
D. K. Spilsbury

6. Natural Materials in Automotive Design 165
Kerry Kerwan and Stewart Coles

7. Rediscovering Natural Materials in Packaging 181
Angela F. Morris

8. Designing Tall Buildings with Natural Materials 199
Richard Harris and Wen-Shao Chang

**9. Emerging Nature Based Materials and Their Use in New
Products** 217
Morwenna Spear

10. Bio-Inspired Design – Enhancing Natural Materials 321
Chris Holland

Index 335

Preface

Experimentation with new materials, improving or creating new manufacturing processes and raising performance expectations are demands that have been faced by all product designers since time immemorial. Today, the criteria must also include an understanding of sustainability and the environmental, social and economic impacts of the materials selected for any design project. Every product design will have different material requirements and different options for making an environmental choice, making material choice a particularly difficult and challenging task in today's environmentally conscious world.

Designers are constantly seeking inspiration, so where better to look than nature? Natural materials have evolved over millennia to a state of perfection and performance that modern science struggles to emulate.

Over recent years, more natural materials are making their way into consumer product design, as are new bio-based materials. The book looks to the future of these innovative materials and how they will influence design. It offers insight to designers of new bio-based materials across a range of different design disciplines, whilst also offering insight to scientists on the process of product design and the needs of a material beyond those traditionally analysed in the laboratory. The book also discusses bio-inspiration and bio-mimicry and how they can influence future material design, products and manufacturing processes.

Historically, science has pursued a premise that nature can be understood fully, its future predicted precisely, and its behaviour controlled at will. However, emerging knowledge indicates that the nature of Earth and biological systems transcends the limits of science, questioning the premise of knowing, prediction and control. This knowledge has led to the recognition that, for civilized human survival, technological society has to adapt to the constraints of these systems.

Natural materials are a supremely credible choice for modern purposes. The plants and animals that produce them have spent millions of years on their research and development, and have reached a degree of technical sophistication that leaves mankind's advanced materials struggling to emulate them. If we accept this fact, and dedicate our research techniques and our cutting-edge technology to the better understanding of natural materials, we will have an opportunity to improve manmade products, but more importantly design innovative new products that will ensure a more sustainable world for ourselves and future generations.

Having said this, there still exists a knowledge gap between pure scientific and academic research into natural materials and the world of modern commerce. The contributors to this book represent a cross-section of designers,

vii

scientists, engineers, academics and business people, with the common aim of helping to bridge this gap by discussing various natural material applications in diverse commercial areas including building, automotive and packaging.

This book aims to bridge the gap between the design community and the scientific knowledge. Split into three parts the book takes the reader on a journey from understanding the principles of design, science and sustainability, through a series of case studies on the use of natural materials and finally to the future with new materials and biomimicry.

Graham A. Ormondroyd
Angela F. Morris
January 2018

Editors

Graham A. Ormondroyd completed his PhD in wood science at Bangor University, Bangor. He has been the head of materials research at the BioComposites Centre for six years and in that time has written many proposals and papers and has undertaken commercial works in all aspects of wood and biomaterials science. Dr. Ormondroyd has over 50 publications including peer-reviewed papers, conference proceedings, book chapters and edited books, and he continues to publish regularly. He is a fellow of the Institute of Materials Minerals and Mining, a member of the International Research Group on Wood Protection, and in 2015 was shortlisted for both the Bangor and the Insider Wales Innovation awards. Dr. Ormondroyd is an editor of two international journals and a reviewer for others.

Angela F. Morris, BA Hons, MA, MinstPkg, FRSA, ran her own creative packaging design businesses for over 30 years and is now CEO and founder of The Wool Packaging Company and creator of the Woolcool® brand. Morris's packaging experience covers a variety of different sectors, from major UK retail high street stores such as BHS and global confectionery companies such as Nestlé and Cadbury's, to water filtration systems for Fairey Industrial Ceramics and automotive components for Rolls-Royce. Working as a packaging consultant and advisor to her clients, she often became an integral part of their businesses, helping companies to market their products successfully, through the introduction of superior and innovative packaging solutions, encouraging the use of natural materials wherever possible.

Morris has lectured and supervised postgraduate students and given talks on design management, innovation and green entrepreneurship at several universities and events across the United Kingdom. Due to the vision, passion, dedication and experience of The Wool Packaging Company founder, Morris was invited by the IOM3 organisation in 2013 to launch the Natural Materials Association. In 2018 The Wool Packaging Company won The Queen's Award for Enterprise in Innovation for its insulated packaging product, Woolcool.

Contributors

Martin P. Ansell
BRE Centre for Innovative
  Construction Materials
University of Bath
Bath, United Kingdom

Wen-Shao Chang
Sheffield School of Architecture
The University of Sheffield
Sheffield, UK

Stewart Coles
Warwick Manufacturing Group
Warwick University
Coventry, United Kingdom

Richard Harris
BRE Centre for Innovative
  Construction Materials
University of Bath
Bath, United Kingdom

Chris Holland
The Natural Materials Group
Department of Materials Science
  and Engineering
The University of Sheffield
Sheffield, United Kingdom

Kerry Kerwan
Warwick Manufacturing Group
Warwick University
Coventry, United Kingdom

Angela F. Morris
The Wool Packaging Company
  Limited
Stone, Staffordshire
United Kingdom

C. Skinner
The BioComposites Centre
Bangor University
Bangor, United Kingdom

Morwenna Spear
The BioComposites Centre
Bangor University
Bangor, United Kingdom

D.K. Spilsbury
The Wool Packaging Company
  Limited
Stone, Staffordshire
United Kingdom

Bruce Wood
School of Engineering and Built
  Environment
Glasgow Caledonian University
Glasgow, United Kingdom

1

An Introduction to the Aesthetics of Design

Bruce Wood

Glasgow Caledonian University

CONTENTS

The Aesthetics of Design 1
The Ten Principles of Good Design 5
Conclusions 13
References 14

The Aesthetics of Design

'The real voyage of discovery consists not in seeking new lands but seeing with new eyes'

Marcel Proust [1]

This quote from the 19th century writer and essayist Marcel Proust is particularly relevant for designers. Designers often create new products and services, which are appropriate and relevant for their time and for the market. The resulting design can be in themselves innovative in terms of technology and manufacturing; however, quite often, designs utilise existing materials and manufacturing technologies in a manner that creates a new product or service that had previously not been imagined. Designers are constantly seeking information on materials and processes that they can incorporate or utilise in a manner that produces a functioning product which brings together function, aesthetics, ergonomics, commercial applicability and many more, that provides a new product for the market place at that time.

Given the importance of design in the commercial success of product services and companies, there have been many authoritative bodies and writers working in this area.

First, if we consider some definitions from the Cox Review [2], the UK Design Council have provided some useful terms to consider:

'Creativity' is the generation of new ideas – new ways either of looking at existing problems or of seeing new opportunities, perhaps by exploiting emerging technologies or changes in markets.

'Innovation' is the successful exploitation of new ideas. It is the process that carries them through to new products, new services and new ways of running the business or even new ways of doing business.

'Design' is what links creativity and innovation. It shapes ideas to become practical and attractive propositions for users or customers. Design may be described as creativity deployed to a specific end.

The UK government's Department of Culture, Media and Sport (DCMS) defines design in the business sector of Creative Industries;

'We define the creative industries as those industries which have their origin in individual creativity, skill and talent and which have a potential for wealth and job creation through the generation and exploitation of intellectual property'.*

Then, design is a process whereby creativity is applied by individuals and teams to solve a problem or create new commercial opportunities. Designers are well educated in this process and utilise many techniques and methodologies with a view to achieve optimum solutions.

In the situation of 'inclusive design' where designers are working with non-designers on a particular project, it is important to ensure that all participants are aware of the general design processes and journey that will be undertaken to achieve optimum solutions.

Again, the Design Council provides a useful methodology for this process and journey. The 'Double Diamond' process provides a useful visualisation. See Figure 1.1 for the process and assists participants to understand where they are in the process journey and why certain activities take place at different times.

Other locations and industry sectors when attempting to visualise the design have developed similarly shaped diagrams and created their own versions relevant to their situations. All the diagrams follow a trend of divergence and convergence, which relates to the essential types of thinking required in order to be creative. Divergent thinking is characterised by certain types of thinking processes or behaviours with a view on creations, options and opportunities, asking very much the 'what if' types of question. A visualisation of divergent thinking can be seen in Figure 1.2, the main point of this process is to create as many ideas, options, opportunities and directions as possible in order to ensure that a wide range of ideas and options is being considered. Ultimately, this part of the process is expansive and relates to 'imagination'; hence, the diagram is an open-ended funnel shape.

Clearly, optimal solutions will not be created by only imagining possibilities; to counter the open-ended nature of the divergent thinking processes,

---

* DCMS http://www.davidparrish.com/creative-industries/.

An Introduction to the Aesthetics of Design

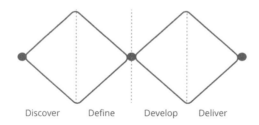

FIGURE 1.1
The Design Council double diamond design process (https://www.designcouncil.org.uk/news- opinion/design-process-what-double-diamond).

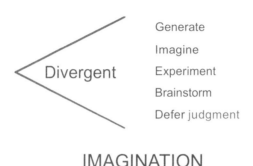

FIGURE 1.2
Divergent thinking characteristics.

the team or individuals must engage in convergent thinking processes. The convergent thinking phase is about applying knowledge and evaluating the output of the divergent thinking phase. Diagrammatically, it is about moving from an open-ended start to a focus. It is here that all of the created ideas are evaluated against appropriate criteria; thus, a selection process(es) is applied using whatever knowledge base is appropriate. Figure 1.3 visualises this phase.

These two phases of divergent and convergent thinking when shown together create the basis for a 'diamond' (see Figure 1.4), which can be applied multiple times to various phases of a project depending on its size and scope.

Imagination and knowledge are in balance and both are necessary, even Albert Einstein had a view on this.

> *Imagination is more important than knowledge. For knowledge is limited, whereas imagination embraces the entire world, stimulating progress, giving birth to evolution. It is, strictly speaking, a real factor in scientific research.*
>
> *Albert Einstein*

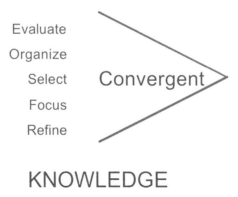

FIGURE 1.3
Convergent thinking characteristics.

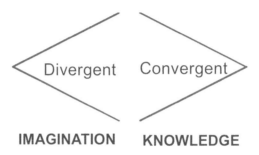

FIGURE 1.4
Divergent convergent thinking diamond.

Referring back to the Design Council's design process double diamond, the process outlines four main stages: Discover, Define, Develop and Deliver. While designers are familiar with this type of process and phases, they are also relevant to designers when they are considering natural materials.

Discover

- This discovery aspect opens up thinking focused on discovering possible solutions.
- In terms of natural materials, what materials are available, what form do they come in, is there a reliable source, how can they be used, what are the performance characteristics, what are the cost implications and many more?

Define

- Following the discovery phase, the results have to be evaluated against a set of criteria; in this sense, defining performance window that any material will have to comply with.

An Introduction to the Aesthetics of Design 5

Develop

- Should there be material options, solutions will have to be developed that will allow material to be used in the product solution that in turn meets the performance and market place requirements.
- This development phase can be time consuming and in itself may require some additional developments in associated areas such as manufacturing.

Deliver

- Given that the previous phases have been successfully completed and there is a solution that satisfies the requirements, the objective is to deliver the solution in a reliable and repeatable manner and thus onto commercialisation.
- Given the possible variability of natural materials versus the more precise nature of the more often used manufacturing materials, these manufacturing materials have been processed in order to provide reliable repeatable performance at a specified cost.

The Ten Principles of Good Design

Back in the late 1970s, Dieter Rams was becoming increasingly concerned by the state of the world around him – 'an impenetrable confusion of forms, colours and noises'. Aware that he was a significant contributor to that world, he asked himself an important question: is my design good design?

As good design cannot be measured in a finite way, he set about expressing the ten most important principles for what he considered was good design (sometimes, they are referred as the 'Ten Commandments').

1. Good design is innovative:
   The possibilities for innovation are not, by any means, exhausted. Technological development is always offering new opportunities for innovative design. But innovative design always develops in tandem with innovative technology, and can never be an end in itself.
2. Good design makes a product useful:
   A product is bought to be used. It has to satisfy certain criteria, not only functional but also psychological and aesthetic. Good design emphasises the usefulness of a product whilst disregarding anything that could possibly detract from it.

3. Good design is aesthetic:
   The aesthetic quality of a product is integral to its usefulness because products we use every day affect our person and our well-being. But only well-executed objects can be beautiful.

4. Good design makes a product understandable:
   It clarifies the product's structure. Better still, it can make the product talk. At best, it is self-explanatory.

5. Good design is unobtrusive:
   Products fulfilling a purpose are like tools. They are neither decorative objects nor works of art. Their design should therefore be both neutral and restrained, to leave room for the user's self-expression.

6. Good design is honest:
   It does not make a product more innovative, powerful or valuable than it really is. It does not attempt to manipulate the consumer with promises that cannot be kept.

7. Good design is long lasting:
   It avoids being fashionable and therefore never appears antiquated. Unlike fashionable design, it lasts many years – even in today's throwaway society.

8. Good design is thorough down to the last detail:
   Nothing must be arbitrary or left to chance. Care and accuracy in the design process show respect towards the user.

9. Good design is environmentally friendly:
   Design makes an important contribution to the preservation of the environment. It conserves resources and minimises physical and visual pollution throughout the lifecycle of the product.

10. Good design is as little design as possible:
    Less, but better – because it concentrates on the essential aspects, and the products are not burdened with non-essentials.
    Back to purity, back to simplicity.
    Dieter Rams – ten principles of good design*

While these 'ten principles of good design' were not written specifically for any industry or belief in particular material use, they are as relevant today as they were when they were devised by Rams. In fact, it could be argued that it is precisely because they were written without any industry sector bias and in general that they have a strong relevance for designers and those working with natural materials.

These principles could be used as the basis for an evaluation tool where design utilising natural materials could be evaluated against other natural materials or more commonly used manufacturing materials. It could be

---

* 10 principles of good design https://www.vitsoe.com/rw/about/good-design.

An Introduction to the Aesthetics of Design

easily imagined that a natural materials design solution would always rate highly in Principle 9 'Good design is environmentally-friendly' but might not rate as highly in Principle 7 'Good design is long lasting'. It was not the original intention of these principles to be an evaluation tool in this manner; however, design solutions must be evaluated before committing to commercialisation to ensure the most effective way to proceed, and elevation tool with a set of criteria based on these principles with an appropriate assessment scheme would be a good starting point.

Case Studies

Case Study 1

In this case study, we will consider the design of a paper pulp lampshade. The designers Van Kotton & Carnduff (VK&C) had a vision to design and manufacture a paper pulp luminaire or lampshade. Paper pulp is the material most frequently used in the packaging industry, and the material has structural properties and a low value, typically supplying food-based products and a significant amount of packaging in the mobile telephone business.

The design comprised of a simple and elegant 'clam' style moulding, which could be folded to create one simple product or joined with another identical component to create a double luminaire. Figure 1.5 shows the initial design concept.

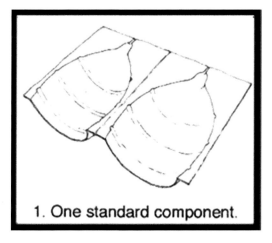

FIGURE 1.5
The initial design concept.

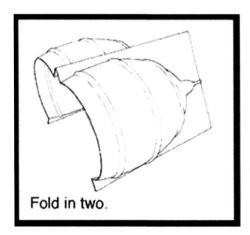

FIGURE 1.6
Fold line.

Figure 1.6 shows the intended fold line to create a single product.
Figure 1.7 shows the intended completed single product folded and fastened together.

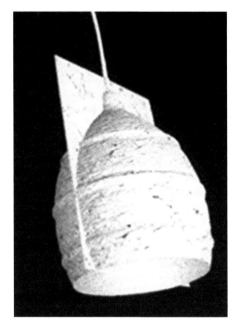

FIGURE 1.7
Complete product visualisation.

As can be seen the designers had considered strengthening ribs being moulded into the product and they had also considered the possibility of adding other possible materials to the paper pulp mix, such as textiles; this is portrayed by the speckle nature of visualisation as outlined in Figure 1.7.

Figure 1.8 explains the method by which two identical paper pulp mouldings would be fastened together to create a double luminaire product.

Figure 1.9 is the visualisation of the double luminaire with similar material.

Fix together.   2. Add another .   3. Fix together.

FIGURE 1.8
Double luminaire construction.

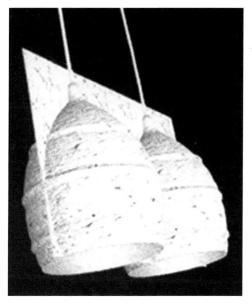

FIGURE 1.9
Completed double luminaire.

Clearly, VK&C were not paper pulp manufacturers and early in the process started working with Universal Pulp Packaging (UPP) who had expertise in this area.

Prototype tooling was ordered, produced and initial samples were manufactured. A number of problems arose from this prototyping exercise, one of the main issues being dimensional stability. This problem was resulting in high degrees of variance of the finished product; therefore, the two mouldings were not fitting together as intended and there was a high degree of variance in the overall form. This variance was clearly visible to the naked eye.

More importantly, this luminaire product required a degree of translucency, this translucency requirement was an aesthetic requirement in that it was intended that 'some light' would shine through the luminaire. This material property had never been asked for paper pulp and was potentially a major blockage for the entire concept. Figure 1.10 shows the initial products of the prototype exercise.

UPP in conjunction with VK&C discussed possible solutions to these problems, UPP solved the dimensional stability issues by creating specialised drying equipment to ensure improved dimensional stability; this worked very successfully and solved this problem.

The translucency issue was again solved jointly by UPP and VK&C; the solution was to develop a new blend of paper pulp material, prior to the moulding process, thus creating a paper pulp product with a high degree of translucency. This new paper pulp blend was jointly owned by UPP and VK&C and jointly lodged with the Intellectual Property Office and achieved the appropriate protection.

FIGURE 1.10
Initial prototype samples.

An Introduction to the Aesthetics of Design 11

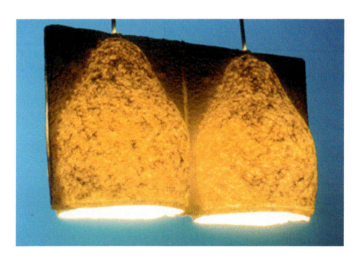

FIGURE 1.11
Completed production paper pulp luminaire.

Figure 1.11 shows the finished product.
The product went on to commercial success, retailing at approximately £20 each; the paper pulp industry does not usually work at this value and is more likely to be selling products at approximately £20 per 1,000 components.

The product went on to win several international awards and was highlighted by several design and eco design publications.

Case Study 2

In this case, the manufacturing company and designers worked together to introduce a naturally occurring material into a market place, which was relatively highly sophisticated and often incorporated high technology components and at high market volumes.

The designer in this case was Brazil based Spazio Uno, and the company was again Brazil based GS Moveis em Fibras Naturais.

Working with natural fibres such as Rush, Rattan and Wicker is part of the tradition of the Santa Felicidade, Curitiba's Italian district. Among the descendants of the pioneering immigrants is Angelo Stival who founded GS Moveis in 1966. The old craftsmanship techniques
have maintained their intrinsic qualities, but design shows little progress or innovation, traditional furniture being the norm.

FIGURE 1.12
Fiber Office chair.

The Fiber Office chair originated from the idea of using natural fibres in setting where this material is still unusual. So, a product that launches brand new possibilities was created. It combines industrialised parts (metal and plastic rotating feet) with a body made of natural fibre.

Figure 1.12 shows the finished product.

Case Study 3

This case is a traditional manufacturing company using wool to weave fabric for the upholstery and interior architecture markets. The Company Bute Fabrics was aware that the market place for their products was relatively mature and was increasingly under market pressure on price. In order to address the pressure on pricing, the company devised a strategy to design a new range of fabrics that would perform better and offer a new range of aesthetics; thus, the product offering could be sold at a higher market value.

Designed by Bute fabrics design team and Jasper Morrison

The Argyll range of Bute fabrics was with the Bute fabric design team working in conjunction with Jasper Morrison.

An Introduction to the Aesthetics of Design   13

FIGURE 1.13
Argyll range by Bute fabrics design team and Jasper Morrison.

The intention was to help drive Bute fabrics' strategy to diversify into more contemporary fabrics whilst using traditional materials and processes (Figure 1.13).

Conclusions

These three case studies demonstrate that natural materials can be incorporated into a wide range of products and market sectors. Although these case studies feature all products that have utilised design in different ways and have incorporated natural materials some in greater depth than others. In all cases, the introduction of natural material needed all the appropriate interested parties to be fully engaged in the process. In all cases, the introduction of contemporary design along with the manufacturing and sales marketing drive of the companies resulted in products that can be sold at higher values.

All three case studies would align with Dieter Rams' 'ten principles of good design'. These case studies demonstrated that design activity applied in a professional and holistic manner including the manufacturing resources from the outset can result in higher value products.

In the case of paper pulp luminaire, the designers' vision for the product would not and could not have been solved without significant input from the expertise of a manufacturer; the result being a new blend of material and a new product selling at a value previously unimaginable for the paper pulp industry. This product vision and concept would have failed, due to lack of translucency, had it not been for the manufacturer (UPP) being prepared to develop new techniques.

These case studies all show to a lesser or greater extent that the manufacturing companies and resources were prepared to go through a design and innovation process themselves and challenge their own practices.

Designers must develop inclusive methods and processes in order to find new solutions to problems and for products. The manufacturing, marketing, research and development and other resources must be prepared to challenge their own previous activities and practises if this is to have positive result.

All parties must be prepared to work with the inherent issues that come along with the incorporation of natural materials, which at times are not aligned with the supply or performance of more commonly used manufacturing materials and processes. There are instances where the compromise of using a natural material may be too much and could adversely affect the resulting products performance or aesthetics; all parties must identify the main issues and resolve them quickly in order to progress.

Natural materials can offer significant product enhancements and if designed appropriately and add value to the resulting products. In this sense, incorporating natural materials into contemporary designs are areal voyage of discovery.

> *'The real voyage of discovery consists not in seeking new lands but seeing with new eyes'*
>
> *Marcel Proust [1]*

References

1. Marcel Proust, 1927, *The Captive, Volume Two of Remembrance of Things Past*, Random House, translated by C. K. Scott Moncrieff, p. 559.
2. Cox, G., 2005, *Cox Review of Creativity in Business*. UK Government Web Archive. http://webarchive.nationalarchives.gov.uk.

2

Selection of Natural Materials Using CES EduPack

Martin P. Ansell
University of Bath

CONTENTS

Introduction to CES EduPack and Materials Selection 15
Natural Organic Materials in the CES EduPack Database............................. 19
Comparison of Mechanical Properties.. 20
Comparison of Thermal Properties 21
Comparison of Durability.. 22
Comparison of Environmental Impact 23
Comparison of Density and Cost.. 25
Discussion and Conclusions.. 26
Acknowledgement 27
References 27

Introduction to CES EduPack and Materials Selection

Cambridge Engineering Selector (CES) EduPack (Granta Design, 2016) is a regularly updated database which provides comprehensive information about the properties of almost 4,000 materials in the form of interactive, visual software. It is intended as a teaching medium for students in the fields of engineering, design, science and sustainable development, but it is also applicable to researchers, industrial scientists and engineers who are able to select engineering materials and compare their properties. The original CES software was developed by Professor Mike Ashby following the publication of his classic text 'Materials Selection in Mechanical Design' (Ashby, 1992), which contained many materials selection charts.

The materials database is accessed at levels 1, 2 and 3 with an increasing number of materials up to 3,907 at level 3. Level 1 provides a list of 69 commonly used materials from A for ABS polymer to Z for zinc alloys. A 'select' function is employed to access some or all of these materials by defining your own subset or choosing all the materials termed the

15

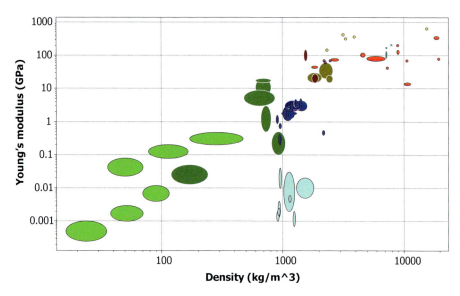

FIGURE 2.1
Young's modulus versus density for all 69 of the level 1 materials.

'Material Universe'. Each material has a description including photographs and properties which are listed under the headings of general, mechanical, thermal, electrical, optical and eco properties. Any combination of properties can be presented in the form of an x-y plot. For example, if all 69 level 1 materials are selected, Young's modulus can be plotted as a function of density (Figure 2.1). The selected axes are logarithmic to more clearly present the wide range of values, but linear axes may be specified. Each material property has a range of values represented by the coloured oval field of values; in this case, the width represents the range of densities and the height represents the range of Young's moduli.

Classes of materials are colour coded (e.g. natural materials are dark green), and the identity of each class is revealed in Figure 2.2. Furthermore, bamboo and cork have been identified and labelled as individual materials.

The results of the investigation of Young's modulus versus density can be significantly expanded by selecting level 3 resulting in Figure 2.3 which includes all of the 3,907 materials in the database. As examples of natural materials, palm and balsa (ochroma spp.) have been labelled. It is obvious that materials selection will almost certainly require a focus on the comparison of a defined subset of materials and the user can pre-select such a subset.

At level 1, natural organic materials include bamboo, cork, leather and wood along and across the grain. For example, cork is described as a natural closed-cell foam, which is waterproof and remarkably stable and a summary of its key features is given, including low density, good thermal and acoustic insulation properties, chemical stability and fire resistance. The chemical

Selection of Natural Materials Using CES EduPack   17

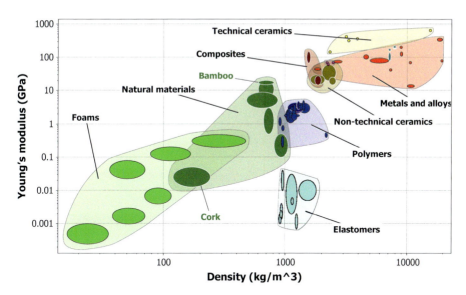

FIGURE 2.2
Young's modulus versus density for all 69 of the level 1 materials with classes of materials identified and labelled.

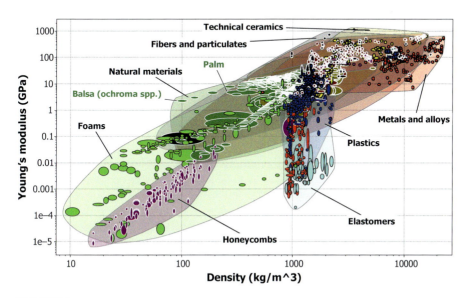

FIGURE 2.3
Young's modulus versus density for 3,851 of the level 3 materials with classes of materials identified and labelled.

composition is given as 40% suberin, 27% lignin, 12% cellulose, 4% friedelin and 17% water.

Properties are listed as follows:

General properties are represented by a range of densities (120–140 kg·m^{-3}) and price (£1.68–8.41 in 2015); the estimated price range being updated annually by Granta Design. Each property is defined by clicking on the term.

Mechanical properties include Young's modulus (0.013–0.05 GPa), yield strength (0.3–1.5 MPa), tensile strength (0.5–1.5 MPa), elongation (tensile strain at failure, 20%–80%), fatigue strength at 10^7 cycles (0.3–1.1 MPa) and fracture toughness (0.05–0.1 MPa·m$^{0.5}$).

Thermal properties include maximum service temperature (117°C–137°C), thermal conductivity (0.035–0.048 W·m^{-1}·°C^{-1}) specific heat capacity (1,900–2,100 J·kg^{-1}·°C^{-1}) and thermal expansion coefficient (130–230 microstrain·°C^{-1}).

Electrical properties are simply specified as poor.

Optical properties are simply stated as opaque.

Eco properties include embodied energy in primary production (3.8–4.2 MJ·kg^{-1}), CO_2 footprint in primary production (0.181–0.200 kg·kg^{-1}) and recyclability (yes or no).

Supporting information is provided indicating typical uses of cork including corks, floats, flooring, insulation and footwear. At level 3, six varieties of cork are described including high- and low-density cork and corkboard with specific gravities of 0.12, 0.16, 0.2 and 0.25. As well as the properties listed above at level 1, there is considerably more property information provided exemplified by data for corkboard with a specific gravity of 0.12 as follows:

Physical properties include relative density and anisotropy ratio.

Mechanical properties are supplemented by compressive modulus (0.00459–0.00561 GPa), compressive strength (0.18–0.22 MPa), flexural modulus (0.0104–0.0128 GPa), shear modulus (0.00531–0.00649 GPa), shear strength (0.807–0.906 MPa), Poisson's ratio (0.05–0.4) and shape factor (3.34).

Thermal properties include glass temperature (77°C–102°C) and minimum service temperature (−52.8°C to −43.2°C).

Electrical properties cover electrical resistivity (10^9–10^{11} μΩ·cm), dielectric constant (6–8), dissipation factor (0.02–0.05) and dielectric strength (1–2 MV·m^{-1}).

Absorption and permeability states water uptake (3.6%–4.4%) after 24 hours.

Durability in broad terms is given for resistance to fresh and salt water, weak and strong acids and alkalis, organic solvents, oxidation at 500°C, UV radiation and fire.

Eco properties are supplemented by water usage (1·kg^{-1}). Processing energy and the CO_2 footprint associated with coarse and fine machining and grinding are quoted as estimated values. The proportion of material recycled (0.1%), heat of combustion (18.5–22.6 MJ·kg^{-1}) and combustion CO_2 (1.56–1.91 kg·kg^{-1}) are included.

Selection of Natural Materials Using CES EduPack   19

Hence, level 1 is seen as a general introduction to the properties of a limited number of generic materials (e.g. cork), whilst level 3 provides considerably more depth in relation to subsets of materials (e.g. high- and low-density cork and corkboards). It should be noted that within CES EduPack, some values for the range of properties are estimates and this is particularly the case for natural materials.

Natural Organic Materials in the CES EduPack Database

Defining your own subset allows natural organic materials from the database to be isolated. Natural fibres include coir, cotton, flax, hemp, jute, kenaf, ramie, silk, sisal and wool. Natural composite materials include corkboards, paper and cardboard. Wood composites include end-grain balsa, glulam, fibreboard, hardboard, insulation board, medium density fibreboard, particleboards, strandboards and plywood. Antler, bone, horn and shell are termed mineralised tissue, and leather is termed soft tissue. Many wood species are accessible with properties longitudinal and transverse to grain. Bamboo, cork and palm are called wood-like.

A plot of Young's modulus versus density is plotted for 520 of these materials (Figure 2.4). Wood and wood-based materials are coloured green, natural fibres are black, leather is purple and mineralised tissue is turquoise.

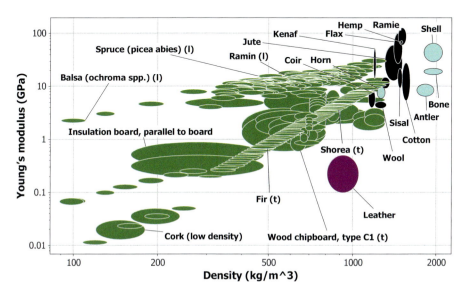

FIGURE 2.4
Young's modulus versus density for 520 natural materials.

Young's modulus of wood in the longitudinal and transverse directions is approximately proportional to density and the wood-based panel products lie between the data fields of the longitudinal (l) and transverse (t) wood properties as a function of density. Natural fibres have a clear advantage over wood and wood-based materials because of their higher density and linear cellular microstructures. In contrast, wood contains a high proportion of transverse cellular elements in the form of medullary ray cells, which reduces the longitudinal Young's modulus. The elastic modulus of the cell wall is ~1,500 kg·m^{-3}, but the structure of low-density wood such as balsa contains a high proportion of air.

Mineralised tissues which have organic precursors (proteins and calcium carbonate in the case of shell) are generally the highest in density. It should be noted that some wood species, including South American hardwoods such as teak, are mineralised with silica.

Comparison of Mechanical Properties

It is instructive to review the mechanical properties of the natural organic and partially organic materials included in Figure 2.4. There is a broad proportionality between Young's modulus and tensile strength of these materials (Figure 2.5).

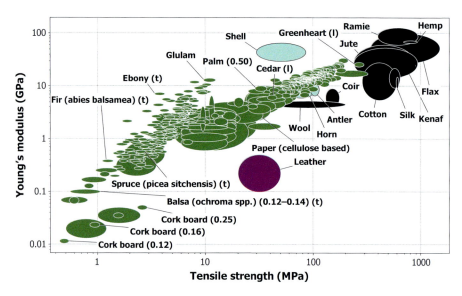

FIGURE 2.5
Young's modulus versus tensile strength for 520 natural materials.

Selection of Natural Materials Using CES EduPack 21

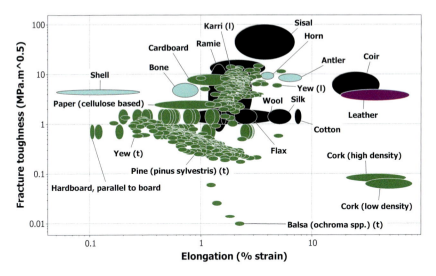

FIGURE 2.6
Fracture toughness versus elongation at failure for 520 natural materials.

The transverse properties of wood species are situated to the left of the diagram, whilst there is an intense clustering of longitudinal properties to the right. The strength of shell is compromised by its brittle mineral content and fracture toughness versus elongation at failure is the subject of Figure 2.6. The fracture toughness is the mode I (crack opening) critical stress intensity factor K_{IC} with units of MPa·m$^{0.5}$ and measures the resistance of materials to the propagation of cracks (Ashby and Jones, 1980).

Sisal is well known as a traditional source of natural fibre for ropes and twine where toughness is a major attribute. Coir has a high elongation at failure at the expense of a lower elastic modulus. Hence, there is a wide selection of fibrous reinforcements for natural fibre composites (NFCs), the choice of which is dictated by the required engineering properties and availability (Ansell and Mwaikambo, 2009). Wood in the transverse (t) orientation is well known to split along radial-longitudinal planes when loaded in tension, whereas crack propagation across the grain is far more difficult and results in complex fracture topographies. Leather has a similar fracture toughness to shell and bone but a much higher elongation at failure, reflecting its traditional applications in drive belts, hinges and clothing, combined with tactility and breathability.

Comparison of Thermal Properties

Natural materials are, broadly speaking, poor conductors of electricity and good thermal insulators. Figure 2.7 plots thermal conductivity against

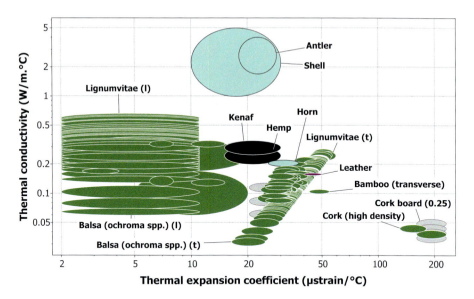

FIGURE 2.7
Thermal conductivity versus thermal expansion coefficient for 520 natural materials.

thermal expansion coefficient and effective thermal insulators include cork and low-density woods such as balsa. In the transverse direction, balsa has an average thermal conductivity of ~0.032 W·m^{-1}·K^{-1} only bettered by a vacuum, air or exotic materials such as aerogels. It should be noted that convective conduction rules out air as an insulator in cavity walls. The ceramic and bony components of shell and antler raise their thermal conductivity. Denser woods are considerably more thermally conductive than low-density woods, and there is a considerable spread in measured values indicated by the width of the property zone. The ratio of thermal conductivity to thermal expansion coefficient reflects materials, which exhibit small thermal distortion. The value of this ratio increases towards the top left of Figure 2.7.

Along the grain, the lowest density balsa possesses an excellent combination of low thermal expansion coefficient coupled with low thermal conductivity.

Comparison of Durability

The CES EduPack contains information on durability including flammability, resistance to acids and alkalis, organic solvents and water. The level of durability is reported in the categories of unacceptable, limited use, acceptable, good, excellent, flammable and non-flammable.

Selection of Natural Materials Using CES EduPack

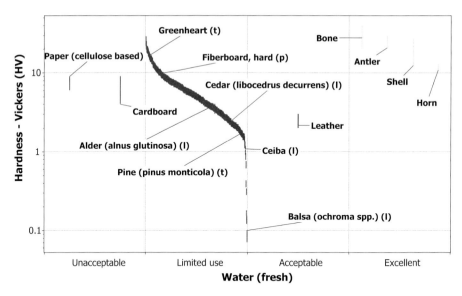

FIGURE 2.8
Vickers hardness versus resistance to submersion in water for 520 natural materials.

Hence, it is not possible to plot durability in terms of parameters such as dissolution rate or ignition time. However, durability can be related to other properties. In Figure 2.8, Vickers hardness is plotted versus resistance to submersion in fresh water. Paper and cardboard have unacceptable durability; but in the limited use range, the natural materials are ranked in order of hardness. Leather falls in the acceptable range, whilst bone, antler, shell and horn are acceptable in declining order of hardness.

Comparison of Environmental Impact

Several environmental impact parameters are included within CES EduPack, and in Figure 2.9, the carbon dioxide generated (kg) in the primary production of one kilogram of the material and released to air is plotted versus the embodied energy for the primary production of the material from feedstock in MJ·kg^{-1}.

These results are derived from life cycle assessment (LCA) databases (e.g. Ecoinvent v2.2, 2016 and Hammond and Jones, ICE database, 2008) together with other published data. This data is contentious in the sense that the results of LCA vary from source to source. However, the information plotted in Figure 2.9 relates to production and does not reflect the levels of embodied CO_2 in the wood-based materials. Bamboo requires the smallest amount of

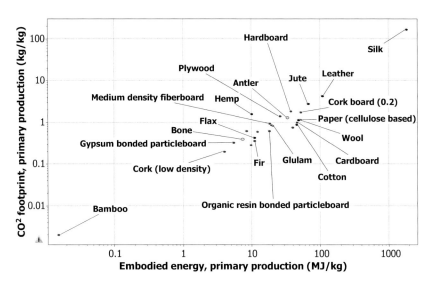

FIGURE 2.9
CO_2 footprint versus embodied energy in primary production for 520 natural materials.

energy and enjoys the lowest CO_2 emissions for production. In contrast, the production of silk has the worst profile. There is a clear linear relationship in the log–log plot between the two parameters. There are apparently few data points on the plot because the property fields overlap.

Retaining CO_2 footprint as a variable and plotting it versus water usage (data derived from factory inputs and outputs in manufacture) a different picture emerges (Figure 2.10). Wool processing is very water intensive

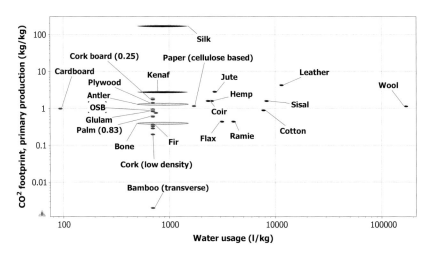

FIGURE 2.10
CO_2 footprint versus water usage in production for 520 natural materials.

(note the log scale), whilst cardboard manufacture is water-lean. Wood-based materials use similar quantities of water per unit mass but natural fibres and leather require considerable more water for processing.

Comparison of Density and Cost

Cost is a key issue when natural materials are selected. In Figure 2.11, density is plotted as a function of price. The wood-based board materials (left of diagram) make very efficient use of the particle- or flake-based feedstock at low cost. The solid wood species (centre) range in density with price being reflected by source (e.g. fir to the left and lignum vitae to the right). The natural fibres vary in price according to source and processing requirements.

At the less expensive end of the spectrum Kenaf, used by Toyota for NFCs, is a tropically-sourced fibre which has superior properties to coir. Flax and hemp are mid-priced and can be grown in temperate Europe and are used by major European car manufacturers (e.g. Mercedes Benz, Volkswagen, Ford) for NFCs. Low-density balsa and cork have a higher price per unit mass (right of diagram); but considering that their density is very low (log scale), the cost of structural balsa cores and cork components is still competitive. Silk and leather have complex and expensive processing stages so their cost per unit mass is high.

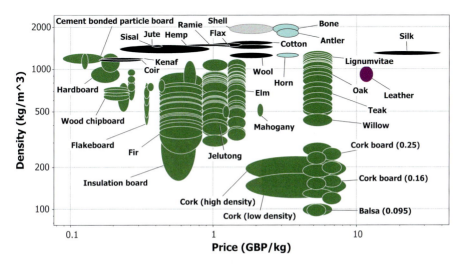

FIGURE 2.11
Density versus price for 520 natural materials.

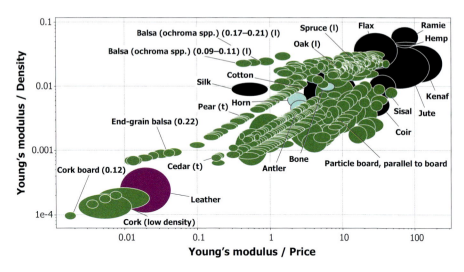

FIGURE 2.12
Young's modulus/density versus Young's modulus/price for 520 natural materials.

A better comparison of density and cost can be made by using CES EduPack's advanced plotting facility. In a structural application, how much stiffness per unit density and stiffness per unit cost is offered by natural materials? In Figure 2.12, specific stiffness (Young's modulus/density) is plotted versus Young's modulus/price.

The two parameters are broadly proportional to each other. The natural fibres offer the greatest stiffness per unit mass and per unit cost. The specific stiffness and price of hemp and kenaf are on a par with glass fibre and the EU End of Life Vehicles Directive (ELVD) EU (End of Life Vehicles, 2016) also favours NFCs in terms of their recyclability. Cork and leather appear to be disadvantaged by their very modest properties but both materials offer significant benefits in terms of thermal insulation (cork) and flexibility and toughness (leather). Balsa has a specific stiffness similar to other wood species. Wool is obscured by the many wood species but lies just below the property fields for horn and antler. Bamboo lies at the lower end of the property zone for flax.

Discussion and Conclusions

This chapter has only scratched the surface of CES EduPack's database at level 3. There is a clear scope for investigating the relationship between the extensive range of property categories (e.g. price, mechanical and thermal properties) and attributes (e.g. shear strength and specific heat capacity).

Only natural materials have been considered here, and there is a clear scope for comparing natural materials with synthetic materials.

Many case studies are included in the textbook 'Materials Engineering, Science, Processing and Design' by Ashby et al. (2010), where 'Exploring Design with CES' is featured.

The selection of natural materials is often driven by the requirement of a combination of properties, for example, thermal insulation, durability and price for construction applications. CES EduPack enables the scientist, engineer or designer to make a sensible selection by judicious comparison of materials via the inspection of data sheets for each material and by graphing property relationships for a shortlist of materials.

Acknowledgement

The preparation of this chapter received support from the European Union's Horizon 2020 research and innovation programme ISOBIO under grant agreement no. 636835. The ISOBIO project is developing innovative strategies to bring bio-based construction materials into the mainstream.

References

Ansell, M.P. and Mwaikambo, L.Y. (2009) The structure of cotton and other plant fibres, *Handbook of Textile Fibre Structure, Volume 2: Natural, regenerated, inorganic and specialist fibres*, Eds. S.J. Eichhorn, J.W.S. Hearle, M. Jaffe and T. Kikutani, Woodhead Publishing, Cambridge, ISBN 978-1-84569-730-3.

Ashby, M.F. (1992) *Materials Selection in Mechanical Design*, Pergamon Press, Oxford, ISBN 0-08-041907-0.

Ashby, M.F. and Jones, D.R.H. (1980) *Engineering Materials*, Pergamon Press, Oxford, pp. 121–128, ISBN: 0-08-026138-8.

Ashby, M., Shercliff, H. and Cebon, D. (2010) *Materials, Engineering, Science, Processing and Design*, 2nd ed., Butterworth-Heinemann (Elsevier), Oxford, ISBN: 978-1-85617-895-2.

EU (End of Life Vehicles). (2016) http://ec.europa.eu/environment/waste/elv/, accessed July 2016.

Ecoinvent v2.2. (2016) http://www.ecoinvent.org/database/database.html, accessed July 2016.

Granta Design. (2016) http://www.grantadesign.com/education/edupack/, accessed July 2016.

Hammond, G. and Jones, C. (2008) Inventory of carbon and energy (ICE) Version 1.6a, http://www.viking-house.co.uk/downloads/ICE%20Version%201.6a.pdf, accessed July 2016.

3

Natural Materials – Composition and Combinations

Morwenna Spear

Bangor Universiity

CONTENTS

Introduction: Natural Materials and Bio-Inspiration 30
Opportunities: The Bio-Based Palette 32
   Principal Organic Components 34
      Polysaccharides 34
      Cellulose 34
      Hemicelluloses and Pectins 41
      Chitin 44
      Proteins 44
      Collagen 45
      Keratin 49
      Fibroin 50
      Elastin 51
      Resilin 51
      Abductin 52
      Other Naturally Occurring Polymers 54
   Principal Inorganic Components – Biominerals 59
      Calcium Carbonate 59
      Hydroxyapatite 60
      Magnetite 61
      Dolomite 62
      Silica 62
      Calcium Oxalate 64
   Combining Ingredients – Recognising Ultrastructure 65
Biocomposites: Combining Bioresins, Biofibres and Green Chemistry 70
   Composites 70
   Man-Made Bio-Derived Organics 72
      Cellulose 72
      Proteins 74
      Chitosan 74
      Bio-Based Polyesters 75

Polylactic Acid (PLA)	75
PHB and PHBV	76
Polybutylene Succinate (PBS)	76
Thermoplastic Starch (TPS)	76
Polyamides	77
Thermoplastic Polyurethanes	77
Bio-Epoxy	78
Unsaturated Polyester Resins (UPE)	79
Acrylate Resins	80
Natural Fibre Composites	80
Fibre Types	83
Bioresins as Matrices	91
Designing with Biocomposite Materials	94
Summary	96
References	96

Where Nature finishes producing its own species man begins, with the help of Nature, to create an infinity of species

Leonardo da Vinci

Introduction: Natural Materials and Bio-Inspiration

There is something compelling about designing with natural materials, something rewarding, tactile and aesthetically pleasing about both the process and the product. So, it is also with designing with the new palette of natural materials, bio-based composites, bioresins, biopolymers and natural fibres. While the additional processing, synthesis, moulding may take two steps towards industrial processes and away from 'nature', there may still be great warmth, texture and naturalness within the finished article, whether it is a flax fibre-carbon fibre hybrid composite or a nanocellulose-filled 3D-printed component (Figure 3.1).

The reason for this pleasantness may stem from the visible natural component (in our first example the flax fibre visible through a transparent matrix); or from the use of this fibre in the strong axis of the material, mimicking the way nature uses the fibre in the plant; or from the elegance and simplicity of the design which harnesses this fibre to form a modern product; or from the functionality brought about by the movement and response of natural materials to environmental stimuli (in our second example, the mobility of the nanocellulose-filled hydrogel material in response to moisture Figure 3.1b, Gladman et al. 2016). These may all be considered to be aspects of bio-inspiration.

Mankind has turned to nature for inspiration in design and use of materials since early in recorded civilisation. How many millennia ago did

Natural Materials – Composition and Combinations

FIGURE 3.1
(a) Monocoque scooter body formed from NFC (see www.vaneko.com), (b) 4D orchid, a 3D-printed cellulose fibril hydrogel which morphs when immersed in water, incorporating the fourth dimension – time (McAlpine 2016).

mankind recognise the benefit of overlapped tiles, shingles or even large leaves, as a roofing material to shed water, reminiscent of fish scales in their optimisation for motion through water? Even in our modern history, we have learnt and forgotten and re-learnt lessons from natural materials. Plywood style cross laminated lay-ups were known to the Egyptians, yet redeveloped into plywood in the 19th century; in this case, an example of harnessing the natural character of the wood, its anisotropy (greater strength in one direction than the others), and utilising it to self-reinforce wooden products in their weaker dimension by cross laminating. Such concepts have been re-used in composites design for glass reinforced plastic (GRP) and carbon fibre composites in the 20th century, and by the end of the century in natural fibre composites (NFCs) as we shall see later.

In more recent decades, we have seen a resurgent interest in the use of natural materials – either due to concerns about resource sustainability, or an affinity for the naturalness of appearance for design and aesthetic values. We have also seen a great deal of research into the nature and structure of bio-based materials – whether plant fibres, or seashells, or animal skin, revealing new approaches to materials science and surface science. The breadth of application of these microstructured materials, and potential to generate surface textures which are either self-cleaning, or ultra-low drag, or highly efficient temporary adhesives seems limitless. These are two aspects of biomimetics, which is the art or science of harnessing nature-based design within materials, and will be considered further in Chapter 9.

Many bio-based materials have properties which are governed by their biogenesis. The strength of wood or bone is far superior to what could be obtained by simply preparing an amorphous mixture of their constituent components. The cellular or fibrillar interior structure permits weight saving while maximising strength in preferred orientations. Some introduction to

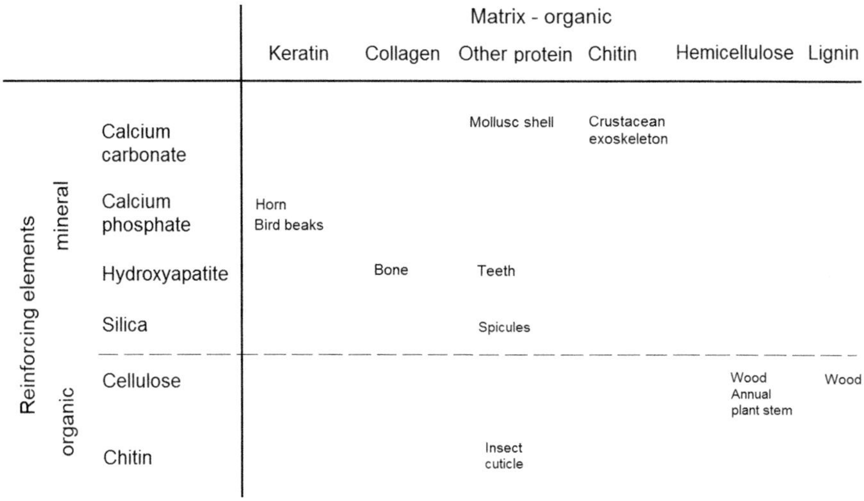

Matrix - organic

Reinforcing elements		Keratin	Collagen	Other protein	Chitin	Hemicellulose	Lignin
mineral	Calcium carbonate			Mollusc shell	Crustacean exoskeleton		
	Calcium phosphate	Horn Bird beaks					
	Hydroxyapatite		Bone	Teeth			
	Silica			Spicules			
organic	Cellulose					Wood Annual plant stem	Wood
	Chitin			Insect cuticle			

FIGURE 3.2

Typical main components of naturally occurring composite materials.

this anisotropy, and its origins in the assembly of the constituent components is given in this chapter. By considering the assembly of materials within biologically produced materials, we can better utilise their constituent elements in man-made analogues.

When we look at natural materials, we discover that many are composite in nature, combining a strong or stiff component with a compliant or more flexible matrix material. The reinforcing elements may be mineral or organic in nature (Figure 3.2), to achieve a range of strength or toughness properties. This chapter aims to introduce the properties and structure of these main ingredients of the natural materials. The application of these concepts to the development of new materials based on bio-derived components will be taken further in Chapter 9, with the aim of inspiring the designer and scientist to greater achievements in the use of nature-based materials.

Opportunities: The Bio-Based Palette

There has been a great deal of interest in different natural materials for a wide range of applications, often as a counterpoint to developments in man-made materials. Printed circuit boards were made with either flax fibre or with glass fibre as the reinforcement. While ceramics pushed new boundaries for toughness and stiffness, scientists looked also at the plate arrangement of nacre in shells. If the 20th century has been described as the 'Materials Age' starting with metallurgy and ceramics, moving into a 'Plastics age' in

Natural Materials – Composition and Combinations   33

the 1950s, and in technology, the 'Silicon Age' with the rise of electronics, semiconductors and computing in the last quarter of the century.

So equally, the period from the 1970s may be regarded as the composites age – with carbon fibre, Kevlar, sandwich composites and many other materials advancing the high strength, lightweight materials in aerospace, motorsport and infrastructure. It could be said that from the 1990s onwards, we entered a 'BioMaterials Age' with the development of biocomposites, biopolymers and the use of renewable or sustainable materials. During this period, ideas from synthetic composites were adopted by bio-based composites manufacture, and renewable alternatives for conventional matrix resins or fibres were sought, leading to both laboratory and market successes. The recognition of depleting reserves of petrochemicals and various mineral resources has led to policy shifts in favour of sustainable and renewable technologies. Note that despite these developments occurring in the same time period as great strides being made in biocompatible materials for surgical use; in this chapter, the term biomaterials will refer to bio-derived materials, not materials for biomedical devices.

There is an extensive palette of biomaterials to choose from. These can be organic, such as cellulose, proteins, waxes or other materials formed by, for example, polymerising bio-derived monomers; or inorganic, such as calcium carbonate, silicate or hydroxyapatite excreted or laid down by animals in bone, or by marine life in exoskeletons (Figure 3.3). The newly emerging field of industrial biotechnology has increased the range of biopolymers

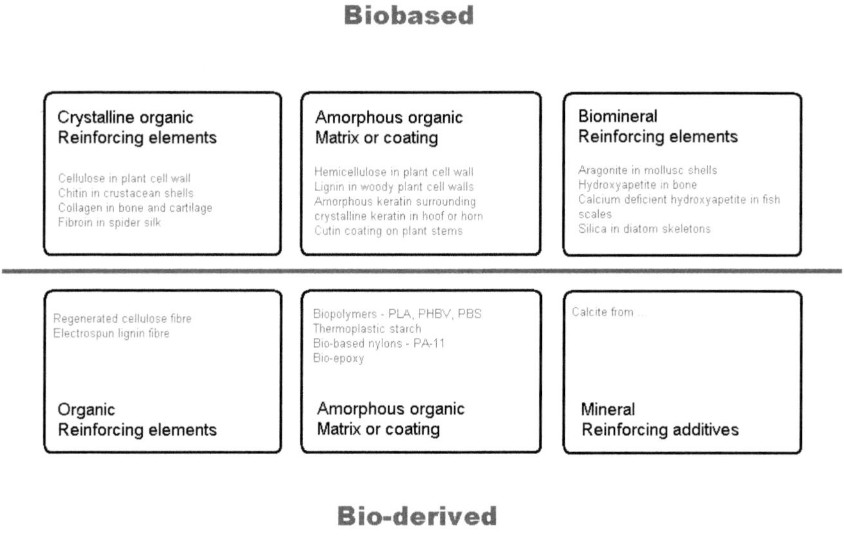

Biobased

Crystalline organic Reinforcing elements	Amorphous organic Matrix or coating	Biomineral Reinforcing elements
Cellulose in plant cell wall Chitin in crustacean shells Collagen in bone and cartilage Fibroin in spider silk	Hemicellulose in plant cell wall Lignin in woody plant cell walls Amorphous keratin surrounding crystalline keratin in hoof or horn Cutin coating on plant stems	Aragonite in mollusc shells Hydroxyapetite in bone Calcium deficient hydroxyapetite in fish scales Silica in diatom skeletons

| Regenerated cellulose fibre
Electrospun lignin fibre

Organic
Reinforcing elements | Biopolymers - PLA, PHBV, PBS
Thermoplastic starch
Bio-based nylons - PA-11
Bio-epoxy
Amorphous organic
Matrix or coating | Calcite from ...

Mineral
Reinforcing additives |

Bio-derived

FIGURE 3.3

Bio-based materials and their bio-derived counterparts. For bio-based, examples of naturally occurring reinforcement roles or matrix roles are given; many are also suited to the equivalent role in biocomposites if suitably extracted and handled.

significantly, with enzymatic breakdown of agricultural and food wastes, followed by industrial or bio-mediated polymerisation of the monomer into new biopolymers such as poly(lactic acid) and poly(hydroxyl butyrate-co-valerate). Polymerisation of vegetable oils has long been known, and the use of fatty acids from plant sources in polyamide production is another example of long known technology which has suddenly flourished with the desire to source bio-based synthetic materials for engineering applications. Let us consider these categories in greater detail.

Principal Organic Components

As can be seen in Figure 3.2, natural materials contain organic components as both rigid reinforcing elements, and as amorphous matrix components. In many cases, the same base polymer may occur in both crystalline and amorphous states, with the crystalline form being embedded in its amorphous equivalent, to combine rigidity with a more visco-elastic shock-absorbing component, to achieve the desired balance of strength and toughness. Examples include the combination of crystalline cellulose and amorphous cellulose seen in the plant cell wall, and the crystalline and amorphous keratin in animal hooves.

Other organic components may be found only with a matrix or adhesive role; for example, lignin has no crystalline order, and is laid down in the wood cell wall late in development, as if to fix the fully formed tissue into a solid structure. Other organic molecules occur combined with biominerals, in particular the proteins which may be found intimately mixed with hydroxyapatite or calcium carbonate in marine creatures. These organic elements clearly introduce toughness to the highly crystalline mineral materials.

To best harness the characteristics of these naturally occurring organic components, a short introduction to the main polysaccharides and proteins is given, alongside other useful organic polymers. Where useful, background information about their crystallinity, long-range order or amorphous state is also included.

Polysaccharides

Plant tissues rely on polysaccharide molecules for the majority of their structural components. Few animals use polysaccharides to this extent within structural elements; however, insects and crustaceans produce chitin, which is a sugar derivative built from glucosamine units.

Cellulose

Cellulose forms the structural element of all plants. It has a cellobiose repeat unit (comprising two chair form β-D-glucopyranose units, with β→1–4 linkages, Figure 3.4). While some rotation is possible around this linkage,

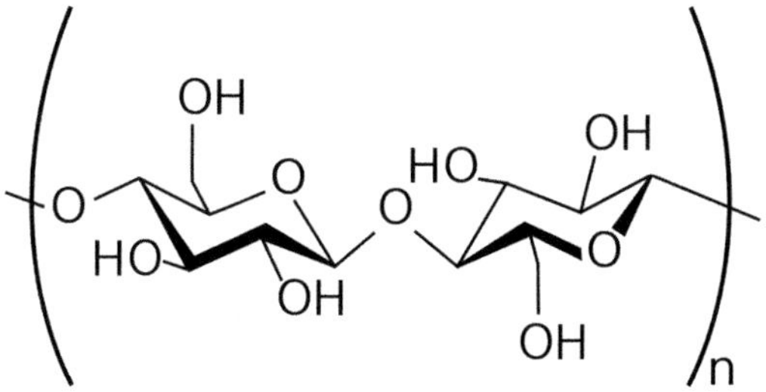

Natural Materials – Composition and Combinations

FIGURE 3.4
Cellobiose unit in which two β-D-glucose monomers are joined between carbon 1 and carbon 4'.

the majority of cellulose in the plant cell wall is assembled into a crystalline form, with only localised regions of disordered amorphous material. In the crystalline state, the two glucose monomers of the cellobiose are oriented oppositely, so that along the cellulose chain each six-membered sugar ring is alternately face up and then face down. The resulting cellulose chain can be considered as ribbon-like – broad and thin, allowing it to be assembled parallel to adjacent chains. Hydrogen bonding between these cellulose chains is strong, especially between those in one layer and the next. Cellulose aggregates of approximately 20–25 nm thickness result, these may often be termed microfibrils. The cellulose aggregates so formed are surrounded by amorphous cellulose (with no crystalline order) and embedded in a matrix of two other polymers: hemicellulose and lignin, which will be discussed later in this chapter.

Within naturally occurring crystalline cellulose, hydrogen bonding holds the extended cellulose chains in the cellulose I conformation. Traditionally, the unit cell for cellulose is accepted as a = 0.835 nm, b = 1.03 nm, c = 0.79 nm and β = 84°, as shown in Figure 3.5 (Meyer and Misch 1937, Frey-Wyssling 1955). In regenerated cellulose (discussed later) and in plant-derived cellulose after processes such as mercerization, the cellulose adopts the energetically favourable cellulose II conformation, which is monoclinic with unit cell a = 0.801 nm, b = 1.036 nm, c = 0.904 nm and β = 62.9 (O'Sullivan 1997). Two other forms, cellulose III and cellulose IV are occasionally identified or discussed; for example, cellulose III can be formed by reaction with ammonia, or heating in the presence of glycerol (O'Sullivan 1997, Bledzki and Gassan 1999).

Recent research has indicated that two naturally occurring forms of cellulose can be identified, cellulose I_α, which is triclinic and cellulose I_β which is monoclinic (Wada et al. 1994, Kataoka and Kondo 1996, Nishino 2004). It is now widely recognised that both cellulose I_α and I_β have the cellulose chains aligned parallel to each other, unlike the anti-parallel system proposed by some of the early models such as Figure 3.5. While the two forms can be generalised as algal type (cellulose I_α) and cotton-ramie type (Cellulose I_β) reflecting two cellulose sources with relatively high proportions of one or the other

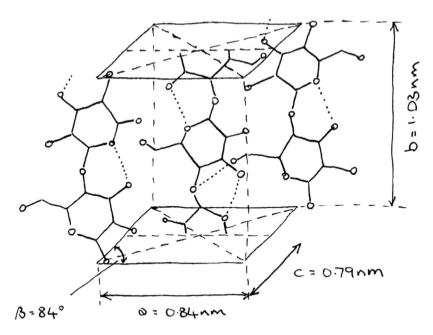

FIGURE 3.5
Unit cell of cellulose, as proposed by Meyer and Misch (1937). Dotted lines indicate intra- and inter-molecular hydrogen bonding. (Adapted from Desch and Dinwoodie 1981.)

form, in practice both forms may be present within the same plant, or even within the same cell wall (Imai and Sugiyama 1998, Nishiyama et al. 2003). Figure 3.6 shows a schematic in which both forms are present, indicating the ease with which one sheet of cellulose chains could be offset sufficiently to switch to the other crystalline form. There is some evidence to suggest that the I_α form is strain induced, with the I_β form relating to cellulose synthesised without imposed strain (Nishiyama et al. 2002); for example, after cell formation has ceased, the later deposited material is a more stable form. Certainly, in wood, both cellulose polymorphs are likely to occur within the same timber; for example, Newman (1999) demonstrated a ratio of 51:49 beta to alpha forms in Radiata pine. It is proposed that slippage between neighbouring sheets of cellulose can account for transition between these two states, without disrupting the hydrogen bonding (Nishiyama et al. 2003).

The cellulose aggregates are found to contain highly crystalline microfibrils, with a 3–4 nm diameter, which can be observed after sonic separation. In the plant cell wall, the microfibrils are most commonly observed aggregated into the larger aggregates, sometime termed macrofibrils, and between plant sources, these are seen to be a range of sizes. Note that an earlier terminology referred to these macrofibrils as microfibrils, composed of elementary fibrils. The microfibrils (or elementary fibrils) are typically 3.5 nm in diameter, but some reports of 5.0–7.0 nm diameter in flax and cotton, and

Natural Materials – Composition and Combinations

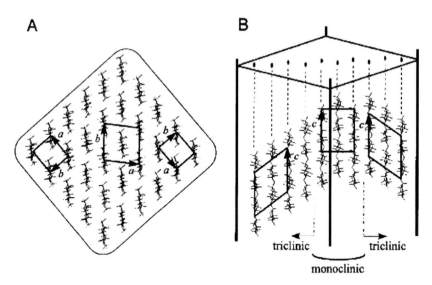

FIGURE 3.6
Packing of cellulose chains proposed by Imai and Sugiyama (1998), showing coexistence of triclinic and monoclinic domains. A relatively small shift of the cellulose chains in one molecular sheet allows this transition. (A) View along the chain direction. (B) View along the [110] (triclinic), and [010] (monoclinic) direction. (Reprinted with permission from *Macromolecules* 31(18):6275–6279, Copyright 1998, American Chemical Society.)

the aggregates typically reported to be 20–25 nm (O'Sullivan 1997). More recently, a wider distribution of dimensions have been recognised, from a single microfibril of 3.5 nm to an upper limit of 30 nm, but most commonly 16–20 nm in size (Salmén and Burgert 2009). The alignment of the cellulose chains longitudinally within the microfibrils, and the efficient hydrogen bonding between chains, results in a high stiffness structural element. The length of the microfibril is considerably greater than its cross section. The degree of polymerisation in wood is in the region of 10,000, and 60%–90% of the cellulose in wood is in the crystalline state (Clemons 2008).

The microfibrils are aligned to provide stiffness in preferential directions within the cell, which varies with cell type and function (Table 3.1). The most frequently reported of which is the softwood tracheid, where the helical alignment of microfibrils within the S2 layer at a low angle to the tracheid long axis provides optimal stiffness for the timber within the trunk of a tree (Mark 1967, Cave 1968, Cave and Walker 1994). In tissue close to the pith, this may be of 20°–30°, falling to 8° or lower in mature wood (Sarén et al. 2001, Sarén et al. 2004). In other strength-providing plant fibres, such as the bast of hemp, flax or ramie, this angle may be as low as 2°–6° (Frey-Wyssling 1952, Preston 1952). Some degree of variability is inevitable, even within one species, as growth rate and agronomy or silviculture play a part in determining microfibril angle. The many complex interactions of microfibril angle with tracheid location within the tree were reviewed by Donaldson (2008). Other

TABLE 3.1
Cellulose Content and Degree of Polymerisation of Cellulose from Different Sources

Fibre	Cellulose Content	Degree of Polymerisation	Fibre Angle (S2) (°)
Flax	71	7,000	10
Ramie	83	6,500	7.5
Hemp	78	9,300	6.2
Jute	61		8
Sisal	67		20
Tracheids	41–51	10,000	8–30

Source: Data from Mukherjee and Satyanarayana (1986), and Bledzki and Gassan (1999).

cell wall layers of the same softwood tracheid provide lateral stability by alignment of the microfibrils at a much shallower angle of winding, as shown in Figure 3.7; and in the primary cell wall, the random alignment allows for the necessary initial rapid growth and expansion of the cell during its formation prior to lignification.

In non-lignified plant cells, such as parenchyma, the microfibril alignment is again random within the primary cell wall, and the combined orthotropic properties of this randomly aligned structure serve to resist cell expansion and maintain turgor pressure (Wainwright 1970, Ennos 2012). Within any

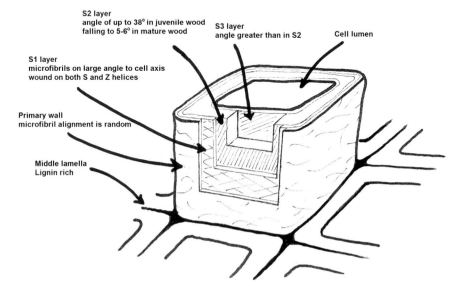

FIGURE 3.7
Schematic of tracheid, with peeled segments revealing microfibril orientation at each cell wall layer.

plant tissue, it is likely that the alignment of microfibrils, whether random or preferential, will relate to some degree to the function of the plant cell. Recent studies of microfibril angle and its effect on cell stress and strain responses have shown that there is a transition between compressive and tensile stresses, depending on microfibril angle, as long as the microfibril stiffness is significantly greater than the matrix material, e.g. 20 or 100 times the stiffness of the matrix are shown in Figure 3.8. The transition from tensile at low microfibril angle to compressive response occurs at angles near 42°–45°, depending on this ratio of stiffness, but a realistic approximation to the plant cellulose microfibrils appears to be 20 (Fratzl et al. 2008). Here, there are two microfibril angles at which swelling stress is zero, near 11° and 43°, as shown in Figure 3.8. Microfibril angle is significantly changed in reaction wood, allowing the tree to control stem angle using the inherent tensile or compressive response of helically aligned cylindrical composites. Tension wood fibres are able to generate stresses of up to 70 MPa (Okuyama et al. 1994), which may relate to the presence of a G-layer with a microfibril angle of 36° (Goswami et al. 2008). Fratzl and co-workers observe that in relation to plant tissue movements, these angles will be relatively stable; the

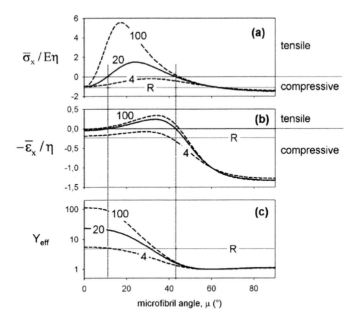

FIGURE 3.8
Mechanical effects related to microfibril angle due to swelling of the cell wall. (a) Stress generated due to swelling when the cell is not allowed to change length, (b) strain generated with no applied stress, and (c) effective Young's Modulus of the cell wall material. Tensile stresses and strains are shown as positive. Data are shown for three values of the fibril's relative Young's modulus, 4, 20 and 100.

Source: Fratzl et al. 2008.

concept of actuators will be discussed in Chapter 9. In terms of bio-inspired design, the fibre tracheid has provided the model for cylindrical composites. In the context of the tracheid – the dominant portion of the tree, many researchers have considered the mechanical optimisation of microfibril alignment for strength (Lichtenegger et al. 1999, Donaldson 2008, Hofstetter and Gamstedt 2009).

There is much interest in cellulose as a raw material for composites or bio-based products. The high stiffness of the microfibril has resulted in great effort to produce these individual crystalline elements for use in polymer matrices as a nanocomposite (Habibi et al. 2010, Klemm et al. 2011). This has been achieved by using plant cellulose, e.g. carrot and sugar beet parenchyma (Dufresne et al. 1997) or from woody fibres, e.g. (Beck-Condanedo et al. 2005), or using cellulose whiskers from tunicin (Anglès and Dufresne 2001). Nanocelluloses have one dimension in the nanometre range, and combine the useful properties of cellulose (e.g. hydrophilicity, suitability for a wide range of chemical modifications, and high crystallinity) with the features of nanoscale materials which arise from their high surface area (Klemm et al. 2011). Three broad groups of nanocellulose can be identified: the microfibrillated cellulose, the nanocrystalline cellulose and bacterial nanocellulose. In addition, microcrystalline cellulose (which is an order of magnitude larger in size and with a lower aspect ratio) can be used for various filler applications; for example, as a filler in thermoplastics, or more commonly in food and pharmaceutical applications.

Microfibrillated cellulose is typically produced from wood pulp or other plant fibres by mechanical pressure after chemical or enzyme treatment. Diameter can be 5–60 nm, and length several micrometers. Nanocrystalline cellulose is produced from a similar range of wood and plant fibres, but by acid hydrolysis, and results in a short fibre length of 100–250 nm but with similar diameter. When produced from algae and tunicates, the length is a much broader range of lengths from 100 nm to several micrometers. The cellulose whiskers of tunicates have a suitably high degree of polymerisation (1,160–2,000) to form stiff fibres of high crystallinity, but the fact that this is lower than found in wood, and bast fibres has allowed readier extraction. The tunicate microfibrils have a cross section diameter of 15 nm. The modulus of tunicate cellulose is very high (143 GPa), which is thought to relate to low amorphous cellulose content (Šturcová et al. 2005).

Bacterial nanocellulose is synthesised directly by bacteria, to produce material with a diameter of between 20 and 100 nm in different networks (Klemm et al. 2011). The cellulose is excreted extracellularly by the culture organism to form a fleece at the air–liquid interface of the culture medium. This biotechnological process offers new opportunities for controlled production of shapes or forms. Bacteria from the genus *Gluconacetobacter*, such as cultures of *Acetobacter xylinium* (Iguchi et al. 2000).

Hemicelluloses and Pectins

Plant tissues also contain other polysaccharide materials, which offer a range of different roles within the architecture of the cell wall. These polysaccharides vary from plant to plant, but can be grouped into two main categories: hemicelluloses and pectins. Both contain a wider range of sugar monomers than cellulose, which are pentoses (five carbon sugars) or hexoses (six carbons). Some of the most common constituents are shown in Figure 3.9. The hexoses in hemicellulose are pyranoses, with a six-membered ring, i.e. bonded by a hemiacetal linkage from carbon 1 to carbon 5 to form the ring. The pentoses such as arabinose may form furanose (bonded between 1 and 4) or the pyranose structures. As the example, monomers are shown in the Fisher projection, and most monomers are in the D-form; the penultimate carbon has the hydroxyl group shown on the right. The exception is L-arabinose, where the penultimate carbon has the hydroxyl on the left.

The balance of the various monomers, e.g. xylose, mannose, glucose and arabinose, and associated uronic acids, e.g. glucuronic acid and galacturonic acid, within the polymer each contribute to the structure, degree of branching, solubility and other properties of the resulting polymer (Figure 3.10). Hemicelluloses tend to exist in various characteristic forms at different locations within the plant cell wall, or within different plants.

The hemicelluloses were initially believed to be precursors of cellulose, which is now known to be incorrect, and a range of distinct compositions based on four main sugar compositions can be defined – the xylans, the mannans, the β-glucans and the xyloglucans (Ebringerova et al. 2005). The pectins are generally considered as a separate group, defined by their high galacturonic acid content, and their ease of extraction from plant tissue by hot water, by weak acids and with chelating agents. Hemicelluloses, in contrast, require aqueous alkaline solutions for their solubilisation.

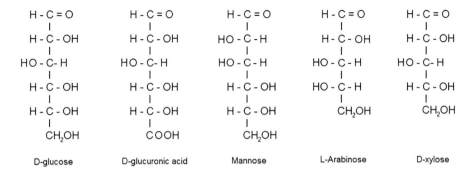

FIGURE 3.9
Fisher projection of some monomers commonly found in hemicellulose: glucose, glucuronic acid, mannose, arabinose and xylose.

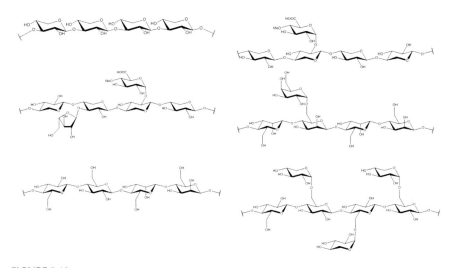

FIGURE 3.10
Primary structures of selected hemicelluloses: (a) β-1–3-xylan, (b) 4-O-methyl glucurono-D-xylan, (c) (L-arabino)-4-O-methyl-D-glucurono-xylan, (d) D-galacto-D-mannan, (e) D-gluco-D-mannan and (f) D-xylo-D-glucan.

Many plants will be found to contain two or more types of hemicelluloses or pectins. The hemicelluloses are typically not crystalline, and may be formed from hexoses (six-membered sugar monomers, such as glucose) and pentoses (five-membered sugar monomers) or in some cases uronic acids. Pectins are very similar in their amorphous nature, and the range of sugar monomers applied. The difference lies in the ability of pectins to gel, due to their high galacturonic acid content. Some plant tissues, such as the parenchyma cells of fruit are high in pectins and hemicellulose, whereas other tissues may contain hemicellulose and lignin (see next section).

Hemicelluloses are typically shorter chained than cellulose, with for example 500–3,000 monomer units. There is a small amount of chain branching present, and the typical monomers include xylose, glucose, mannose and galactose (Gibson 2012, Ebringerova et al. 2005). The hemicelluloses vary widely, depending on the plant tissue. For example, galactoglucomannans, xyloglucans and glucuronoarabinoxylans occur in softwoods. Monocotyledons might contain a greater quantity of mixed linkage glucans. Some hemicelluloses, e.g. the xyloglucans, are structured in a way which allows them to align closely with the cellulosic microfibrils, whereas have a greater degree of chain branching, or create locations onto which other monomer units may attach, such as ferulic acid or acetyl groups onto galacturonoarabinoxylans (Ebringerova et al. 2005, Sorieul et al. 2016).

In both softwoods and hardwoods, the hemicellulose content may vary considerably, typically ranging from 23% to 31% in softwoods, and potentially up to 38% in hardwoods. Within the hardwoods, the proportion

Natural Materials – Composition and Combinations 43

of glucuronoxylans dominates (Table 3.2). Softwood hemicelluloses typically contain a greater proportional of additional monomers arabinose or galactose, and in softwoods, the galactoglucomannans are the dominant fraction (Table 3.3).

Pectins are found in the tissue of many plants, and a wide range of structures and variations are likely to exist. The majority of characterisation has focused on crops of economic importance, and three main structures can be recognised – homogalacturonan, rhamnogalacturonan I and rhamnogalacturonan II. Despite the similarity of name, rhamnogalacturonan I and II have significant differences in their structure, with the main backbone of rhamnogalacturonan I being an alternating galacturonic acid rhamnose polymer, while rhamnogalacturonan II has a backbone of galacturonic acid alone. Homogalacturonan, as implied by the name, contains only galacturonic acid, while the other two contain significant chain branching. For greater detail, a comprehensive review by Ridley et al. (2001) provides structural and functional insights. One key role of pectin in the plant is binding the cells together. In this capacity, the gel-forming nature of pectin appears to contribute to the formation of a strong yet flexible bond.

TABLE 3.2

Typical Composition of Five Hardwoods

	Cellulose	Lignin	Glucurono-xylan	Gluco-mannan	Pectin, Starch etc.
Acer rubrum	45	24	25	4	2
Betula papyrifera	42	19	35	3	1
Fagus grandifolia	45	22	26	3	4
Populus tremuloides	48	21	24	3	4
Ulmus procera	51	24	19	4	2

Source: Timell (1967).

TABLE 3.3

Typical Composition of Five Softwoods

	Cellulose	Lignin	Glucurono-arabinoxylan	Galactogluco-mannan	Pectin, Starch etc.
Abies balsamea	42	29	9	18	2
Picea glauca	41	27	13	18	1
Pinus strobus	41	29	9	18	3
Tsuga canadensis	41	33	7	16	3
Thuja occidentalis	41	31	14	12	2

Source: Timell (1967).

Chitin

Chitin is a naturally occurring polysaccharide. Chitin is a polysaccharide with *N*-acetyl glucoseamine repeat unit ($C_8H_{13}O_5N$). It has a six-membered ring with the acetyl amine unit at the second carbon. The monomer is linked by β-(1–4) linkages in the same manner as cellulose. As a result, chitin is structurally very similar to cellulose, but with the secondary hydroxyl group of the alpha carbon being substituted with an acetamide group. Like cellulose, chitin is also commonly found in a crystalline state, with three known polymorphs, α-, β- and γ-chitin.

Crustacean exoskeletons and insect carapaces are formed from chitin. Chitin is also found in cephalopod beaks (squid, octopus) and forms the organic component of molluscs (Chen et al. 2012). It is also found in the fungal cell wall and bacteria. In the fungal wall, chitin forms a layer beneath a manoprotein exterior, and outside the cytoplasmic membrane. The chitin is intermixed with glucan, and forms a strong tough structure. In arthropods, the chitin is frequently modified, including sclerotin to form hard stiff exoskeletons, e.g. in insect carapace. Sclerotin is a tanned protein-based matrix material, cross-linked by phenolic compounds or in some cases by di-sulphide linkages. The harder nature of sclerotinized chitin is harnessed, for example, in the biting mouthparts of scorpions and beetles. In molluscs, the chitin is present alongside calcium carbonate (Peters 1972), producing a hard stiff composite, to which the chitin has provided the toughness (Feng et al. 2000).

Proteins

The family of structural proteins contains both strong, stiff, and flexible, elastic materials, both of which are high molecular weight polymers, not to be confused with other protein based biological polymers such as hormones, antibodies and enzymes. The strong, stiff group of proteins includes collagen, keratin and fibroin, and typically has a density in the region of 1.20 g/cm^3; whereas the elastomeric proteins – resilin, elastin, cartilage and abductin – are slightly lower in density, at 1.15 g/cm^3 (Wegst and Ashby 2004).

All proteins are formed from amino acid monomers, such as Glycine (Gly), Cysteine (Cys), Proline (Pro) and Hydroxyproline (Hyp). Several of the common amino acids referred to in this text are illustrated in Figure 3.11. All amino acids contain an amino group (NH_2) and a carboxylic acid group (COOH). The general formula for an amino acid is $–NH_2CHRCOOH$, in which R varies, it is this functional group which determines the nature of the amino acid, for example, in glycine R is H, making this the smallest amino acid; it is also non-polar. Alanine (Ala) is the next non-polar amino acid with R = CH_3, while serine (Ser) has R = CH_2OH, making this an uncharged polar molecule. Asparagine (Asn) and Glutamine (Gln) are also uncharged polar molecules, of slightly longer chain length.

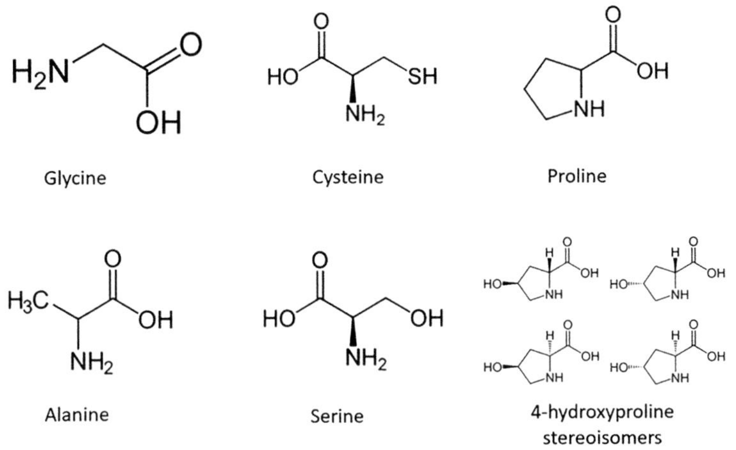

Natural Materials – Composition and Combinations

FIGURE 3.11
Structure of six amino acids commonly found in structural proteins.

Acidic and basic functional groups also occur; for example, in aspartic acid (Asp) and lysine (Lys), respectively. Many of the structural proteins have hydrophobic side groups, including Alanine (Ala), Valine (Val), Leucine (Leu) and Isoleucine (Ile), which are all aliphatic; and Phenylalanine (Phe) and Tyrosine (Tyr) which have aromatic and phenolic functional groups.

Many of the common amino acids are primary amines, and all except glycine are chiral molecules. When combined into polypeptides, the structure of the monomers, and the sequence in which they have been combined, contributes to the structure of the material. Peptide bonds are formed by a condensation reaction between the amine group on one amino acid and the carboxylic acid group of the other amino acid, eliminating water. Collagen and keratin are best representatives of the strong protein materials, while elastin, resilin, abductin and fibroin are elastic, resilient materials.

Collagen

In mammals, the most abundant protein is collagen. It occurs in cartilage, bone, teeth, skin and blood vessels. It is also found in fish scales, and the muscle of other vertebrates. In bone, fish scales and the dentine of teeth, the collagen exists in a biomineralised state with hydroxyapatite, as introduced in Section 2.2. Collagen is not a single protein, but a class containing 29 different types of collagen. Five types of collagen occur in the human body, where it makes up more than half of the total protein (Currey 2002, Chen et al. 2012).

Collagens are characterised by the presence of hydroxyproline, which is not found in most other proteins. Proline and hydroxyproline form about 20% of the total amino acid residues in collagens. Collagen contains amino acid sequences of the form Gly-X-Y, where X and Y may be proline and hydroxyproline respectively, but could be any amino acid (Canty and Kadler 2002). (i.e. Gly-Pro-Y or Gly-X-Hyp; X or Y = other amino acid). Here, the small nature of glycine allows it to tuck into the centre of a tight tropocollagen triple helix in which three strands of collagen polymer twist together. Hydrogen bonding occurs between strands within the triple helix, increasing its stability. The occurrence of proline and hydroxyproline, which are imino acids (where the nitrogen atom of the side chain is associated into a five-membered ring structure) reduces the amount of chain rotation and results in a relatively inflexible macromolecule.

The tropocollagen helix length is approximately 300 nm, and in fibrillar collagens, these are organised into arrays, with staggered gaps which create bands that are visible on the collagen fibrils under transmission electron spectroscopy. The tropocollagen chains are aligned with a stagger of approximately 67 mm, or one d-period (Figure 3.12c, Orgel et al. 2011,

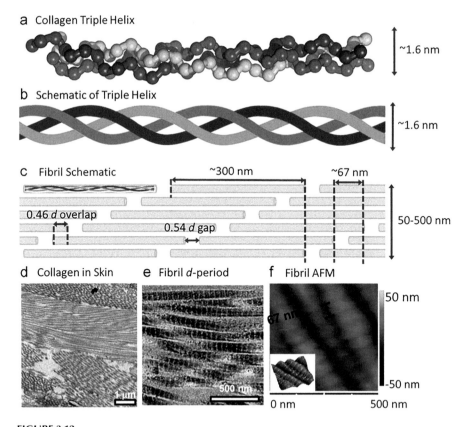

FIGURE 3.12
Schematic showing formation of the collagen triple helix, and the banding which is visible under transmission electron microscope and nanoindentation.

Source: **Sherman et al. 2015.**

Natural Materials – Composition and Combinations 47

Sherman et al. 2015). Within this stagger, there is a gap (36 mm, or 0.54 d) between the end of one molecule and the start of the next, and an overlap of 31 mm (or 0.46 d) between the fibril and its nearest neighbour. Smith (1968) proposed a collagen filament of five chains cross section with one d-period as the stagger, as an arrangement to maximise the number of quarter staggered tropocollagen chains. This has been observed by Orgel et al. (2006), who also noted the right-handed helix of these fibrils, within which the tropocolagen helical structure was also discerned. The resultant collagen fibrils are micrometers to tens of micrometers long (Figure 3.12d,e). Fibrils are frequently aligned preferentially to provide strength, for example longitudinally in tendon and in antler tissue.

Several other forms of collagen are known, including the fibril associated collagens with triple helixes (FACIT collagens), the beaded filament collagens, basement membrane collagens, short chain collagens and transmembrane collagens (Sherman et al. 2015). The FACIT collagens interact with the fibrillar collagens. They have collagenous domains, but these are interrupted by non-helical domains, and thus may provide some interaction with matrix regions. Other FACIT collagens are not known to interact with collagen fibrils, but share the interrupted domain structure.

Tendons are formed from collagen, and can vary in Young's modulus by several orders of magnitude, from 30 to 80 MPa (rabbit patellar tendon) to 1.10 GPa (cross-linked rat-tail tendon) according to data compiled by Sherman et al. (2015). The strength depends on the origin, and the age of the animal which the tendon was taken from. Model studies have demonstrated the influence of cross-linking in determining the modulus and the tensile yield stress of tendon (Buehler 2006a,b, 2008), and tendon exposed to cross-linking agents has been shown to have higher modulus (Gentleman et al. 2003). In cartilage, which is a collagen II based material, the density of cross-linking between fibrils has a strong influence on the material properties (Chen et al. 2017).

Tendon comprises tens of millions of collagen fibrils, each hundreds of microns long, which are synthesised parallel to the tendon long axis (Canty and Kadler 2002). The collagen fibrils are the principal source of mechanical strength, but each fibril also may contain other (non-collagen) surface associated proteins and small leucine-rich proteoglycans which are thought to limit lateral fusion between fibrils. The fibrils are associated into bundles, or fibres, which are further organised into orthogonal layers (e.g. in the cornea), basket weaves (as in skin and bone) or parallel arrays (as in tendon) (Canty and Kadler 2002). The modulus of skin is much lower, from zero up to 50 MPa (Yang et al. 2015), reflecting this different arrangement of the fibres. The modulus of corneal tissue is 0.2–1.0 MPa (Orssengo and Pye 1999), where fibrils are not organised into fibres, but instead are arranged in a highly ordered manner to ensure transparency.

Strain in tendon is reversible when loaded up to 4%, but irreversible changes occur if extended beyond this, with breakage occurring at 8%–10% strain (Rigby et al. 1959). Several mechanisms are at work in collagen when under applied tensile load, including straightening of originally curved fibres, increased alignment with the axis of load, elastic stretching, and sliding of the collagen fibrils (Yang et al. 2015). Simulations by Gaultieri et al. (2009) indicated that for low pulling rates, the straightening of the tropocollagen molecule occurred up to 8% strain. Collagen stress strain behaviour has long been known to be highly dependent on strain rate, with fast loading resulting in near linear elastic behaviour. Slow strain rates show an initially low modulus which increases to a linear region after an initial toe region, giving a J-shaped curve (Wainwright et al. 1976, Gentleman et al. 2003). The deformation behaviour in the toe, heel and linear regions of the stress-strain graph relate to stretching of macroscopic crimp, followed by stretching of microscopic kinks (predominantly in the gap regions), then sliding of the molecules over one another (Fratzl et al. 1997, Puxkandl et al. 2002).

At high strain rates, the modulus increases, e.g. when pulled at 100 m/s Young's modulus is 15 GPa; Gaultieri et al. (2009) related this to different molecular uncoiling mechanisms seen at slow intermediate and high strain rates. Thus, in action in running and jumping animals, the tendon provides efficient load transfer (Wainwright et al. 1976, Screen 2008). Greater understanding of strain rate and creep in tendon is of great interest in preventing injury.

Tendon has also been shown to exhibit creep behaviour in long-term loading and relaxation experiments (Hall 1951, Wainwright et al. 1976). A hysteresis effect of between 5% and 10% is reported for gross tendon samples, with the energy being dissipated as heat (Ker 1981). The non-collagenous components, such as the proteoglycan matrix material, which is present in tendon influences the sliding behaviour of collagen fibres (Puxkandl et al. 2002). Viscoelasticity has been indicated to relate to friction between the fibres, and between the fibres and matrix components. The main proteoglycan type in tendon is decorin, a glucoseaminoglycan chain of dermatan sulphate. Studies where the decorin content has been reduced by enzymes showed reduced viscoelasticity (Millesi et al. 1995).

Hydration is also a significant factor in determining response to stress. Within the body, hydrated collagen contains a significant quantity of water, which is capable of hydrogen bonding with the collagen. Gaultieri et al. (2012) calculated that a decrease of chain spacing from 1.6 to 1.1 nm could be expected on drying collagen. In the hydrated state, the slippage of collagen molecules against one another may be facilitated by the slippage of hydrogen bonds with the water within the structure. As collagen dries, intermolecular bonding directly between collagen fibres allows less slippage, increasing modulus. Skin shows the stiffening effect to a great extent, with the modulus of wet skin being very low (near zero) but for dry skin is nearer 60 MPa (Yang et al. 2015). It has been suggested that the viscous component of the viscoelastic behaviour decreases with increased strain rate, and that at high rates of strain water molecules present within the collagen may allow a hydroplaning effect permitting the sub-fibrillar subunits to slip under tension (Silver et al. 2002).

Natural Materials – Composition and Combinations 49

Various models, including viscoelastic behaviour, and structure-based models of extension, are reviewed by Sherman et al. (2015).

Keratin

Keratin is the second most abundant animal protein. In mammals, keratin forms the main component of hair, nails, hooves and horns, as well as the epidermis of skin. Reptiles and birds have β-keratin in their claws, feathers, beaks and scales. The tensile strength of keratin tissues differs; for example, hairs and wool are typically an order of magnitude stronger in tension than hoof or nail, relating to the greater and lesser degree of alignment of the α-keratin dimers; however, the mode of loading for hooves and nails is typically compression and bending. The β-keratin of claws and beaks is tougher than α-keratin. Approximately, 30 different keratin variants are known in mammals (McKittrick et al. 2012).

Like collagen, keratin also contains relatively high quantities of small amino acids glycine and alanine, which make it possible to construct α-helices. Keratin is also characterised by approximately 22% cysteine; this monomer contains a thiol group (-SH), enabling it to form disulphide bonds with neighbouring cysteine residues. These disulphide bonds provide stability, and resistance to solubility in water, especially in the crystalline regions. Keratin can be found in an α- and a β-form. Pairs of α-keratins tend to form coiled-coil dimers, in which two α-helices are held together by sulphide linkages, these dimer helices are 45 nm in length. The dimers aggregate into protofibrils, in which the coiled-coil dimers are aligned head to tail to form two staggered rows. These protofibrils are approximately 2 nm in diameter, and further aggregate into intermediate filaments, of approximately 7 nm diameter, and at a spacing of 10 nm embedded in amorphous keratin matrix (McKittrick et al. 2012). The intermediate filaments are differently arranged in different keratinous materials, and their alignment influences mechanical properties. Skin, for example, has loosely packed bundles of intermediate filaments to give high elasticity, while hard α-keratins have ordered arrays of intermediate filaments embedded in amorphous α-keratin matrix (McKittrick et al. 2012). Within the dimers, the chemistry of the two chains, and in particular their terminal domains, has a strong influence on the properties of the resulting α-keratin, which is believed to relate to their specific functions in vivo (Parry and North 1998). In textile fibres, the amorphous keratin contains two proteins, one rich in Cys, the other rich in Gly and Tyr residues, and can be seen as an elastomer (Chapman 1969).

Beta-keratin is arranged in a pleated structure, with antiparallel chains (McKittrick et al. 2012). The so-called β-sheet has a crimp of 0.70 nm in length. Hydrogen bonding holds the molecules within the sheet in lateral relationship to each other. It has been shown that if the α-form is stretched, it will transform to a β-form (Frazer et al. 1971). The β-form of keratin is unique to reptiles and birds, and is found in their claws, feathers, beaks and scales, while they also produce α-keratin in other tissues, e.g. skin. Evolutionary genetics studies have shown that β-keratinogenesis has been important in the evolution of feathers and adaptation to ecological niches (Greenwold et al. 2014).

Fibroin

Fibroin has gained great interest in materials science, as the main component of spider silk, and the main component of the silk moth *Bombyx mori*. Fibroin is highly crystalline, although the crystal unit cell differs between organisms, relating to the composition of the polypeptide. Early work on *Bombyx mori* silk showed that about 86% of the amino acids were accounted for by glycine, alanine and serine, with the number of glycine residues being approximately equal to the combined quantity of alanine and serine. Many other amino acids were identified in minor quantities. Studies of *Antheraea pernyi* moth silk and *Araneus diadematus* spider silks revealed the same three main amino acids but in different proportions. While all fibroin crystals have a and b spacings of 0.944 and 0.695 nm, the c-axis spacing differs, depending on the size of amino acid side chains, from 0.93 nm for *Bombyx mori* to 1.06 nm for *Antheraea pernyi* and 1.57 nm for *Nephila senegalensis* (Wainwright et al. 1976).

Spider silk fibroin has a distinctive amino acid sequence of –(Gly-Ser-Gly-Ala-Gly-Ala-)$_n$ which takes up an anti-parallel β-sheet arrangement. Spider silk is produced for different functions, and has markedly different properties, between the strong stiff radiating frame of the web, and the dragline (both are formed by the major ampullate gland), and the more elastic-catching spiral (viscid spiral) of the web (Goseline et al. 1999, Chen et al. 2012). Two types of proteins have been identified in the dragline silk of *Nephila clavipes* and *Argiope auranta*, and shown to have different protein conformations (Brooks et al. 2005). MaSp1 contains segments of two protein types – a glycine-rich (Gly-Gly-X) repeat and polyalanine domains; the MaSp2 type also contains the polyalanine segments, but there is a higher proline content in the glyine-rich portions, with a Gly-Pro-Gly-X-X repeat. The combination of these two types of protein segment allows sheets of polyalanine to form highly aligned segments, with amorphous regions of the glycine-rich protein providing elasticity and flexibility (Keten and Buehler 2010). The catching spiral silk has a higher proline content, replacing much of the alanine and serine (Chen et al. 2012). The presence of proline induces a kink in the protein chain, reducing crystallinity, maintaining the amorphous properties required for this filament within the web (Szent-Györgyi and Cohen 1957).

In the web frame, the β-sheet fibroin forms nanocrystals within a semi-amorphous phase, which make up the silk fibrils (of which several are co-aligned within the core of the spider silk fibre) (Keten and Buehler 2010). The crystalline regions contribute the stiffness to the silk, which can reach 1.2 GPa maximum stress, and have Young's modulus of approximately 10 GPa. In the catching spiral, the Young's modulus is nearer 0.003 GPa

Natural Materials – Composition and Combinations 51

(Goseline et al. 1999). In nanoindentation tests, the modulus was also shown to differ between the two silk types, with values of 7.49 and 3.90 GPa reported for the radial and the catching spiral silks of *Argiope aemula*, respectively (Das et al. 2017).

Strain rate increases Young's modulus of the major ampullate (web frame) silk, but also increases the strain to failure (Denny 1976) – this is not normally observed in synthetic materials, and has prompted research into high performance fibres. Dynamic mechanical analysis (DMA) of spider silks by Das et al. (2017) showed a strong change in loss modulus at low frequencies (below 50 Hz), changing from 0.07 to 0.30 GPa for the radial silk from 1 to 50 Hz, and a slower increase at frequencies above 50 Hz, to a maximum of 0.31 GPa at 100 Hz. The spiral silks showed a very similar trend over a slightly higher range of values (0.08 to 0.35 GPa). This non-elastic component is likely to assist in energy damping during prey capture. The low frequency section of the DMA frequency scan also showed an increase in the storage modulus, which was greatest in the radial silks. The increase in stiffness with increasing frequency is proposed to be due to an increase in hydrogen bonding interactions between crystalline portions of the silks. In addition, the study revealed that spider silks perform most efficiently in humid conditions, with increased energy damping capacity during prey capture (Das et al. 2017).

Elastin

Elastin is found in the skin and arterial walls of vertebrates, as well as the nuchae ligaments of ungulates. Elastin can have several different morphologies, as rope-like networks in skin or lungs, as concentric sheets in blood vessels, or as three-dimensional cellular structures in cartilage (Pasquali-Ronchetti and Baccarani-Contri 1997). The elastin itself is a highly cross-linked amorphous protein, which is highly insoluble due to the cross-linking at lysine residues (Vhrovski and Weiss 1998). This matrix also contains highly hydrophobic amino acids glycine and proline, making it one of the most hydrophobic proteins known.

Elastin is found in virtually all vertebrates, but not known in invertebrates. The elastic fibres of connective tissue are largely composed of elastic, with up to 10% microfibrils of five different glycoproteins embedded within this matrix (Vhrovski and Weiss 1998). The high extensibility of elastin relates to its amorphous, randomly coiled polypeptide chains. The system recoils spontaneously to regain maximum entropy. Elastin can undergo 200% strain before the chains are fully straightened (Aaron and Gosline 1981).

Resilin

Resilin is predominantly found in insects, for example, in wing joints and cuticle structures. It is also the material responsible for a flea jumping, and is part of the structure allowing cicadas to produce sounds. It is known to be highly resilient, and have excellent fatigue properties. It has been reported to be possible to strain to 300% and to return to its original shape on unloading (Weis-Fogh 1961a,b).

Resilin is a highly efficient elastic protein, comprising randomly coiled chains which are cross-linked by di-tyrosine and tri-tyrosine linkages. While resilin is insoluble, it performs as an elastomer when swollen by water or other polar solvents. In the dry state, it becomes hard and brittle. The elastic efficiency is likely due to the complete freedom of the long segments (approximately 40–60 units) between cross links to uncoil under load and resume their coiled state instantaneously (Elliott et al. 1965). The combination of resilin and adjacent segments of the chitin-based exoskeleton is used to store energy required for jumping in insects such as the froghopper. Large amounts of energy must be stored and released in a very short space of time for propulsion, Burrows et al. (2008) reported that the resilin itself could store only 1%–2% of the energy required; however, the stiffness of the cuticle allows large quantities of energy to be stored in small deformation, while the rapid return of stored energy within the resilin component restores the body to its original shape after jumping.

Abductin

Abductin has been isolated from the pad at the junction of a bivalve shell. It has excellent elastic resilience, and extensibility, necessary for the swimming and feeding activities of the mollusc. The constriction and release of the internal triangular hinge ligament (ITHL) of a scallop is used to propel the mollusc through the water. It is the only natural elastomer to be known to work excellently in compression (Denny and Miller 2006). Abductin from ITHL has been found to have similar properties to elastin and resilin (Kelly and Rice 1967), while Alexander (1966) found Young's modulus to be 4 MPa, which is greater than 0.6–1.8 MPa for elastin and resilin. The dominant amino acid in abductin from ITHL is glycine.

TABLE 3.4

Typical Stiffness and Strength of Bio-Based Materials and Related Tissues in Which They Occur (m.c. = moisture content)

	Young's Modulus (MPa)	Failure Stress (MPa)	Strain at Failure	Sources
Biopolymers and tissues				
Cellulose (longitudinal)	150,000	18	0.024	Meyers and Chawla (2009)
Cellulose (longitudinal)	134,000			Salmén (2004)
Cellulose (longitudinal)	120,000–140,000	750–1,080		Gibson (2012)

(Continued)

Natural Materials – Composition and Combinations 53

TABLE 3.4 (*Continued*)

Typical Stiffness and Strength of Bio-Based Materials and Related Tissues in Which They Occur (m.c. = moisture content)

	Young's Modulus (MPa)	Failure Stress (MPa)	Strain at Failure	Sources
Biopolymers and tissues				
Cellulose (tunicate)	143,000			Šturcová et al. (2005)
Hemicellulose	2,000			Salmén (2004)
Hemicellulose (xylan and glucomannan, dry)	5,000–8,000			Cousins (1978)
Hemicellulose (xylan and glucomannan 70% m.c.)	10			Cousins (1978)
Chitin	41,000			Nishino et al. (1999)
Lignin (Kraft)	2,500–4,000 decreases with m.c.	25–75		Cousins (1976), Gibson (2012)
Lignin (Periodate)	3,100–6,700 decreases with m.c.			Cousins (1976)
Collagen (longitudinal) fibril	1,000	50–100	0.09	Meyers and Chawla (2009)
Collagen (longitudinal) molecular	2,400–9,000			Sherman et al. (2015)
Tendon	2,900	20–40	0.8-1.4	Sasaki and Odajima (1966)
Tendon collagen	1,500	1,500	0.12	Goseline et al. (1999)
Tendon		50–100		Elliot (1965), Wainwright et al. (1976)
Tendon	1,000			Rigby et al. (1959), Sherman et al. (2015)
Tendon fibres	50–250			Gentleman et al. (2003), Sherman et al. (2015)
Tendon fibres	960–1,570			Sherman et al. (2015)
Tendon		up to 250		Sherman et al. (2015)

(*Continued*)

TABLE 3.4 (*Continued*)

Typical Stiffness and Strength of Bio-Based Materials and Related Tissues in Which They Occur (m.c. = moisture content)

	Young's Modulus (MPa)	Failure Stress (MPa)	Strain at Failure	Sources
Biopolymers and tissues				
Skin	zero to 50	15		Sherman et al. (2015), Yang et al. (2015)
Keratin (wool, dry)	5,600			McKittrick et al. (2012)
Keratin (wool, 65% r.h.)	4,500			McKittrick et al. (2012)
Keratin (pig hair)	3,930–8,080		6–15	Mohan et al. (2014)
Fibroin	10,000	70	0.09	Meyers and Chawla (2009)
Spider silk (*Araneus diadematus*, webframe and dragline)	10,000	1,100	0.27	Goseline et al. (1999)
Spider silk (*Araneus diadematus*, viscid, catching spiral)	3	500	2.7	Goseline et al. (1999)
Cocoon silk (*Bombyx mori*)	7	600	0.18	Goseline et al. (1999)
Abductin	1–4			Meyers and Chawla (2009)
Elastin	0.3–1.5		>2	Meyers and Chawla (2009)
Resilin	0.6–1.8	3	>3	Weis-Fogh (1960)

Other Naturally Occurring Polymers

Lignin

Lignin is an amorphous polymer, abundant in wood, and present in many other plant tissues. The name is derived from *lignum*, the Latin word for wood. Lignin is based on three hydroxycinnamyl alcohol monomers with different levels of methoxylation: hydroxycinnamyl alcohol; coniferyl alcohol and sinapyl alcohol, as shown in Figure 3.13. Once combined into the polymer, these phenyl propane structures are referred to as 4-hydroxy phenol, guaiacyl and syringyl units, respectively.

The composition, and abundance of these phenyl propane repeat units varies depending on the plant type, with, for example, coniferyl alcohol (or guaiacyl units, Figure 3.13b) being abundant in the lignin of softwoods, with a small quantity of hydroxyphenyl units, whereas hardwoods

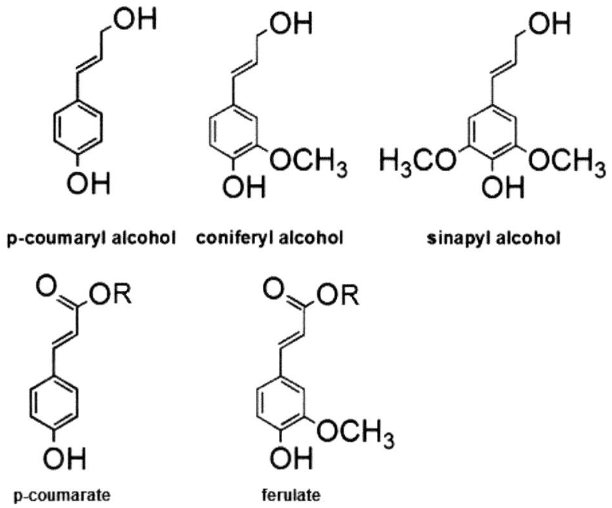

Natural Materials – Composition and Combinations

FIGURE 3.13

(a) p-Hydroxy phenyl, (b) guaiacyl and (c) syringyl monomers (d) coumaric acid ester (e) ferulic acid ester, where R signifies an oligomer, often of hemicellulose.

have a mixture of syringyl and guaiacyl units, and a small proportion of hydroxyphenyl. Grasses tend to have some syringyl and guaiacyl units, but proportionally more p-hydroxyphenyl units. The grass lignins may also contain esters of hydroxycinnamic acids as pendant groups on the polymer, and this is becoming recognised in other plants (Boerjan et al. 2003). The most common additional monomers are p-coumarate and ferulate, which are closely related in structure to the coumaryl alcohol and coniferyl alcohol, respectively, and acetate. The bonding of hydroxycinnamic acids is predominantly through the benzyl position, as an ether linkage to the next lignin monomer (Lam et al. 2001) or by a mixture of ether and ester linkage (Sun et al. 2002). The ester at the γ-carbon provides a linkage to hemicellulose within the grass plant cell wall (Scalbert et al. 1985, Lam et al. 2001).

There are other differences between lignins from different sources, for example, in the type of cross-linking occurring within the lignin (Table 3.5). This is to some extent related to the lignin monomer composition, the location within fibre, vessel or parenchyma tissue, and the deposition processes. The level of guaiacyl and syringyl units, for example, influences the susceptibility to chemical degradation, due to differences in the number of β-5, 5-5 and 5-O-4 bonds formed. The carbon-5 position is unavailable in syringyl units, resulting in fewer 5-5 bonds in hardwoods lignins (2.3% for beech, whereas spruce has 9.5%) and fewer β-5 bonds (6% in beech, 9%–15% in spruce). Instead, beech lignins have a greater proportion of β-1 linkages, whereas the 1 position is barely used in softwood lignins (see Table 3.5). As a result, coniferous lignins show greater resistance to pulping chemicals. This

56　　　　　　　　　　　　　　　　　　　　　*Designing with Natural Materials*

TABLE 3.5

Types and Frequencies of Inter Unit Linkages in Softwood and Hardwood Lignins, Indicating the Number of Linkages per Hundred C_9 Monomer Units (Brackets Indicate Lignin Types Not Reported, But Present in Low Quantity)

Linkages	Spruce Lignin (Erikson et al. 1973)	Beech Lignin (Nimz 1974)	Poplar Lignin (Stewart et al. 2009)
β-O-4	49–51	65	87.4
α-O-4	6–8		
β-5	9–15	6	2
β-1	2	15	2.1
5-5	9.5	2.3	(0.3)
4-O-5	3.5	1.5	(0.3)
β-β	2	5.5	7.9

altered susceptibility to degradation has economic implications in pulping, or industrial biotechnology, whether by fungi and microbial action, or by industrial biotechnology processes. In addition, lignin may cross link with hemicellulose during deposition (Figure 3.14).

As lignin cannot be removed from plant tissue except by cleavage of bonds, it is not possible to fully quantify the molecular weight. However, empirical calculations have indicated that a weight average molecular weight of up to 20,000 may be representative of milled wood lignin of spruce. Pulping chemistry alters the character of lignins collected from industrial processes, which can strongly impact their suitability for use in development of new products; for example, lignosulfonates are harvested from the sulphite

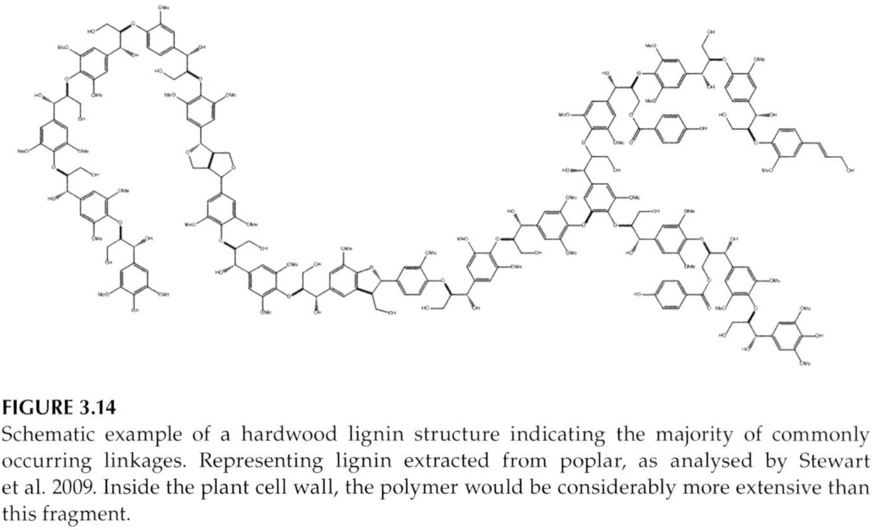

FIGURE 3.14
Schematic example of a hardwood lignin structure indicating the majority of commonly occurring linkages. Representing lignin extracted from poplar, as analysed by Stewart et al. 2009. Inside the plant cell wall, the polymer would be considerably more extensive than this fragment.

Natural Materials – Composition and Combinations

pulping process, whereas kraft lignin can be harvested from sulphate of kraft pulping, but this may also contain inorganic salts and contaminants. Organosolv pulping, using acidified alcohols, avoids the introduction of sulphur to the lignin and has been the subject of recent interest for green chemistry approaches to the use of lignin as a platform chemical.

Within the woody tissue of the plant, the lignin is frequently likened to the glue which binds plant cells into the structure. The bond created by lignin is stiffer than that provided in the growing plant by pectins, which hold the cells of less-woody plant tissues in place. Lignification of the xylem (wood) begins in cell corners, and at the middle lamella between the primary walls of adjacent tracheids, as part of programmed cell death. Later, it diffuses to locations further into the cell wall (Donaldson 2001, Boerjan et al. 2003). After lignification, the primary wall and middle lamella region are lignin rich and referred to as the compound middle lamella. Lignin content of the compound middle lamella is therefore high, but lignin also occurs dispersed within the cell wall, and is present in all layers of the plant cell wall at lower concentrations. Within the cell wall, the role of the lignin appears to be to encrust itself around and between lamellae of hemicellulose and cellulose into the cell wall structure, and each distinct region of the cell wall may have distinct lignin content relating to the requirement for this lignin to act as a matrix material within the composite structure of the cell wall.

The majority of the knowledge about lignin has been derived by scientists studying the pulping chemistry to prepare paper fibre from trees and other biomass (e.g. Hatakeyama and Hatakeyama 2010), or plant scientists studying metabolic pathways and plant form and function (Boerjan et al. 2003). However, in recent years, the interest has transferred to the use of enzymatic processes in degrading lignin as an initial step, allowing polysaccharide-rich residue to be used in fermentation for bioethanol, or generation of other platform chemicals for green chemistry, biopolymer production or bioresins (Hatakeyama and Hatakeyama 2010).

Due to the cross-linking and physical dispersion of the lignin throughout the cell wall, pure lignin is relatively difficult to harvest as a source material. However, lignin is produced as a by-product of the pulp and paper industry, and lignosulphonates, kraft lignin and hydrolysis lignin may be used as starting materials for biopolymers. Each of these lignins will have undergone some degree of chemical reaction during extraction from the plant cell wall, changing the typical functional groups present in the extracted lignin material. However, lignin-based polymers have been developed, for example, utilising the hydroxyl functionality (both aromatic and aliphatic may be present) to derive polyurethanes, polycaprolactones, or epoxy systems (Hatakeyama and Hatakeyama 2010). Other research has sought to use the phenolic nature of lignin as a potential replacement in phenol formaldehyde resin systems. The presence of the aliphatic side chains reduces the reactivity of lignin, compared to phenol, but researchers have, for example, looked into controlled depolymerisation methods which could increase the available hydroxyl units to increase reactivity (Pizzi 2013, Ghorbani et al. 2016). One thermally mouldable polymer filled with wood flour is available commercially, as Arboform (Nägele et al. 2002), and can be used in as an alternative to other plastics in various products.

Cutin and Suberin

One frequently overlooked biopolymer which occurs within plants is cutin. Cutin is a structural component of the plant cuticle, attached to the outer part of the epidermal cell wall in the aerial parts of plants. It is prominent and thick in the stems, leaves and fruit of many plants, e.g. ivy leaves and apples, forming a protective surface. On leaves it may range in thickness from 0.5 to 14 µm, with an area coverage of 20–600 µg/cm^2, whereas in fruits with a well-developed cuticle it may reach 1.5 mg/cm^2 (Martin and Juniper 1970, Holloway and Baker 1970). Grape cuticles may be 1–4 µm thick, tomato 4–10 µm and apple 30 µm (Velásquez et al. 2011). The below-ground parts of most plants utilise a similar but chemically distinct biopolymer, suberin. This is also a biopolyester, and derives its name from the cork oak (*Quercus suber*) in which it is found in the bark, and provides the water-repellent properties associated with the use of cork as a stopper for bottles. In fact, suberin is also found in the wound periderm of most plants, and many other locations (Kolatukkudy 1980).

The cutin is composed of long aliphatic monomers, typically 16 carbon hydroxylated fatty acids, 18 carbon hydroxy or epoxy fatty acids are also present (Kolatukkudy 1980). Lime cutin, for example, has been shown to be formed from 10,16-dihydroxyhexadecanoic acid, 10-hydroxyhexadecanoic acid or hexadecanoic acid (Kolattukudy 1984, Fang et al. 2001). Cutin composition alters with age, and trihydroxyoctadecenoic acid is reported to be dominant in thicker membranes (Holloway and Baker 1970). There is considerable cross-linking, which appears to be in the form of ester linkages. Plant waxes are frequently associated with the cutin; these are often alkanes, often with chain length from 29 to 37. The wax also contains esters, primary and secondary alcohols or fatty acids, depending on the fruit (Eglington and Hamilton 1967, Holloway and Baker 1970). The waxes do, however, often contribute significantly to the thickness of the protective layer. In apples, for example, the wax coverage may be 30–40 times heavier than the wax on leaves, which further contribute to low surface energy and protection of the underlying fruit or leaf tissue.

The suberin composition has two distinct components, the first is a polyester similar to cutin, based on longer fatty acids (length 20–32). The fatty acids are hydroxylated in a similar manner to those in cutin. The second is a phenolic component, which may have some similarity to lignin however containing a lower degree of substitution on the phenol propane units (Kolatukkudy 1980). The cross-linking between the two domains is proposed to occur through the γ-position of the phenyl propane unit forming an ester linkage with the fatty acid of the aliphatic component.

Natural Materials – Composition and Combinations 59

Both cutin and suberin have roles in controlling the passage of water or gases through the surface of the plant tissue, root or fruit, and the protection from diseases. Cutin has interest for the materials scientist in the field of tribology and surface science, as this, in combination with plant waxes, provides much of the water shedding behaviour of plant leaves due to low surface energy. Plant epicarps have low surface energy, ranging from 37 to $44\,mJ/m^2$ (Velásquez et al. 2011). A flat sheet of plant wax may provide a surface energy sufficient to reduce the contact angle with water to 74° (carnauba wax, smooth, Cheng et al. 2006), and to 84° for the surface of a sweet pepper fruit (epicuticular waxes show only the pattern of underlying epidermal cells, Khayet and Fernandez 2012). However, it has long been recognised that wax on plant tissue is often in rods, tubules, platelets or other highly textured states (Velásquez et al. 2011). This effect will be considered further in Chapter 9, where, for example, the lotus leaf effect achieves a contact angle of over 150°, meaning a water droplet is readily shed from the surface.

Principal Inorganic Components – Biominerals

Many natural materials with high stiffness and hardness incorporate ceramic components, or biominerals. Bone, antler and shell are all examples in which a protein component has been reinforced by the deposition of a mineral element, such as hydroxyapatite, calcite or aragonite. The density of materials with a biomineral component varies from 1.8 to 3.0 g/cm^3, i.e. greater than the solely organic cellulosic or proteinaceous materials. The moduli of these bioceramics tend to be lower than engineering ceramics, but tensile strengths are approximately the same, while toughness can be considerably greater, e.g. by a factor of ten (Wegst and Ashby 2004).

Many of the main biominerals are introduced in the following section. It is worth noting that the form of the mineral may be influenced by the presence of organic molecules during deposition, leading to altered crystalline structures, which may further enhance mechanical performance, compared to crystals formed in their pure inorganic state (Meyers et al. 2008). Understanding the nature of interaction between the organic and mineral component, in particular the control exerted by the organism in deposition of crystals of a defined shape and structure, can lead to advances in materials science. In addition, incorporation of organic polymer within the developing crystal may contribute to further improvement of toughness and resistance to crack propagation.

Calcium Carbonate

Calcium carbonate ($CaCO_3$) occurs in biomaterials in several forms. The most common of these are calcite, aragonite, vaterite and the amorphous form. Within different crustacean and mollusc shells different forms of calcium carbonate occur, sometimes of more than one type, imparting different mechanical properties to specific layers within the structure. Birds eggs, on the other hand, are predominantly calcite, as are the spicules of calcarea sea sponges and the spines and teeth of sea urchins. Corals tend to be aragonite, and arthropod exoskeletons are of the amorphous form.

Calcite is the rhombohedral polymorph of calcium carbonate, and is most stable, while aragonite is orthorhombic. Vaterite (hexagonal) is less commonly seen in biomaterials, as it is relatively unstable. In herring, for example, normal mineralisation of the fish otolith is aragonite, but deformed structures with vaterite were shown to be deficient in sodium, strontium and potassium, and can be related to impaired functioning of the inner ear of the fish (Tomas and Geffen 2003). Similarly, low quality pearls with poor lustre were shown to contain vaterite, whereas high quality pearls are almost exclusively aragonite in structure (Ma and Lee 2006). Aragonite is also present in the mother of pearl or nacre found within many mollusc shells (Jacob et al. 2008, Chen et al. 2012). The controlling effect of collagen and the ion concentration of magnesium and calcium has been demonstrated to influence the deposition of calcium carbonate as aragonite or as vaterite in vitro (Jiao et al. 2006) and this may shed light on the control of deposition exhibited in molluscs and sea urchins.

In addition to the crystalline order of this inorganic material, the calcite and aragonite both have tendencies to form tiles, lamellae or foliated structures, specific to their crystallinity, and their deposition mechanism within the organism. Various efforts to classify these structures have been undertaken, (Currey and Taylor 1974, Kobayashi and Samata 2006). Each structure offers advantages to the strength, toughness or other properties of the mollusc shell, as will be considered further in Chapter 9.

Hydroxyapatite

Hydroxyapatite ($Ca_{10}(PO_4)_6(OH)_2$) is a calcium mineral which is commonly seen in bone and teeth. The formula above is often used to indicate that one hexagonal crystalline unit cell comprises two entities of $Ca_5(PO_4)_3(OH)$. In fact, in both teeth and bone the hydroxyapatite is impure. In bone there is 4%–6% replacement of carbonate groups for phosphate groups, making it more a carbonate apatite, and in dental enamel, a carbonated form which is deficient in calcium is seen.

The carbonate substitution in hydroxyapatite of bone occurs more near the edges of the bone, and tends to reduce the crystallinity in these regions (Currey 2002). The shape of the hydroxyapatite crystals in bone is thought to be plate-like, although discussion continues. The work of Landis et al. (1996) shows platelet-shaped domains of 4–6 nm thick, with widths of 30–45 nm, and lengths of 100 nm. It appears that later these platelets fuse sideways and lengthways to form sword-like sheets, which retain the thickness of approximately 5 nm. It is proposed that mineralisation occurs within the collagen hole and overlap zones within the fibrils, as previously described for Type I collagen. Modelling studies have shown that the charged side

Natural Materials – Composition and Combinations 61

chains of the collagen polypeptide are orientated into these hole regions, and may preferentially nucleate the hydroxyapatite (Xu et al. 2015).

Hydroxyapatite is also seen in the dentine and enamel of teeth, and in osteoderms. Fish scales typically contain calcium deficient hydroxyapatite alongside the collagen component.

Magnetite

Magnetite (iron (II,III) oxide, Fe_3O_4) is found in various organisms, ranging from magnetotactic bacteria and chitons to honeybees and homing pigeons (Mizota and Maeda 1986, Liu et al. 2011, Prozorov 2015). While this is a relatively minor component of the organism, the magnetoreception which is conferred to the organism, allowing navigation or prompting behaviour, has triggered great interest within biology (Kirschvink and Gould 1981, Desoil et al. 2005). From a materials point of view, the potential of biocompatible magnetic nanoparticles for use in medicine has also prompted the study of the magnetite structure and the deposition processes within these organisms (Liu et al. 2011). With a density of 5.1 g/cm^3, magnetite is the most dense biogenic material known; it also has a high electrical conductivity (Kirschvink and Gould 1981).

One source of magnetite is the radulae of chiton (a marine mollusc). The radula is used by the organism to scrape algae from rocks, so the hardness and abrasion resistance which this material confers to the cusp of the radular teeth is a useful attribute. Within the teeth, the magnetite is deposited alongside other minerals, lepidocrocite (γ-FeOOH), maghemite (γ-Fe_2O_3) and hydroxyapatite, and the level of mineralisation reflects the maturity of the teeth (Lowenstam 1967, Mizota and Maeda 1986). The biosynthesis of magnetite in chiton radular teeth was described by Kirschvink and Gould (1981) beginning with the formation of an iron storage protein, ferritin, containing a 60Å subunit of the mineral ferrihydrite ($5Fe_2O_3 \cdot 9H_2O$). This organic material builds up into a honeycomb-like structure of compartments, approximately the size of the later formed mineral grains (0.1 µm diameter). The crystalline ferrihydrite is later converted to magnetite, during which one third of the Fe^{3+} ions are reduced to Fe^{2+}, with loss of water and repacking to form a distorted cubic structure. High resolution transmission electron microscopy studies have revealed that the magnetite crystals within the cusp of the tooth has a rectangular shape, 100–300 nm wide by 500–1,000 nm in length, which are aligned parallel with the surface of the cusp (Liu et al. 2011).

Within magnetotactic bacteria, nanometer-sized particles of magnetite (or of iron sulphide, greigite, Fe_3S_4) can be found in short chains, with each particle encapsulated in a phospholipid membrane. The membrane also contains fatty acids and proteins. The magnetosome chains effectively form small compass needles within the bacterium cell, and allows the organism to passively align itself and locate the oxic–anoxic boundary within its liquid environment (Prozorov 2015). There is interest in the magnetosome

62   *Designing with Natural Materials*

deposition, and the uniformity and relatively high level of crystalline perfection of the magnetite crystals, which is superior to magnetic nanoparticles produced by synthetic means.

Dolomite

Dolomite $(CaMg(CO_3)_2)$ is found in sea urchin spicules and teeth. The magnesium substitution of calcite increases the hardness of this otherwise relatively soft mineral. The sea urchin tooth structure contains several different elements, with a hard and tough fibre composite material (the stone) being encased in a set of plate-like calcite elements. The stone comprises needles of high Mg-calcite (approximately 1 μm cross section), and is supported by a lamellar needle complex which grades into the keel of the structure. Magnesium distribution is non-uniform throughout the tooth, and is even heterogeneous within a single component. Single crystals can be observed to have both low and high levels of Mg substitution.

The composition and hardness of different regions of the sea urchin tooth was studied using nanoindentation, and compared with geologically occurring calcite and dolomite. The stone can have up to 5.2% weight magnesium, and average hardness of 5.7 GPa, and highest observed value of 8.8 GPa. By comparison, the primary plates contain between 1% and 2% magnesium and were found to have a hardness of 4.6 GPa (sagittal cut), 5.0 GPa (coronal cut). The hardness of dolomite was recorded as 7.3 GPa, and geologic calcite was 3.0 GPa (Goetz et al. 2014). Their study also attempted to indicate the change in hardness, moving from the older to younger material within the tooth; clear differences were seen between the regions of the tooth structure and a general decrease in hardness from the tip to the plumula.

Sea urchin spicules are also formed from this magnesium substituted calcite, with higher levels of substitution (up to 12%). There is also a very low content of included organic material (approximately 0.1% weight). These organic macromolecules inform the shape of the spicule, providing a spongy structure in which the crystallites are perfectly aligned (Magdans and Gies 2004). It has been shown that the degree of magnesium substitution is related to the water temperature, with sea urchins grown in higher temperatures having a greater proportion of magnesium (Magdans and Gies 2004, Chen et al. 2012). In addition, composition studies by Magdans and Gies indicated that the accumulation of magnesium ions in the spine base provides strength to the spine, as the smaller ion size of Mg^{2+} than Ca^{2+} alters the crack propagation behaviour of the material.

Silica

Hydrated silica $(SiO_2 \cdot nH_2O)$ is found in diatom skeletons and sea sponge spicules. In the diatoms, the silica structures (frustules) may take amazing intricate patterns (Round et al. 1990), with a porous valve face-controlling uptake of particles from the environment (Hale and Mitchell 2001). The structures and forms have provided stimulus for design and architecture (Kooistra and Pohl 2015). Sea sponge skeletons further demonstrate the potential of silica to construct intricate and mechanically efficient structures (Figure 3.15). These intricate basket-like structures are built up using fine filaments which have concentric layers of silica with a small proportion of organic material providing interlayers (Woesz et al. 2006). The hexactinellida and demospongiae classes of sea sponges construct their skeletons from silica, while the calcarea, another class, use calcium carbonate. Spongin, a collagenous protein, may also be used to form the skeleton of the demospongiae, either with the silica deposits or without – in this case, the spongin provides greater flexibility and the sponge has a different form.

The skeleton provided by the spicules of the hexactinellid sponges is unique in having a three-axis symmetry, and its remarkable intricate structure led to many studies dating back to the 18th and 19th centuries. Effectively, a basket-like structure is formed of longitudinally and circumferentially aligned spicules, with additional diagonally aligned spicules providing a secondary structure (Weaver et al. 2007). Modern researchers have been interested due to the mechanical properties and high optical transmission of the silicate material (Sarikaya et al. 2001).

Weaver et al. (2007) combined observations of *Euplectella aspergillum* from scanning electron microscope and atomic force microscopy to show that the spicule consists of a central proteinaceous filament onto which the silica is deposited. This central element has a square cross-section, surrounded by a central axial cylinder of silica, around which concentric rings or lamellae of consolidated silica nanoparticles are deposited outwards. In the demospongiae, the central proteinaceous filament is hexagonal or triangular in cross section (Garrone 1969, Simpson et al. 1985). The filament has been demonstrated to catalyse the hydrolysis and polycondensation process of silicon alkoxides (in vitro); so, the filament is thought to provide a template

FIGURE 3.15
Sea sponge skeleton from a spicule of *Euplectella aspergillum*.

Source: **Woesz et al. 2006.**

for silica deposition (Shimizu et al. 1998, Cha et al. 1999). It has also been shown that while the demospongiae are limited in spicule length to use the length of original filament, the hexactinellida are able to extend the central filament and therefore the spicule (Weaver et al. 2007). This may contribute to the remarkable lengths of structure seen in these organisms.

The layers of silica in the spicule are typically 0.1–2.0 µm thick, and separated by thin layers (5–10 nm) of organic material (Woesz et al. 2006). The laminated architecture of the spicule cortex contributes to its strength and toughness, with cracks being arrested and deflected along the organic layers. In addition, the outermost layers of the spicule cortex are thinner than the central ones, meaning that the region which is under greatest strain when bending load is applied have the greatest number of crack arresting interfaces (Weaver et al. 2007).

The silicate structures of sea sponge spicules, and the silicate cages excreted by diatoms are less stiff than calcite, allowing greater deformation per unit load. Young's Modulus of silica in the sea sponge spicule is approximately 40 GPa (Woesz et al. 2006, Chen et al. 2012), while for the frustule of a diatom it is 22.4 GPa (Hamm et al. 2003). For calcite, Young's Modulus is nearer 76 GPa (Chen et al. 2012). Each diatom species has a biosilica wall with splits or pores which can be 10–1,000 nm in size. A great number of diatoms have been identified, with a wide range of structures and researchers continue to identify new species.

Silica is also produced by many grasses and other plants as phytoliths. These are thought to lend structural support, and to make the plants more difficult to digest, giving the plant tissue a prickly or grainy texture. Simply comparing the texture on a blade of grass, to run your finger one way along the leaf gives a rougher texture than the other – due to the small scale-like silicate structures on the surface.

Water, and soil water at pH values below 9, contains monosilicic acid ($Si(OH)_4$), and this monomer solute is widely distributed over the Earth's surface (Lewin and Reimann 1969, Iler 1979). When taken up by plants, the monosilicic acid increases in concentration, and a polymerised silica gel of $(SiO_2)_n$ is formed. This hydrated silicate is an amorphous polymer. Within the plant, this may solidify either inside or between cells, forming the solid silica bodies known as phytoliths (Epstein 1994, Snyder et al. 2007). A large range of morphotypes (shapes) can be seen, relating to the plant tissue type, and cells in which the phytolith was formed (Ball et al. 1999, Mulholland and Rapp 1992). These are distinct for plant species, and as they are returned to the soil after the plant dies and decomposes, can be used to identify ancient vegetation cover in paleoarchaeology (Strömberg 2004, Mercader et al. 2009).

Calcium Oxalate

Some plants form calcium oxalate (CaC_2O_4) structures rather than silicate-based phytoliths. Cacti, for example, use calcium oxalate as a reserve of carbon dioxide, for use during the day when they close their stomata to prevent

Natural Materials – Composition and Combinations 65

excess loss of water. In baobab trees, the calcium oxalate has been suggested to provide structural support to the tree. It also occurs in various other species of plant across many genera, and it believed to relate to detoxification of calcium ions within the plant metabolism.

In animals, calcium oxalate crystal formation is associated with kidney stones, in which the monohydrate of calcium oxalate is the most common hydrate type. Calcium oxalate adopts a large variety of crystalline forms and hydration types, and studies comparing the effects of concentration, buffering, pH and temperature alter calcium oxalate crystal growth and nucleation (Thongboonkerd et al. 2006).

Combining Ingredients – Recognising Ultrastructure

Let us now look at how these ingredients combine within natural materials. The strength and stiffness of each component contributes to the properties seen in the bio-based material, whether this is wood, bone or mollusc shell. Some excellent work in the field by Ashby and co-workers has demonstrated the vast range of properties which can be attained using the bio-based ingredients introduced above. A series of charts such as Figure 3.16 have been developed which shows the similarities and differences in strength or modulus or density for the many materials formed within nature using these individual building blocks.

For the elastic modulus-density chart (Figure 3.16), it is clear that broad groupings of natural ceramics or natural cellular materials tend to form clusters with similar levels of stiffness and density, but also that within each group, some materials have significantly greater stiffness or greater density, than others. Hydroxyapatite has a greater stiffness than dentine, although dentine contains approximately 45% hydroxyapatite. Within the cellular materials, balsa wood tested in parallel with its grain ('Balsa (//)') has a greater stiffness than balsa tested perpendicular to its grain ('Balsa (⊥)', i.e. balsa wood is anisotropic.

When strength data is plotted against density, different regions of overlap are seen between these same materials. Cellulose is still a high performance material; however, silk, which has a relatively low modulus (Figure 3.17), has a tensile strength which can rival cellulose. Some members of the natural elastomers group now extend into high strength values (skin, cartilage, elastin) despite relatively low stiffness, while others (artery, muscle) remain relatively weak. Such comparison charts are very useful in understanding the relative performance of natural materials, and their suitability for their role within the living organism. Many more features can be noted, as capably reviewed by Wegst and Ashby (2004).

Figures 3.16 and 3.17 remind us that the data in the tables presented in Sections 2.1 and 2.2 are only the beginning of the story (Tables 3.4 and 3.6). While you could be forgiven for assuming that the engineer is now equipped to harness all of the benefits of biomaterials, using simple engineering and

66   *Designing with Natural Materials*

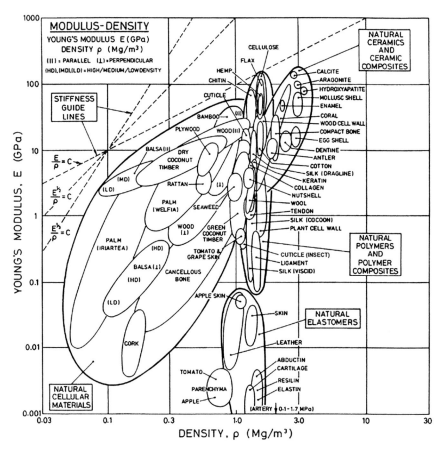

FIGURE 3.16
A material property chart for natural materials, plotting Young's modulus against density.

Source: **Wegst and Ashby 2004.**

design principles, this would be to overlook one significant factor which has been hinted at, but not specifically discussed in the paragraphs introducing the individual materials. It is not sufficient simply to identify and copy the composition of the material. The mechanical properties of biological materials are frequently anisotropic, with strength optimised in different directions within the material, thus saving weight or conserving the resources and energy of the organism in its growth. To achieve this, the microstructure, or ultrastructure (smaller than that which is visible using a microscope), of these natural materials governs properties in each plane of loading, and ensures that the material is efficient and suitable for its purpose within the plant or organism. Nature can teach us much about composite materials.

If we look at the properties of some biological materials at the macro scale (Table 3.7), we see that the properties vary enormously. In some cases, they

Natural Materials – Composition and Combinations 67

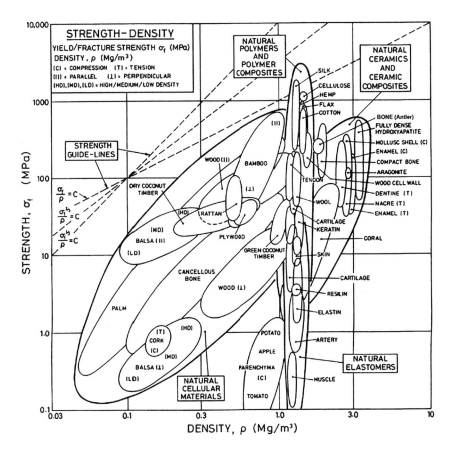

FIGURE 3.17
A material property chart for natural materials, plotting strength against density.

Source: Wegst and Ashby 2004.

obey the rule of mixtures, by which the properties of a composite with fibrous reinforcement can be estimated based on volume fraction within the long axis of the sample (this will be discussed in Section 3.3). Well-investigated examples are nacre and wood fibres, both of which will be covered shortly. Even these examples rapidly show the influence of other factors, such as the scale effect – a single softwood fibre has a different tensile strength to a matchstick-sized piece of the same wood. Similarly, the mechanical properties of nacre indicated in the table span a large range (30–120 GPa) due to the measurements being conducted at different length scales – from macroscopic, through microscopic, and down to techniques such as nanoindentation. These differences are demonstrating a change in properties on moving to a different hierarchical level in the microstructure.

In addition, on moving from one region of a material to another the properties may change by an order of magnitude, as the composition and the

68 *Designing with Natural Materials*

TABLE 3.6

Typical Stiffness and Strength of Biominerals and Selected Man-Made or Mineral Reference Materials

	Young's Modulus (MPa)	Strain at Failure (MPa)	Hardness (MPa)	Sources
Biominerals				
Aragonite ($CaCO_3$)	70,000	170		Jackson et al. (1988)
Hydroxyapatite (from enamel)	65,000			Marshall et al. (2001)
Dentin (~45% hydroxyapatite)	20,000			Marshall et al. (2001)
Hydroxyapatite (bone)	6,700–31,600	63–180	178–785	Currey and Brear (1990)
Radular teeth (Fe_3O_4) from chiton	90,000–125,000		9,000–12,000	Weaver et al. (2010)
Dolomite ($CaMg(CO_3)_2$) from sea urchin teeth	64,000–137,000		4,600–8,800	Goetz et al. (2014)
Silica ($SiO_2 \cdot nH_2O$) in diatom frustules	600–2,800		33–116	Subhash et al. (2005)
Minerals				
Hydroxyapatite ($Ca_{10}(PO_4)_6(OH)_2$) (compacted)	115,000	96–124 (dry) 66–115 (wet)		De With et al. (1981)
Magnetite (Fe_3O_4)	175,000		10,500	Weaver et al. (2010)
Silicon carbide (SiC)	410,000–480,000	3,500	20,000–30,000	Chen et al. (2012), Weaver et al. (2010)
Zirconia (ZrO_2)	140,000–240,000	250	10,000–15,000	Chen et al. (2012), Weaver et al. (2010)
Dolomite ($CaMg(CO_3)_2$)	111,000		7,300	Goetz et al. (2014)
Calcite ($CaCO_3$)	70,000		3,000	Goetz et al. (2014)
Magnesite ($MgCO_3$)	145,000			Goetz et al. (2014)

Note: Hardness values are from different methods and may not be fully comparable between sources.

size and shape of any reinforcing elements changes. Values in Table 3.7 show examples for different regions within cortical bone and different orientations of test in trabecular bone. It is equally true for regions within antler, regions within the tree trunk or within the plant stem. Differences between flax fibres with different levels of separation into fibre ultimates, or different preparation techniques also show a large variability. We will revisit the concepts of hierarchy and anisotropy in Chapter 9.

One additional point based on Table 3.7 is the importance of testing conditions. Many biological materials show different properties in the wet state than the dry state – both wood and bone are examples. In many cases, bio-based materials require tests to be carried out at specified conditions,

Natural Materials – Composition and Combinations

TABLE 3.7

Typical Stiffness and Strength of Bio-Based Composite Materials and Selected Synthetic or Inorganic Materials for Reference

	Young's Modulus (MPa)	Failure Stress (MPa)	Strain at Failure	Sources
Mineral Biocomposites				
Trabecular bone	800–14,000	1–100		Currey (2002)
Cortical bone	6,000–20,000	30–150	0.03	Currey (2002)
Trabecular bone, transverse	13,400 dry			Rho et al. (1997)
Trabecular bone, longitudinal	22,700			Rho et al. (1997), Roy et al. (1996)
Cortical bone, osteon lamellae	22,500 dry 18,000 wet			Bechtle et al. (2010), Rho et al. (1997)
Cortical bone, interstitial lamella	25,800 dry 18,00 wet			Bechtle et al. (2010), Rho et al. (1997)
Antler bone	5,500–7,600			Bechtle et al. (2010)
Nacre (abalone)	40,000–70,000	160		Currey and Taylor (1974)
Nacre	30,000–120,000	170–230		Bechtle et al. (2010), Meyers et al. (2008)
Nacre tablet (nano indentation)	79,000	2,500		Barthelat et al. (2006)
Nacre interface (nano indentation)	2,840			Barthelat et al. (2006)
Organic Biocomposites				
Maple	16,500 dry 14,600 12% mc			Schneider and Phillips (1991)
Softwood	7,000–17,000			Hoffstetter et al. (2005)
Southern pine sulfate pulp fibre	5,000–10,000	350–700		Meyers et al. (2008)
Tracheids wet		82		Klauditz (1957)
Tracheids dry		120		Klauditz (1957)
Flax fibre	26,000–107,000	550–1,500		Ivens et al. (1997), Van den Oever et al. (2000)
Flax fibre	50,000–70,000	750–850		Van den Oever et al. (2000)
Flax ultimates	60,000–80,000	1,500		Van den Oever et al. (2000)
Steam exploded flax	43,500	270	5.5	Hoffstetter et al. (2005)
Jute	20,000–55,000	200–773		Gassan and Bledzki (1999), Ivens et al. (1997)

(Continued)

70 *Designing with Natural Materials*

TABLE 3.7 (*Continued*)

Typical Stiffness and Strength of Bio-Based Composite Materials and Selected Synthetic or Inorganic Materials for Reference

	Young's Modulus (MPa)	Failure Stress (MPa)	Strain at Failure	Sources
Hemp, field retted	19,100	270	0.8	Eichhorn et al. (2001)
Man-made				
Kevlar	130,000–200,000	3,000–4,000	0.03	Hoffstetter et al. (2005), Wegst and Ashby (2004)
E-glass fibre	76,000	2,000		Hoffstetter et al. (2005)
Carbon fibre	23,000	3,400		Hoffstetter et al. (2005)
Rayon	9,400–32,200	220–710	3.9–17.2	Eichhorn et al. (2001)
Rubber	0.01–0.1	15	8.5	Chen et al. (2012)

such as after conditioning at a standard temperature and relative humidity (wood, biopolymers), or after oven drying at a prescribed temperature. Note that even apparently 'dry' materials such as bone show differences, so interaction with water is not prevented by biomineralisation.

Various of the trends in data for natural fibres will be considered further in the next section. In addition, the benefit of combining strong or stiff natural fibres with a flexible compliant matrix can clearly be recognised, and NFCs will be introduced in the next section.

Other aspects such as strength and toughness in biomineral composites like nacre and bone will be considered in Chapter 9. When you look at the two constituent components of nacre, the stiffness value of the hydroxy-apatite tablet is almost thirty times larger than the stiffness of the polymer interlayer. It is no surprise that results between different samples, species and laboratories show so much variability. Bone shows differences between bone type, sample location within the bone, and test orientation (anisotropy).

Biocomposites: Combining Bioresins, Biofibres and Green Chemistry

Composites

Composite materials can achieve many things by combining the different mechanical properties and attributes of their component materials. A fibre composite can be formed to achieve strength in a single direction (by aligning fibres unidirectionally, Figure 3.18a); or to achieve strength in two directions of a plane (by adopting a criss-cross lay up, or a woven fibre mat) resulting in an orthotropic material. The technique can be extended to complex shapes with fibre reinforcement, for example, forming the conical profile of a wind turbine casing. Techniques such as filament winding allow continuous fibres or fibre rovings to be wound around a form or rotating mandrel. Continuous fibres or yarns are required for this, and the technique was developed for synthetic fibres; however, progress has been made to allow long natural fibre rovings of sufficient integrity to be used in the process. The engineer selects the angle of orientation for each layer of fibre in order to provide the required strength longitudinally and in the hoop direction.

Other composites may use short fibres or particles of various aspect ratios to alter the mechanical properties of the matrix (Figure 3.18c). The ratio of length to diameter (if the particle is cylindrical) has a great influence on the level of strength imparted to the composite, due to load transfer mechanisms between the particle and the matrix. Similarly, if plate-like reinforcing particles are used, the geometry will provide a stiffening effect which is greatest in the plane of the platelets, and to a lesser degree perpendicular to their alignment. These two examples both assume relatively uniform alignment of the reinforcing elements in one dimension.

These high performance materials are easily formed using glass, carbon or Kevlar fibres, and can also be produced using natural fibres, as will be discussed in Section 3.3. The short fibre or particulate reinforcement approach has also been undertaken, with the development of wood plastic composites (WPCs). Here, the observation that particle size and geometry has a strong influence on resulting composite properties is worth consideration, and WPC products have been developed with varying levels of control. Some manufacturers or researchers have used closely controlled wood flour mesh sizes, while others have used relatively heterogeneous feedstocks with minimal sieving or process control. As a result, a wide range of properties may be seen, both in the scientific literature and in the marketplace. This subject will not be considered further in this text; however, good

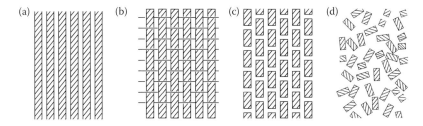

FIGURE 3.18
Schematic of the reinforcing element in various types of composite. (a) unidirectional continuous fibre; (b) typical woven roving with unidirectional fibre in the 0° direction, and minor fibres in the weft to stabilise the fabric; (c) short parallel aligned discontinuous fibres and (d) randomly aligned fibres or particles.

introductory texts and reviews can be found which address manufacture and processing (Klyosov 2007), the effect of fibre and matrix choice (Bledzki and Gassan 1999), the potential for improvement of compatibility between fibre and matrix (Spear et al. 2015) and the mechanical properties and long-term performance of WPC materials (Klosov 2007, Sobczak et al. 2012).

Man-Made Bio-Derived Organics

Before considering the NFCs and other biocomposites which have been developed, it is necessary also to consider a growing family of bio-based polymers and resins which can be used as matrix materials, or in some cases to form bio-derived fibres. Some of these bio-based materials have been in use for many years (e.g. cellulose acetate and rayon), whereas others have emerged as a result of research into sustainable and renewable materials as alternatives to crude oil based polymers.

Cellulose

The production of celluloses by dissolution of plant fibre, and spinning fibres or forming films from the cellulose solution has been well explored, and the use of cellulose acetate as a packaging material (cellophane) pre-dates modern interest in renewable materials by a significant margin. There is some modern ongoing interest in cellulose acetate for biopolymer production, such as Eastman's (Kingsport, USA) Tenite™ product. A recent example is the production of electric guitar bodies using Tenite in conjunction with regenerated cellulose fibre (Thrive™ by Weyerhaeuser (Seattle, USA)).

The initial work with nitrocellulose and cellulose acetate for fibres was largely replaced by cuprammonium rayon and by viscose in 1899 and 1905, respectively. While rayon is frequently used as a name for all varieties of regenerated cellulose, viscose refers specifically to the fibres developed by Charles Cross and Edward Bevan, who used a three-step process to react cellulose with strong base, then convert to xanthate using carbon disulphide and finally convert back to cellulose during spinning. Their work was the platform on which Courtaulds Fibres was established in 1905. The viscose product underwent further development in the 1950s, when high wet modulus rayon was developed by S. Tachikawa (Japan). The higher crystallinity of this fibre leads to lower deformation when wet and the product is sometimes termed 'Modal'.

Later, in 1972, a new solvent N-methylmorpholine N-oxide, was used to develop lyocell, which uses a dry jet – wet spinning process (Woodings 1995). The amine oxide solvent was preferred for its environmentally benign nature, and the lyocell fibres were shown to be stronger than any other regenerated cellulose fibres, especially when wet. In addition, many other benefits in ease of processing, blending, dyeing and finishing were observed, as well as stability in washing and comfort when worn. Originally known as 'Newcell' but the developers at Enka Fibers (USA), it was further developed as 'Tencel'

Natural Materials – Composition and Combinations

by Courtaulds (UK) and finally transferred to Lenzing AG (Austria) who maintained the TENCEL® brand name.

The regenerated cellulose fibres have a different degree of crystallinity to natural fibres, which is determined by the processing parameters used in manufacture. In addition, the cellulose tends to adopt a cellulose II crystalline structure, which is monoclinic, in regenerated fibres. The chains in cellulose II may be anti-parallel aligned, resulting from chain folding during precipitation from solution. The stiffness of all regenerated celluloses tends to be lower than that of the naturally occurring cellulose I found in bast and leaf fibres. A theoretical modulus for the crystalline regions of the fibre (cellulose II) is 98 GPa (Eichhorn et al. 2005), although literature values from 77 to 163 GPa have also been reported.

Mechanical properties of regenerated celluloses are affected by degree of polymerisation, crystal structure and degree of crystallinity, the orientation of chains in both the crystalline and non-crystalline regions relative to the fibre axis, and void content (Bledzki and Gassan 1999). Several of these parameters are influenced by the fibre-drawing ratio during production. Different drawing ratios in lyocell fibres were compared by Kong and Eichhorn (2005) and revealed that high draw ratio fibres were of lower diameter, and that the skin of the fibres had higher orientation than the core (Kong et al. 2007). A similar change in orientation level between skin and core is also known in viscose fibres, and limits the mechanical properties of the fibre. The modulus of elasticity of regenerated cellulose is generally lower than that of natural fibres with their inherent cellulose I conformation; however, it is still significant (9–32 GPa, Table 3.8). The fibre modulus is affected by the crystal orientation of the fibres, with high draw ratios (e.g. 8.9) leading to high crystallite alignment with the fibre axis and Young's modulus of approximately 26 GPa, compared to 11 GPa for a lower draw ratio of 0.7 (Kong et al. 2007).

TABLE 3.8

Mechanical Properties of Cellulose Fibres, Tested in Tension at 23°C and 50% r.h

	Fibre Diameter (μm)	E (GPa)	Stress at Break (MPa)	Strain at Break (%)
Cordenka EHM	9.4	32.2	710	3.9
Cordenka 1,840	12.9	16.9	660	12.7
Enka viscose	18.0	9.4	220	17.2
Cordenka 700	13.1	20.0	660	10.7
Lyocell	12.3	15.2	540	7.0
Steam exploded flax	17.1	43.5	270	5.5
Field retted hemp	31.2	19.1	270	0.8

Source: Eichhorn et al. (2001).

Proteins

Proteins are abundant macromolecules, synthesised by a wide range of organisms for their stiffness and strength, or their elasticity, which are governed by composition and molecular organisation. Many proteins have been introduced and discussed above, for their contribution to high strength materials and highly elastic materials in natural organisms. There is great potential for protein-based polymers and adhesives, and this is increasingly being recognised in development of bio-based materials.

Adhesives based on proteins such as casein (milk protein) and animal proteins have been known since the Egyptian era, with some examples of laminated materials surviving until this day. Adhesives based on soy, wheat, zein (corn) and whey are all likely to be considered for modern products as the search for bio-derived materials continues. For example, soy protein adhesives are already available to the wood-based panels industry, for particleboard and medium density fibreboard (MDF) production.

Another possibility is to use the protein macromolecules as the basis for biopolymers in applications where thermoplastics would routinely be used. As the strength and stiffness of proteins relates to their molecular and supermolecular structure, and the formation of structures such as triple helices, or beta sheets, it is essential to break down this order if thermoplastic materials are to be produced. Some proteins based on agricultural residues or by-products from the food processing industry have been found suitable for this. Keather keratin, egg albumin, lactalbumin, wheat gluten and corn zein are all candidates, with different levels of higher order structure or intermolecular bonding (Athamneh et al. 2008).

These proteins can often be plasticised to act as a formable polymer. Glycerol is one candidate plasticiser due to its bio-derived nature (as a by-product of biodiesel manufacture). Athamneh et al. (2008) defined the critical plasticisation point for each protein source they compared, using the reduction in modulus of elasticity once critical plasticisation was exceeded. Similar analysis using amide data from Fourier-transform infrared spectroscopy allowed trends to be observed correlating the ease of plasticisation with the degree of cross-linking due to cysteine residues. For example, feather keratin and lactalbumin with high cysteine required greater proportions of glycerol for plasticisation. The plasticisation of proteins not only allows their use as thermoplastics, but also offers scope for production of composites using protein as a matrix material in composites. Other proteins including bloodmeal and chicken feathers have been investigated, for example, a tensile strength of 69 MPa and modulus of 7.7 GPa was seen (Reddy and Yang 2011).

Chitosan

Chitin can be easily extracted from waste seafood shells, and converted to chitosan. Chitosan is prepared by deacetylation of chitin, usually using sodium hydroxide or enzyme reactions, and often has 60%–100%

Natural Materials – Composition and Combinations 75

deacetylation of the acetylamine units, compared to chitin. As a result, it has amine groups on the alpha carbons instead of acetamide. The chitosan retains a similar macromolecular skeletal orientation to cellulose I. Unlike crystalline cellulose, chitosan is soluble in acidic aqueous solutions, which allows handling for spray applications or film casting. The tensile modulus of chitosan was found to be 65 GPa in a study by Nishino et al. (1999), whereas the modulus of α-chitin was 41 GPa. Again, these moduli are lower than the equivalent value for cellulose I, which is 138 GPa. Differences in structure, namely the contracted state which the chitin skeleton adopts, explain the greater extensibility of this molecule compared to its close relative cellulose (Nishino et al. 1999).

One of the areas of interest for chitosan is as a scaffold or antibacterial matrix for cell culture (Maeda et al. 2008, Gallezot 2012). It may be suitable for biomedical materials, due to its biocompatibility. Chitosan also has antimicrobial properties. Chitosan has good film forming properties, and there is interest in using this material in packaging due to its anti-microbial properties (Elsabee and Abdou 2013). The presence of amino and hydroxyl units on the polymer makes it suitable for modification reactions.

Bio-Based Polyesters

The upsurge of interest in renewable materials and green chemistry has led to many biopolymers being investigated in the move to diversify from petrochemical polymer technologies. Several dominant products are now widely known, including polylactic acid (PLA), the polyhydroxy alkanoates (PHA) which include polyhydroxybutyrate (PHB), polyhydroxy butyrate-valerate (PHBV). There are many options for synthesis of polyesters, by reaction of diacids such as succinic acid and adipic acid, with diols such as isosorbide, 2,3-butanediol, or 1,3-propanediol. Molecules with three or more hydroxyl groups or acid groups give the option of introducing chain branching, e.g. glycerol, sorbitol or tartaric acid. Many of these monomers are available from fermentation reactions of biomass, and several breakthroughs have been made to allow platform chemicals to be formed by fermentation, this includes as bio-succinic acid and 1,3-propanediol which are useful in biopolyester synthesis among other reaction platforms (Gallezot 2012).

Polylactic Acid (PLA)

PLA is a thermoplastic polyester, which can be formed from lactic acid by polycondensation, using ring-opening polymerisation reactions via the lactide (Södergård and Stolt 2002). The lactic acid can be produced by fermenting corn, and the biopolymer can be biodegraded, depending on composting conditions. PLA has been the fastest growing new biopolymer in the market (Spear et al. 2015). The lactide polymerisation route is increasingly well understood, and control of the lactide stereoforms has allowed properties to be tailored for higher performance. Poly-L-lactic acid (PLLA), based on the L,L-lactide, has potential for high performance applications, and its toughness can be controlled by precise control of D,L-lactide content.

The tensile strength of semicrystalline PLA (50–70 MPa) and flexural strength (100 MPa) compares well with polypropylene (Södergård and Stolt 2002). By controlling polymer isomer composition, the strength of PLLA can be increased to 15.5–150 MPa. Melt spun fibres of PLLA show even higher tensile strengths, 0.39–1.8 GPa, and 6.5–9.3 GPa tensile modulus (Eling et al. 1982). Grijpma and Pennings (1994) showed that introduction of a small quantity of D-lactide increased the impact strength of PLA and altered crystallinity. Toughness can be increased by co-polymerisation with other polyesters. Mechanical properties of all polymers relate to some extent to other factors such as their molecular weight and polydispersity, so selecting the grade which will give the desired performance is important. As with most of the biopolyesters, the hygroscopic nature of PLA means that care must be taken to minimise moisture uptake prior to or during manufacture, to avoid chain hydrolysis under the high processing temperatures.

PHB and PHBV

The polyhydroxy alkanoates (PHAs) are also polyester-based polymers. Two of the most studied are polyhydroxy butyrate (PHB) and Polyhydroxy butyrate-valerate (PHBV). The tensile strength of PHB can be in the region of 40 MPa, making it suitable for many simple applications; however, the polymer tends to readily form large crystals, leading to brittleness, which can be overcome by use of the PHBV copolymer, and by introducing nucleating agents to reduced crystal size.

Polybutylene Succinate (PBS)

Polybutylene succinate is also a polyester, formed by reaction of succinic acid with butane-1,3-diol. The biodegradability of the polymer was its initial interest; however, recent commercialisation activity to develop bio-based succinic acid as a feedstock has allowed this polymer to be formed from bio-based materials rather than traditional sources using fermentation technologies.

Thermoplastic Starch (TPS)

Another biopolymer which has successfully established a market during the 1990s and early 21st century is thermoplastic starch. As the name suggests, it incorporates a plasticiser (often glycerol) to increase formability. Starch from natural sources – maize (Zein), potatoes, wheat and many others – is blended with plasticiser in a high shear mixer or extruder to break down the crystalline domains, allowing the starch chains to interact in a plastic state.

While the current product is found typically in disposable applications due to its high biodegradability, there is scope for future development of this material.

Polyamides

Various polyamides (nylons) based on castor oil and other sources have been in existence for decades; however, the growing market interest in bio-derived polymers has opened new opportunities for this polymer. PA-10 and PA-11 can be formed using castor oil, while a PA-6,10 with 60% renewable content is also in existence. Rennovia launched a PA-6,6 product using bio-based adipic acid. This sector is undergoing rapid growth and change to meet market desire for high performance biopolymer grades.

Polyamides typically have a processing window which is higher than the polyolefins (PE, PP etc.) and the PLA or PHB/PHBV biopolymers. The tensile strength is higher, allowing components and composites to have higher heat deflection temperatures, to be used in more demanding environments. The bio-derived polyamides have lower melting points than the majority of synthetic polyamides, increasing interest in their use in NFCs as matrices, due to working temperatures which are within the acceptable range for natural fibre reinforcements.

Thermoplastic Polyurethanes

Bio-based polyols from various sources are relatively easy to produce for laboratory work, and increasingly produced commercially. These offer a suitable starting material for reaction with isocyanates to form polyurethane polymers. A review of novel polyols derived from vegetable oils by Lligadas et al. (2010) presents many of the strategies which seek to ensure sufficient sites for cross-linking are formed. Development of bio-derived di- or poly-isocyanates has also been undertaken. Today, polyurethanes with partial or near full bio-derived status can be achieved using various combinations. One example presented by Calvo-Correas and co-workers (2015) used a macrodiol derived from castor oil and a diisocyanate from the amino acid lysine. In addition, a chain extended based on corn sugar was used to alter the polymer behaviour. Various of the combinations investigated gave a thermoplastic which showed shape memory behaviour.

Polyurethane based on bio-succinic acid (Biosuccinium, from Reverdia) is proposed to be used within the Desmopan® brand thermoplastic polyurethane range by Covestro (formerly Bayer MaterialScience, Leverkusen, Germany). Biosuccinium has been in production since 2012, using yeast fermentation at low pH. The commercialisation of polyurethane based on this platform chemical was proposed in 2015. Up to 65% bio-based component is possible in the Desmopan polyurethane, for use in trainers and other foams (Black 2015).

A range of bio-derived monomers can be used depending on the desired chemistry and mechanical properties of the resulting polyurethane. Bio-based polyurethane foams have been developed and used within the automotive industry for several years now. One product used by Ford is based on 20% lignin polyols.

Bio-Epoxy

While synthetic epoxy resins dominate the market, in particular, those based on epichlorohydrin and bisphenol A (BPA) known as diglycidyl ether of bisphenol-A (DGEBA), there is steady interest in developing epoxy-based resins using bio-based monomers. One motivation is the need to reduce the emissions of BPA to the environment due to health concerns. The other main driver is a desire to utilise renewable feedstocks and green chemistry. A range of epoxy resins have therefore been developed based on various bio-based sources.

Epoxy resins are characterised by the oxirane ring structures (also termed epoxide or ethoxyline units). These can be generated on various bio-based oligomers where there is a high hydroxyl content, such as glycerol, sugars and phenolics such as tannins. The initial density of functionalisable groups has an effect on the degree of cross-linking which will be attained in the final resin. One of the dominant technologies is the epoxidation of vegetable oils, which are triglycerides of fatty acids. The fatty acid components are typically 16–18 carbons long, with relatively few locations for epoxidation, resulting in relatively low cross-link density. The mechanical properties may therefore be lower than traditional synthetic epoxy resins based on bisphenol A and epichlorohydrin (Niedermann et al. 2015). As a result, some systems have used hybrids of epoxidised vegetable oil with the synthetic DGEBA.

Other approaches have included investigation of epoxy resins derived from polysaccharides where there is greater potential for functionalisation (Sachinvala et al. 1998), or other monomers such as cardanol and rosin. The sucrose-based system developed by Satchinvala et al. was cross-linked using diethylene triamine. It was demonstrated that the sugar-based epoxies could be synthesized with a wide range of glass transition temperatures, ranging from 16°C up to 134°C, offering suitability for different mechanical applications. A different sugar-epoxy system was used by Niedermann et al. (2015), and compared with synthetic DGEBA systems. When used to form composites with jute fibre, the sugar-based epoxy showed potential in a wide range of aircraft interior applications. Another approach has been to functionalise on end of fatty acids from vegetable oil with an epoxide, and the unsaturations in the centre of the chain with anhydride, to allow self-cross-linking. An example was proposed by Huang et al. (2015) based on tung oil.

Various bio-epoxy resins are commercially available, e.g. Super Sap® by Entropy Resins Inc (Bay City, USA). (20%–50% bio), Epicerol® by Solvay

(Brussel, Belgium), SR GreenPoxy 56 by Sicomin (over 50% bio-based carbon) (Chateauneuf les Martigues, France) , and products by Eco Green Resins LLC, DSM and Alpas srl. The Super Sap product has been built on experimentation with sugar-based (sorbitol) and vegetable oil esters based technologies, as well as other bio-based chemicals. The stated aim is to create a product which is a drop in replacement for traditional resins, and the resin has been demonstrated in a range of sporting goods including snowboards and surfboards (Black 2015). Surf Clear EVO is a clear epoxy specifically for the surfboard market from Sicomin. Tailoring the resin formulation to match the intended manufacturing processes is important, now that the potential of bio-based epoxies has been well demonstrated.

Cardolite, a cashew nut shell liquid (CNSL)-based epoxy has been used in brake linings, brake pads and clutch plates for many years. Elmira offer an epoxidised CNSL product named Coral incorporating glass fibre or carbon fibre into prepregs for composites. Fully bio-based versions using flax and cellulose fibres have also been developed (Black 2015).

Unsaturated Polyester Resins (UPE)

Unsaturated polyester resins for thermosetting composites have also been developed with up to 100% bio-content using vegetable oils and some of the newly emerging bio-based platform chemicals such as 1,3-propanediol. As with many of the thermosetting bio-based polymers, the bio-content depends on the choice of monomers combined to form the oligomer, and the selected hardener system. Several products based on unsaturated polyester are available for biocomposite production, including Envirez® 1807 from Ashland (18% bio), Palapreg Eco P by DSM (55% bio), Envirolite by Reichhold (20% and 25% bio) and EkoTek by AOC resins (which has 42% renewable and/or recycled content) (Black 2015). Ongoing developments in green chemistry and fermentation technologies for generating industrial quantities of platform chemicals are likely to increase the options available in polyesters for composites manufacture.

Polyesters based on maleic anhydride reacted epoxidised vegetable oils have also been developed. This chemistry is slightly different to the products listed above, but provide properties comparable to the unsaturated polyesters. A range of products has been launched by Dixie Chemical Co. Inc., based on linseed oil (maleiniated acrylated epoxied linseed oil [MAELO]) and soybean oil (maleiniated acrylated epoxied soybean oil [MAESO]). In systems which contain a reactive diluent (such as methacrylated fatty acids), the bio-content can be 85%, whereas for more traditional diluents, the bio-content is nearer 65%. The MAELO and MAESO products have been developed into sheet moulding compound and bulk moulding compound by Premix, now part of A. Schulman (Black 2015).

Acrylate Resins

Functionalisation of vegetable oils can also use acrylate chemistry to form resins (Mosiewicki and Arangueren 2013). Various researchers have looked at grafting acrylate functional groups onto oils after epoxidation, most commonly this is done to soybean oil, forming acrylated epoxidised soybean oil (AESO). In particular, Wool and co-workers have demonstrated many biocomposites based on this platform (Khot et al. 2001, O'Donnell et al. 2004). AESO can also be used as a starting material for coatings, adhesives, lubricants and plasticisers, as well as thermosetting resins.

The development of a bioresin tractor panel for John Deere (Figure 3.22) gained and Innovation in Real Materials award in 1998 (Bassett et al. 2016). The AESO resin technology used by an early commercial form of Envirez has continued to be developed. Another AESO product named Tribest resin was available from Cognis, Germany, until the company was bought by BASF. Tribest is based on acrylate functionalised soy oil. It was used with hemp reinforcement by students from the Royal Institute of Technology (KTH), Sweden, to manufacture a concept car for the Shell eco-marathon in 2009. Tribest S350–01 EXP from Cognis performed well with flax reinforcement in laboratory tests (Spārniņš et al. 2012).

Table 3.9 shows the maximum bio-based content for a range of commercially available products, including thermoplastic biopolymers and all resin functionalities. The sector is rapidly developing, and in some cases, bio-based content is still increasing as additional monomers become commercially available.

Natural Fibre Composites

One area of biomaterials which has been well advanced is the development of biocomposites or NFCs. Here the length and strength of natural fibres such as flax and hemp can be harnessed in composites in a manner similar to the GRP or carbon fibre composites, which have become widely recognised in moulded panels, sports cars and boat hulls. Many other forms of biocomposite exist – in fact, any composite with a bio-based reinforcement or bio-based matrix could be considered to be such, but the NFCs have been the most iconic. Initial developments centred on natural fibres in conventional unsaturated polyester or epoxy matrices, but relatively quickly bio-derived or partially bio-based equivalent resins were developed, thus increasing the bio-based content of the material as a whole.

The high strength of the main natural fibres has already been shown in Table 3.7 (Section 2.3), and the reasons for their superior properties will be further explored in Chapter 9. Young's modulus of hemp (30–90 GPa) and flax (27–80 GPa) fibre compares relatively well with E-glass fibre (73 GPa); however, in tensile strength, hemp (580–1,100 MPa) and flax (343–1,035 MPa) are lower performance than E-glass (2,000–3,500 MPa). However, when

Natural Materials – Composition and Combinations

TABLE 3.9

Commercially Available Bio-Based Polymers, Excluding Bio-Polyethylene and Related Technologies

Polymer	Bio-Based Polymer Fraction	Bio-Based Content (%)	Status
Cellulosics	Cellulose	Up to 100	Commercial
Thermoplastic starch (TPS) and starch derivatives	Starch	Up to 100	Commercial
Polyhydroxy alkanoates (PHA)	All	100	Commercial
Polylactic acid (PLA)	L-lactic acid	Up to 100	Commercial
Polytrimethylene terephthalate (PTT)	1,3-Propanediol	30	Commercial
Polyethylene terephthalate (bioPET)	Ethylene glycol	30	Commercial
Polyethylene terephthalate (bioPET)	Ethylene glycol Para xylene	100	Expected
Polyethylene furanoate (PEF)	Ethylene glycol, Furan dicarboxylic acid	100	Expected 2020
Polyacrylic acid (PAA)	3-hydroxypropionic acid	100	Expected
Polyethylene (bioPE)	Ethanol	100	Commercial
Polybutylene succinate (PBS)	Succinic acid	Up to 100	Commercial
Polyamide 11 (PA11)	Ricinoleic acid	100	Commercial
Polyamide 6,10	Ricinoleic acid	66	Commercial
Polyethers	1,3-propanediol, fatty acids	Up to 100	Commercial
Polyisoprene	Isoprene	100	Expected
Aliphatic polyesters	1,3-propanediol, fatty acids	Up to 100	Commercial
Polyurethane (PUR)	Fatty acids	Up to 70	Commercial
Epoxy resins	Triglycerides, Sugars	Up to 100	Commercial
Furan resins	Xylose	Up to 100	Commercial
Acrylate resins	Soya oil	Up to approximately 25	Commercial

Source: Derived from Zini and Scandola (2011) with revisions.

comparing the specific strength (strength per weight), the natural fibres offer competitive properties, typically 20–50 GPa·cm^3/g, whereas the specific modulus for glass fibre is approximately 29 GPa·cm^3/g, and for aramid fibre 45–48 GPa·cm^3/g (Biagiotti et al. 2004).

The majority of NFCs are formed with cellulosic natural fibres. The strength values used above are generalised, and represent a very wide range of properties. The strength of natural fibres in practice depends on many factors, beginning with the agronomy and the plant variety, but being influenced by all the processing steps employed in harvest and conversion. Separation of the fibre from the stem (there is only 15%–25% bast fibre) involves the removal of a large volume of pith and non-load bearing tissue. This is typically done by retting to break down pectin-based bonding between plant cells. Traditionally, retting was conducted by leaving the stems of flax in the field, or in ponds, to allow microbial enzymes to take effect (Biagiotti et al. 2004). There has been a shift away from ponding, due to the quantity of water consumed and generation of fermentation waste, although the fibre from this method was considered of high quality. Modern processing may use enzymes which have been selected for their efficiency when used in industrial vats under tighter control of temperature and pH, and higher throughput. The strength of dew-retted fibre may vary to a greater extent than the enzyme-retted fibre, and requires specialist knowledge by the farmer in monitoring the retting progress in the field to gather the fibre crop at the correct time. There has been steady progress in plant selection, agronomy and processing for fibre composites, which have led to high quality fibre production targeting NFC manufacture.

Products such as linoleum used natural fibres in a polymer matrix (based on linseed oil) long before the widespread development of high tech composites based on glass or carbon fibre. Similarly, early printed circuit boards used either glass fibre or flax fibre as the reinforcing element to provide stability in the panel. Despite these examples, it has taken a relatively long time to develop and perfect composites which utilise natural fibres to their full advantage.

Some examples of NFC materials are shown in Figure 3.19. As you can see, the majority of applications are in marine or automotive applications, where manufacturers are already familiar with glass fibre and carbon fibre composite materials. In this chapter, most discussion is directed towards natural fibre in thermosetting matrix composites, due to their good utilisation of the fibre length and strength, compared with other technologies such as WPCs where wood flour or relatively short fibre are used. However, advances in handling long fibres for use in thermoplastic composites must also be noted, such as the direct long fibre thermoplastic process (D-LFT) developed for the automotive industry in Germany by the Fraunhofer Institute for Chemical Technology. This allows direct feed of long fibres into either extrusion or injection moulding systems, overcoming many issues in fibre length reduction and orientation which exist if pre-chopped fibre is blended through the traditional methods. The result is a high level of control to engineer the mechanical properties of the component, and the ability to combine flax fibre with other continuous fibres such as glass rovings.

There has been perennial interest in developing NFCs for canoes and boat hulls; however, moisture uptake by the natural fibre is a concern in this application. Successful examples such as the Flax Cat (by Technical University of Delft and NPSP) showed the possibility but did not lead to products in the market.

Natural Materials – Composition and Combinations 83

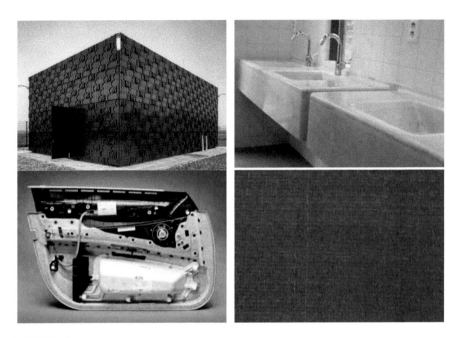

FIGURE 3.19
(a) Moulded NFC panel façade formed using NaBasCo 5010, combining hemp and bioresin. (b) Wash basin of NFC moulded with hemp flax and sisal in polyester resin, NaBasCo by NPSP (bio-based.org). (c) Car panel moulded in NFC. (d) Composite panel formed using flax 2 × 2 twill with a PLA matrix.

Source: **Biotex, Composites Evolution.**

Fibre Types

Various plant fibres have suitable length and strength to be considered for composite manufacture. In addition to hemp and flax previously discussed, other bast fibres, such as kenaf and ramie are grown for NFC manufacture. All bast fibres are harvested from the bast, which is the outer portion of the stem, just beneath the epidermis (Figure 3.20). The bast fibres are longer than the other plant cells and have very thick cell walls. There are also various leaf fibres which possess similar strength values, and have sufficient length to allow handling for composite manufacture. These include sisal, pineapple leaf fibre and curauá. All these plants have long leaves of similar morphology; in sisal, the leaf may be over a metre in length. Leaf fibres are also retted to separate them from the other plant tissue (in which the cell size and shape is not suitable). In sisal, the mechanical fibres are generally located on the outer surfaces of the leaf, while additional fibres may be obtained from a line inside the leaf, known as ribbon fibres; these are associated with the vascular bundles (Li et al. 2000).

The tensile modulus of the long plant fibres used in engineering differs from the reported value for cellulose in the same way that was discussed

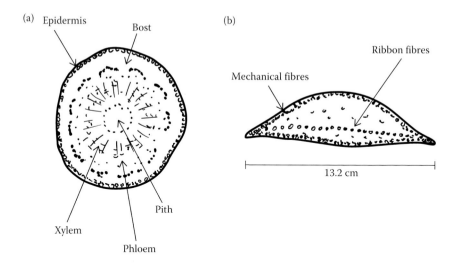

FIGURE 3.20
(a) Schematic showing the location of bast fibre within a transverse section of the plant stem.
(b) Location of fibres within a cross section of a sisal leaf.

for wood. In flax fibre, or jute fibre, the majority of the bast fibre wall is the S2 layer, with a very high proportion of cellulose (approximately 70%) and the microfibrils are aligned at approximately 10° to the fibre axis (Mukherjee and Satyanarayana 1986, Bos et al. 2002, Charlet et al. 2010). The cellulose within the S2 layer is bound by approximately 15% hemicellulose and 2% pectin, with approximately 2% lignin and some minor constituents, e.g. wax and water soluble matter (Baley 2002, Batra 1998).

An empirical study by Mukherjee and Satyanarayana (1986) reviewed the spiral angle of many bast, leaf and fruit fibres, and reported this alongside aspect ratio, cross sectional area and ultimate tensile strength for technical fibres. The data revealed a clear trend between microfibril angle and elongation, with high angles correlating with high strain at break. While degree of fibre separation into the textile fibres varies with processing technology, the cross sectional area data and cell length data shown in Table 3.10 can give a useful guide to some of the less well-known fibre types. This clearly illustrates the potential of flax, hemp and ramie fibre on the basis of their aspect ratio. The trend for the low cellulose content fibres (coir, talipot and palmyrah) to have low tensile strengths is also apparent.

Continuing research has increased the pool of data on fibre performance since this early study. Some companies have developed enzyme retting to a high level, while others continue to use field-retted fibres. Some compiled data on strength properties comparing dew-retted and enzyme-retted materials is shown in Table 3.11; in addition, two compatibiliser treatments, silane and acrylic acid are shown to further increase the tensile strength of the single fibres. Many researchers have reported a beneficial effect of alkali

Natural Materials – Composition and Combinations

TABLE 3.10

Structural and Strength Properties Data Compiled by Mukherjee and Satyanarayana (1986)

Fibre	Cellulose Fraction (%)	Microfibril Angle (°)	Cross Section Area (mm²)	Cell Length (mm)	Aspect Ratio	UTS (MN/m)	Strain at Break (%)
Sisal	67	20.0	1.10	2.20	100	580	4.3
Pineapple	82	15.0	0.26	4.50	450	650	2.4
Banana	65	12.0	0.39	3.30	150	540	3.0
Coir	43	45.0	1.20	0.75	35	140	15.0
Ramie	83	7.5	0.03	154.00	3,500	870	1.2
Hemp	78	6.2	0.06	23.00	906	690	1.6
Jute	61	8.0	0.12	2.30	110	550	1.5
Flax	71	10.0	0.12	20.00	1,687	780	2.4
Sunn hemp	69	9.5	0.71	8.00	284	440	5.5
Sansevieria	75	22.0	0.48	2.00	166	584	4.0
Maur hemp	64	22.5	0.66	1.90	142	500	5.2
Pita floja	72	20.2	0.44	1.75	145	600	3.6
M. eneste	73	12.0	0.29	4.30	138	550	3.8
Talipot	50	24.5	0.40	1.15	47	210	3.1
Palmyrah	40	42.5	1.60	1.30	43	190	11.0

TABLE 3.11

Simplified Data for Single Fibre Tests on Flax

	Treatment	Load Rate (mm/min)	Average Strength at 5 mm Gauge Length (MPa)	Sources
Water retted, scutched	none	0.1	645	Van de Weyenberg et al. (2000)
Dew retted, scotched	none	0.1	562	Van de Weyenberg et al. (2000)
Dew retted, hackled	none	0.1	732	Van de Weyenberg et al. (2000)
Green, bundle	none	1	678	Zafeiropoulos et al. (2000)
Dew retted, bundle	none	1	906	Zafeiropoulos et al. (2000)
Enzyme retted	None	0.5	1,149	Joffe et al. (2003)
Enzyme retted	Acrylic acid	0.5	1,439	Joffe et al. (2003)
Enzyme retted	Acrylic acid	0.5	1,369	Joffe et al. (2003)
Enzyme retted	Silane	0.5	1,380	Joffe et al. (2003)
Enzyme retted	Silane	0.5	1,000	Joffe et al. (2003)

All data are for single fibre unless stated as a bundle.

treatment, or mercerisation, on the strength of NFCs; for example, Van de Weyenberg et al. (2006) showed an increase in tensile strength and modulus longitudinally in unidirectional flax epoxy composites with sodium hydroxide treatment, and an increase of over 100% in the same properties for transversely loaded samples.

In addition, the development of yarns and products specifically for NFC production has continued, with efforts to govern level of twist to avoid problems relating to coiling or uncoiling of the yarn with associated extension in length when under tensile load in the composite. Values for textile and composites grade fibres are reported in Table 3.11.

The distinctive feature of bast and leaf fibres is their well-developed secondary cell wall, in which the cellulose is well aligned with the fibre axis giving excellent strength and stiffness. The technical fibre used in composite manufacture typically comprises many individual plant cells (or 'ultimates') which have diameters of 10–25 μm, compared to the 50–100 μm of the technical fibre. Lengths of fibre ultimates in flax are between 2 and 5 cm, while the technical fibre is typically tens of centimetres in length (Bos and Donald 1999). The degree of separation and fibre diameter is controlled by the processing – decortication, scutching, hackling, carding etc – which follow the retting. Preparation of leaf fibres follows a similar process, with retted fibre being decorticated or scutched, washed, degummed and dried. Further information about the industrial production of fibres for NFCs can be found in reviews by Van de Weyenberg et al. 2003, Biagiotti et al. 2004, Yan et al. 2014a. Reviews by Summerscales et al. (2010a,b) cover aspects of fibre production, composite manufacture and properties prediction.

The bast fibre production processes generate two main products: fibre sliver (long fibres) and tow (short fibres and fragments of fibre bundles). The pith is also removed early in the process, and this 'shiv' (or for hemp, 'hurd') may be used in chip-based composites or for animal bedding, but has very short fibre and low bulk density governing this use in alternative products. Some flax processing techniques seek only to create tow, for example, for paper manufacture, however scutching followed by hackling is required to generate the long technical fibres required in long NFC.

The incorporation of flax or hemp fibre in thermosetting matrices has been demonstrated in a wide range of products. Many are flat panels or moulded curved panels produced by hand lay-up, compression moulding or vacuum infusion. Resin transfer moulding may also be used. To study unidirectional composites, simple panels of aligned fibres or yarns have increasingly often been produced by filament winding. Many examples are discussed by Ho et al. (2011) and Yan et al. (2014a).

In hand lay-up, a mat of natural fibre can be laid into a prepared mould in which the gel coat and initial matrix coat have already been applied, stippled to embed the fibre mat into the matrix and coated with additional resin prior to application of additional coats of fibre and matrix according to lay-up design. The vacuum bagging system, and resin transfer moulding systems,

Natural Materials – Composition and Combinations 87

rely on drawing or pushing resin through a stack of fibre mats placed within a mould or bag, with benefits in the process simplicity, speed and reproducibility. Flat sheets and simple designs can be achieved given sufficient consideration of design to ensure flow of resin throughout the form. The advantage of resin transfer moulding and other closed mould techniques is the better containment of solvent and reagent vapours during production.

As with traditional composites manufacture, ensuring good wet-out of the fibre by the matrix is essential, and in the NFCs, this is all the more relevant due to the cellular nature of the fibre, in which air may remain in the cell lumen, or trapped between partially separated fibres. Permeability of the fabric and viscosity of the resin (and pot life prior to increase in viscosity) are therefore important to minimise defects in the formed product. Good results have been demonstrated nonetheless, and some typical properties for thermoset flax composites produced using a range of manufacturing techniques are shown in Table 3.12. Considerably, more data is reviewed by Shah (2013). Several introductory texts relating to their use in NFCs or biocomposites provide greater depth than is possible here (Bledzki and Gassan 1999, Yan et al. 2014a, Spear et al. 2015).

Many different mats of fibre can be considered, with early studies using either textiles (e.g. burlap/hessian) or mats specifically prepared for composites manufacture. In early work, non-woven mats formed by a needle punching process were studied, to achieve a quasi-random fibre alignment. Air-laid mats have been developed in which the needle punching process can be avoided (it was shown to induce weak points in the composite, where the fibre alignment perpendicular to the plane by the needle action allowed strain to develop). Air-laid mats often have a bidirectional alignment, both diagonal to the machine direction, with the angle being governed by feed rate. In some air-laid mats, a small quantity of binder may be added to avoid introducing the needle-punch marks.

In addition, weaves of natural fibre have been developed specifically for composites manufacture. Examples from one company include unidirectional rovings, hopsack and twill (Figure 3.21). Mats and yarns have also been formed in which synthetic fibre is intermixed with the natural fibre (Baghaei et al. 2013). This offers potential for thermoplastic matrix composites, assisting the dispersion of fibre and matrix, which is dependent on thermal transfer and flow of the polymer once molten. To best achieve the level of reinforcement predicted for natural fibre reinforcements, the long fibres should be aligned with minimal twist or deviation from the length direction of the weave. These natural fibre mats and weaves have shown good results in laboratory tests, prototypes and commercial production (Phillips and Lessard 2011, Di Landro and Janszen 2014). These differ from textile cloths in their avoidance of excessive twist in the yarns (which can induce defects in the composite during loading). Lay-up of natural fibre mats can still be optimised, and unidirectional or bidirectional fabrics are selected as appropriate. Considerations such as the level of compaction (depending on number of

TABLE 3.12

Mechanical Properties for Flax Epoxy Composites with Selected Weave Architectures and Preparation Methods

		Tensile Strength (MPa)	Flexural Strength (MPa)	Tensile Modulus (GPa)	Flexural Modulus (GPa)	Sources
Flax (unidirectional 40% V_f) epoxy longitudinal	Autoclave cure	133	218	28	17.7	Van de Weyenberg et al. (2003)
Flax (unidirectional 40% V_f) epoxy transverse	Autoclave cure	4.5	8	2.7	0.36	Van de Weyenberg et al. (2003)
Arctic flax (unidirectional 42% V_f) epoxy	RTM	280		35		Oksman (2001)
Arctic flax (unidirectional 47% V_f) epoxy	RTM	279		39		Oksman (2001)
Flax (unidirectional 32% V_f) epoxy	RTM	132		15		Oksman (2001)
Flax (UD 40% V_f) epoxy	RTM	258.8		20.2		Duc et al. (2014a)
Flax (twill 40% V_f) epoxy	RTM	85.1		7.9		Duc et al. (2014a)
Flax (1/1 weave 34% V_f) epoxy	Hand lay-up	64.28		5.95		Liu and Hughes (2008)
Flax (1/6 weave 34% V_f) epoxy	Hand lay-up	79.73		6.85		Liu and Hughes (2008)
Flax (various knit 18% V_f) epoxy	Hand lay-up		43–57		1.3–2.1	Muralhidar et al. (2012)
Flax (various weave 27% V_f)	Hand lay-up		78–100		2.1–2.9	Muralhidar et al. (2012)

plies and thickness of the moulded piece) and the permeability are important (Francucci et al. 2010, 2011, Di Landro and Janszen 2014). Research has also addressed the yarns for filament winding and other processes where a continuous strand of natural fibres is required.

Further opportunities exist with yarns and unidirectional tapes which can be utilised in filament winding systems (Ma et al. 2016), where greater precision of the fibre alignment on each layer of the composite is possible. The tensile properties of yarns removed from a woven flax mat were tested by Yan et al. (2014b) and are considerably lower than the tensile properties of single fibres, e.g. 136.3–154.8 MPa tensile strength and 15.2–17.3 GPa tensile modulus compared to 1129 ± 390 MPa and 59.1 ± 17.5 GPa as reported for monofilament flax fibre in the literature (Charlet et al. 2009). This lower

Natural Materials – Composition and Combinations

(a) 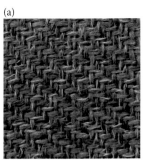 (b) 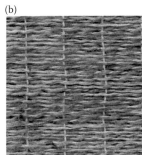 (c)

FIGURE 3.21
Examples of natural fibre weaves for composites manufacture: (a) twill, (b) unidirectional and (c) hopsack, as manufactured by Composites Evolution.

strength in the yarn can be explained by considering the interaction of multiple filaments, all of which are angled with some degree of twist relative to the test axis, allowing a complex set of interactions and localised movements when loaded. This is considered in greater detail elsewhere, but may in fact contribute significantly to beneficial properties such as vibration damping in service (Duc et al. 2014b). Shah (2014) developed materials selection charts for both unidirectional and nonwoven natural fibres in composites.

Pultrusion has also been considered for unidirectional applications, and combined wool-hemp pultruded rods with tensile modulus values of 113–123 GPa have been demonstrated (Peng et al. 2011). The development of various feed mechanisms allowing reels of natural fibre to be handled into pultrusion or extrusion machines despite the non-continuous nature of the natural fibre represents a significant breakthrough for large-scale use of NFCs in future. Both pultrusion and filament winding have potential for working with regenerated cellulose fibres in addition to natural fibres. In fact, the regenerated cellulose may allow greater flexibility in processing conditions due to its availability in controlled fibre diameters and continuous lengths. Excellent toughness has been shown in unnotched Charpy impact test of viscose reinforced epoxy composites produced by pultrusion (Mader et al. 2012). This relates to the remarkable elongation at break values of the regenerated cellulose fibres.

Experimental data frequently reports simple flat section unidirectional composites formed by this process, however, to produce a tubular section offers efficient design and lightweight. One example was developed by Yan and Chouw (2013) using woven flax fibre in epoxy matrix, with the weft direction of the fabric aligned with the longitudinal axis of the tube. The energy absorption and crashworthiness were well demonstrated, and have potential in the automotive industry. Further, it is possible to adjust the angle of the plant fibres so that they provide helically aligned reinforcement, the unidirectional tapes offer great potential in developing a structure which shows bio-inspiration with a bio-derived fibre. The parallel between fibre composites and the microfibrillar structure of the plant cell wall is frequently commented upon. This will be discussed in Chapter 9.

Another feature of natural fibres is their affinity for water. This means that fibre may typically contain moisture when stored in dry conditions. If allowed to stand in a damp environment, the moisture content would be considerably higher and spoilage might result. However, in practical terms, the small amount of moisture (10%–12% in air-dry fibre) may contribute to development of voids within the matrix of the composite if it is driven off during the cure process. Temperatures for curing NFCs are typically up to 80°C depending on resin, with many thermosetting systems relying on processes such as UV cure which are conducted at ambient conditions.

Natural fibres contain voids, relating to the central lumena of the plant cells. It can also be difficult to eliminate small air bubbles forming between fibres during infusion of the matrix. Madsen and co-workers studied this effect, and its origins from the fibre and the processing method. They proposed a revised rule of mixtures for NFCs in which porosity term significantly improved prediction of the composite properties (Madsen and Lilholt 2003). Conventionally, the rule of mixtures allows tensile modulus of a unidirectional composite with continuous fibres to be related to the modulus of the fibre and the matrix by the volume fraction of fibre (V_f) and matrix (V_m), where the sum of V_f and V_m is one (Hull and Clyne 1996). Thus, the longitudinal tensile modulus of the composite (E_{c1}) can be calculated as follows:

$$E_{c1} = V_f \cdot E_f + (1 - V_f) \cdot E_m$$

Transverse tensile modulus can also be calculated:

$$E_{c2} = \frac{E_f \cdot E_m}{(1 - V_f) \cdot E_f + V_f \cdot E_m}$$

However, in the presence of pores, an additional term V_p must be introduced to account for volume fraction of pores (Madsen and Lilholt 2003). The weight fraction of fibre and matrix still obeys the simple relationship, summing to unity. Thus, they worked with the volume of fibre, matrix and pores (v_f, v_m and v_p); the mass of the composite (m); the weight fraction of fibre and of matrix (W_f and W_m), which can all be measured or calculated during manufacture, and the density of fibre and matrix (ρ_f, ρ_m) in order to calculate the volume of pores.

$$v_c = v_f + v_m + v_p$$

$$v_c = \frac{\dfrac{m \cdot W_f}{\rho_f} + \dfrac{m \cdot W_m}{\rho_m}}{1 - V_p}$$

Natural Materials – Composition and Combinations 91

From this true value of composite volume, it is then possible to derive the true volume fraction of the fibre and the matrix. In addition, the volume of pores was assumed to be a function of the fibre weight fraction, and two terms relating to the process-related pore formation and fibre structure related pore formation were used to best model the behaviour of NFCs. Full derivation of these terms is given in the literature, and validation for model data from different NFCs with unsaturated polyester and with polypropylene as matrices (Madsen and Lilholt 2003, Madsen et al. 2007, Madsen et al. 2009).

In addition, the revised rule of mixtures addressed the anisotropy of fibre properties, as it is recognised that fibre tensile strength is typically several times greater than transverse strength; yet, the transverse strength is rarely measured (Madsen and Lilholt 2003). An anisotropy factor (f_a), defined as the ratio of the ratio between transverse and axial properties, was also added. The resulting revised rule of mixtures is as follows:

$$E_{c1} = \left(V_f \cdot E_f + \left(1 - V_f \right) \cdot E_m \right) \cdot \left(1 - V_p \right)^2$$

$$E_{c2} = \frac{\left(E_f \cdot f_a \right) \cdot E_m}{\left(1 - V_f \right) \cdot \left(E_f \cdot f_a \right) + V_f \cdot E_m} \cdot \left(1 - V_p \right)^2$$

The model described above presented a significant step forward in modelling the mechanical properties of NFCs, and has application in other composite materials. Pores can be associated with the fibre lumen, with the fibre–matrix interface (i.e. fibre outer surface), with the matrix, or related to poor impregnation (i.e. restricted flow due to the proximity of two or more fibres). Predictions of each pore type can be made according to Madsen et al. (2007). The observation also informed the ongoing development of yarn types with greater suitability to use in NFCs. Twist degree and linear density are key determining factors in suitability for composite manufacture, and increases in linear density have been observed to introduce difficulties for full impregnation of the resin into the yarn (Ma et al. 2016).

Bioresins as Matrices

The development of biocomposites may rest on the use of biofibres as reinforcing elements, or the use of bioresins as the matrix material, or in an ideal world, the combination of both bio-based elements to form a 100% biocomposite. Many of the polymers found in biological materials may be harnessed as matrix polymers, although in many cases, they may be modified to a greater or lesser extent during extraction, or to aid processing, or to prevent problems in service. For many years, cellulose acetate and other reconstituted celluloses have been available, and there is some interest in using these as matrix polymer for microfibril nanocomposites. Chitin, extracted from source materials such as crab shells or prawn shells, can be deacetylated and used in the form of chitosan. And lignin is altered significantly by the extraction process chosen, but may be re-formed into a thermoplastic state, such as in the Arboform product, or altered to produce thermosetting resins using phenolic resin chemistry. In all three examples, the regenerated product is not the same as the parent polymer, but has gained processability. For example, in regenerated cellulose, the presence of the acetate units reduces the incidence of hydrogen bonding between chains and limits crystalline development, ensuring an amorphous transparent polymer is formed.

Another example is the use of thermoplastic starch, in which the crystalline order of the starch is broken down by addition of plasticiser and the effect of mechanical action. This enhances the thermoplastic character to allow the starch to be formed into useful commodities, from food trays to biodegradable bags. In the case of the thermoformed food trays, containers and moulded components wood flour has been successfully incorporated into the blend to increase rigidity, and form a particulate composite. Typical properties are shown in Table 3.13. Many of the biopolymers, for example, PBS, PLA, PHBV etc. (as introduced in Section 3.2) also have potential as thermoplastic matrices for wood flour composites with a very high bio-based component (Spear et al. 2015).

Thermoplastic biopolymers form ideal matrix systems for many short-term applications, due to their ease of forming, the significant progress in range of additives and processing aids which allow them to be used to replace traditional fossil-based polymers in moulded or extruded materials. Here, the bio-based nature may lead to relatively simple recycling or compositing depending on product type and end of life recovery, or low carbon footprint due to the renewable nature of the initial feedstock (Mohanty et al. 2002, Miller 2013). However, the greater interest in biopolymers from a bio-inspiration point of view may be derived when the matrix is selected for the unique attributes which has led to its development in nature, and these are designed into the composite system. For example, recent work with

TABLE 3.13

Physical Properties of Some Commonly Available Biopolymers or Biodegradable Polymers

	MOE (GPa)	Tensile Strength (MPa)	T_m (°C)	T_g (°C)	Density (g/cm³)
PLA	1.50–3.61	21–60	150–162	45–60	1.21–1.25
PLLA	2.8–4.4	15.5–150	145–186	53–64	
PHB	0.93	21–40	161–182	−10 to 15	1.18–1.26
PHBV	2.38	25.9	153–162	−1 to 21	1.25
PCL	0.34	19.3–21.6	58–65	−60 to −65	1.11–1.15
PBS	0.5	40–60	90–120	−45 to −10	1.2

Source: Data from: Daniels et al. (1990), Van de Velde and Kiekens (2002), Bhattacharya et al. (2005), Zhang and Sun (2005), Lee and Wang (2006), Bledzki and Jaszkiewicz (2010), Yan et al. (2014a).

hemicellulose gels which harnesses the water retaining character to provide shape morphing behaviour, which can be harnessed in actuators, as will be discussed in Chapter 9.

Thermosetting matrices may also have a medium to high bio-based component, offering a reduction in carbon footprint for an engineering composite material compared to traditional epoxy or polyester matrix systems. Examples are sugar based and vegetable oil/fatty acid based epoxies. In developing bio-derived thermoset resins, one aim has been to offer a drop in replacement, as far as possible, so that existing processing and moulding techniques can be easily used with the new resins. Control of viscosity, ease of wetting the fibre mat, compatibility with existing processing aids and release agents is required. Good examples have been demonstrated in large mouldings for agricultural machinery (Figure 3.22), or for boat hulls. These one-off examples may form the output of large research projects, serving as proof that the bioresin, or the biocomposite, can perform at the level intended in service, and allowing properties and longer term effects such as wear, fatigue or the effects of weathering to be studied in addition to demonstrating the potential of the material.

FIGURE 3.22
Composite panel manufactured using Envirez bio-based thermoset derived from soybean oil for John Deere combine harvester.

In practice, since the various demonstrations by John Deere, various smaller companies have taken up the day-to-day use of the resins. Successful products include boat hulls, skis, snowboards and surfboards, as the bio-derived nature of the product gained the interest a group of environmentally motivated consumers, creating a niche market. Meanwhile, research into the use of long fibre biocomposites in high performance applications such as wind turbine blades and aircraft interior panels has continued.

Various bio-derived thermosetting resins have been developed and demonstrated in NFCs, and a selected few are compared in Table 3.14. Several of these fall outside the traditional groupings of epoxies, polyesters or polyurethanes, but utilise a combination of bio-based platform chemicals and reactions to form a useful resin formulation for cross-linking in situ. One example is the lactic acid based thermoset investigated by Åkesson et al. (2011), in which methacrylated star-shaped oligomers were formed based on a central pentaerythritol molecule, which provides four reactive groups to create the star shape.

As can be seen in the table, a range of strength properties have been attained using many available formulations (Table 3.14). This depends partially on the properties of the matrix, but also strongly influenced by the architecture of the fibre reinforcement, as discussed above. As progress in the field of bio-derived resins considers apace, this is a snapshot of the potential, and it is likely that steady refinements of the resins and improvements in handling and preparation methods will lead to continued advances.

One further aspect for consideration in biocomposites production is frequently overlooked, as the bio-derived matrix materials may also have potential as foam forming materials. The sugar-based bioresins, for example, found early use in foams for surfboard blanks. The same foams could equally provide a bio-based core material for sandwich panels and other lightweight composite assemblies. In large boat hulls, for example, the fibre composite product is used in two skins, either side of a lightweight foam core, such sandwich structures are not necessary in shorter boat hulls or canoe mouldings, where single skin fibre composite is sufficiently stiff over the span of the moulding.

Designing with Biocomposite Materials

It is clear that great progress has been made in controlling natural fibre quality and performance, and in the development of bio-derived matrices. The long natural fibres can offers the same advantages in designing high performance composites as synthetic materials, but incorporate a sustainable raw material.

What of this design potential? Having prepared a weave or mat of natural fibre of known properties, the engineer may incorporate it into his design software, select fibre orientation for each layer to balance stresses within the component, control fibre volume fraction and select an appropriate matrix to

Natural Materials – Composition and Combinations

TABLE 3.14

Example Properties for NFC with UPE and with Epoxy, Different Manufacturing Techniques

Fibre	Resin	Technique	Tensile Strength (MPa)	Flexural Strength (MPa)	Tensile Modulus (GPa)	Flexural Modulus (GPa)	Sources
Flax (plain weave)	Bioepoxy	RTM	63		5.87		Di Landro and Janszen (2014)
Flax (plain weave)	AESO thermoset	Compression moulding	280	250	32	25	Adekunle et al. (2011)
Flax (twill)	AESO thermoset	Compression moulding	60–80	100–120	8–11	8–10	Adekunle et al. (2011)
Flax (UD yarn)	Soy protein resin	Winding and compression moulding	246.0	81.7	3.8	4.7	Huang and Netravali (2007)
Flax (weave)	Soy protein resin	Compression moulding	54.6–68.7	22.6–35.8	994–1,123	753–960	Huang and Netravali (2007)
Flax weave 0° MMSO bio-thermoset	MMSO bio-thermoset	Compression moulding	136.9		14.7		Adekunle et al. (2012)
Flax weave 90° MMSO bio-thermoset	MMSO bio-thermoset	Compression moulding	17.5		3.2		Adekunle et al. (2012)
Flax non-woven	MMSO bio-thermoset	Compression moulding	43.6		5.4		Adekunle et al. (2012)
Flax (weave-non-woven hybrid)	MSO bio-thermoset	Compression moulding	50–140	up to 180	6–14	up to 17	Adekunle et al. (2012)
Flax (weave non-woven hybrid)	MMSO bio-thermoset	Compression moulding	50–150	up to 195	7–16	up to 20	Adekunle et al. (2012)
Flax 70% W_f	Thermoset lactic acid resin	Compression moulding	62	96	9	7.5	Åkesson et al. (2011)
Flax short fibre 30% W_f	PLA	Injection moulding	54		6.3		Bax and Müssig (2008)

achieve the product properties required. The availability of this data enables him to use this in a thermosetting matrix in the same manner as any other composite fibre.

The same design constraints apply to this process as would in GRP or carbon fibre, with wettability of the fibre surfaces, and maximum diffusion of matrix into the fibre being important. Various sizing agents may be considered for this; in UPE or epoxy-based systems, silane-based sizes can be chosen, as might be used in GRP. Control of load transfer at the interface of the fibre and matrix has a great influence on the properties of the resulting material.

The NFC may form one layer in a larger composite element. Foam cores or other elements may be introduced to form a sandwich structure, achieving a lightweight structure. Design, not only of the fibre lay up within the sheet of composite but also in the interaction of opposing contoured sections may introduce additional constraints on design, for example, if chosen for an aerofoil or wind turbine blade. Here, jointing may also become an issue, as strain in the joint between opposing faces is significant.

Summary

Nature provides many examples of design optimisation, as well as a wide range of constituent components which are suitable for use in developing modern materials. The main groups of natural material have been introduced, at a fundamental component level, and here many levels of hierarchy have been introduced; for example, the construction of triple helices from polypeptides to combine into fibrils and ultimately create a tissue which we know as collagen.

It has also been shown that natural materials have scope for use within high-tech materials, to achieve a wide range of properties. The best-known examples are the long NFCs. In Chapter 9, some more applications of bio-based materials within components or products demonstrating bio-inspired design will be considered.

References

Aaron B.B. and Gosline J.M. (1981) Elastin as a random-network elastomer: a mechanical and optical analysis of single elastin fibers. *Biopolymers* **20**(6):1247–1260.

Adekunle K., Cho S.-W., Patzelt C., Blomfeldt T. and Skrifvars M. (2011) Impact and flexural properties of flax fabrics and lyocell-fiber-reinforced bio-based thermoset. *Journal of Reinforced Plastics and Composites* **30**(8):685–697.

Adekunle K., Cjp S.-W., Ketzscher R. and Skrifvars M. (2012) Mechanical properties of natural fiber hybrid composites based on renewable thermoset resins derived from soybean oil, for use in technical applications. *Journal of Applied Polymer Science* **124**:4530–4541.

Åkesson D., Skrifvars M., Seppälä J. and Turunen M. (2011) Thermoset lactic acid-based resin as a matrix for flax fibers. *Journal of Applied Polymer Science* **119**:3044–3099.

Alexander R.M. (1966) Rubber-like properties of the inner hinge ligament of Pectinidae. *Journal of Experimental Biology* **44**:119–130.

Anglès M.N. and Dufresne A. (2001) Plasticized starch/tunicin whiskers nanocomposite materials. 2. Mechanical behavior. *Macromolecules* **34**(9):2921–2931.

Athamneh A.I., Griffin M., Whaley M. and Barone J.R. (2008) Conformational changes and molecular mobility in plasticized proteins. *Biomacromolecules* **9**:3181–3187.

Baghaei B., Skrifvars M. and Berglin L. (2013) Manufacture and characterization of thermoplastic composites made from PLA/hemp co-wrapped hybrid yarn prepregs. *Composites Part A* **50**:93–101.

Baley C. (2002) Analysis of the flax fibres tensile behaviour and analysis of the tensile stiffness increase. *Composites Part A: Applied Science and Manufacturing* **33**(7):939–948.

Ball T.B., Gardner J.S. and Anderson N. (1999) Identifying inflorescence phytoliths from selected species of wheat (*Triticum monoccum*, *T. dicoccon*, *T. dicoccoides*, and *T. aestivum*) and barley (*Hordeum vulgare* and *H. spontaneum*). *American Journal of Botany* **86**(11):1615–1623.

Barthelat F., Li C.M., Comi C. and Espinosa H.D. (2006) Mechanical properties of nacre constituents and their impact on mechanical performance. *Journal of Materials Research* **21**:1977–1987.

Bassett A., LaSacala J. and Stanzione J.F. (2016) Richard P. Wool's contributions to sustainable polymers from 2000 to 2015. *Journal of Applied Polymer Science* **133**:4381, 14 pp.

Batra SK. (1998) Other long vegetable fibres. In: *Handbook of Fibre Science and Technology, Fibre Chemistry*, M. Lewin and E.M. Pearce (eds.) Marcel Decker, New York, pp. 505–575.

Bax B. and Müssig J. (2008) Impact and tensile properties of PLA/Cordenka and PLA/flax composites. *Composites Science and Technology* **68**:1601–1607.

Bechtle S., Ang S.F. and Schneider G.A. (2010) On the mechanical properties of hierarchically structured biological materials. *Biomaterials* **31**:6378–6385.

Beck-Candanedo S., Roman M. and Gray D.G. (2005) Effect of reaction conditions on the properties and behavior of wood cellulose nanocrystal suspension. *Biomacromolecules* **6**(2):1048–1054.

Bhattacharya M., Reis R.L., Correlo V. and Boesel L. (2005) Material properties of biodegradable polymers. In: *Biodegradable Polymers for Industrial Applications*, R. Smith (ed.) CRC Woodhead Publishing Ltd, pp. 336–356.

Biagiotti J., Puglia D. and Kenny J.M. (2004) A review on natural fibre-based composites—part I: Structure, processing and properties of vegetable fibres. *Journal of Natural Fibres* **1**(2):37–68.

Black. (2015) http://www.compositesworld.com/articles/green-resins-closer-to-maturity [Accessed 11/4/17].

Bledzki A.K. and Gassan J. (1999) Composites reinforced with cellulose based fibres. *Progress in Polymer Science* **24**(2):221–274.

Bledzki A.K. and Jaszkiewicz A. (2010) Mechanical performance of biocomposites based on PLA and PHBV reinforced with natural fibres—A comparative study to PP. *Composites Science and Technology* **70**(12):1687–1696.

Boerjan W., Ralph J. and Baucher M. (2003) Lignin biosynthesis. *Annual Review of Plant Biology* **54**:519–546.

Bos H.L. and Donald A.M. (1999) In situ ESEM study of the deformation of elementary flax fibres. *Journal of Materials Science* **34**:3029–3034.

Bos H.L., van den Oever M.J.A. and Peters O.C.J.J. (2002) Tensile and compressive properties of flax fibres for natural fibre reinforced composites. *Journal of Materials Science* **37**(8):1683–1692.

Brooks A.E., Steinkraus H.B., Nelson S.R. and Lewis R.V. (2005) An investigation of the divergence of major ampullate silk fibers from *Nephila clavipes* and *Argiope aurantia*. *Biomacromolecules* **6**(6):3095–3099.

Buehler M.J. (2006a) Atomistic and continuum modeling of mechanical properties of collagen: elasticity, fracture and self assembly. *Journal of Materials Research* **21**(8):1947–1961.

Buehler M.J. (2006b) Nature designs tough collagen: explaining the nanostructure of collagen fibrils. *Proceedings of the National Academy of Sciences USA* **103**(33):12285–12290.

Buehler M.J. (2008) Nanomechanics of collagen fibrils under varying cross-link densities: atomistic and continuum studies. *Journal of the Mechanical Behaviour of Biomedical Materials* **1**(1):59–67.

Burrows M., Shaw S.R. and Sutton G.P. (2008) Resilin and chitinous cuticle form a composite structure for energy storage in jumping by froghopper insects. *BMC Biology* **6**:41, 16 pp.

Calvo-Correas T., Santamaria-Echart A., Saralegi A., Martin L., Valea A., Corcuera M.A. and Eceiza A. (2015) Thermally-responsive biopolyurethanes from a biobased diisocyanate. *European Polymer Journal* **70**:173–185.

Canty E.G. and Kadler K.E. (2002) Collagen fibril biosynthesis in tendon: a review and recent insights. *Comparative Biochemistry and Physiology Part A* **133**:979–985.

Cave I.D. (1968) The anisotropic elasticity of the plant cell wall. *Wood Science and Technology* **2**:268–278.

Cave I.D. and Walker J.C.F. (1994) Stiffness of wood in fast-grown plantation softwoods: the influence of microfibril angle. *Forest Products Journal* **44**:43–48.

Cha J.N., Shimizu K., Zhou Y., Christiahnsen S.C., Chmelka B.F., Stucky G.D. and Morse D.E. (1999) Silicatein filaments and subunits from a marine sponge direct the polymerization of silica and silicones in vitro. *Proceedings of the National Acadamy of Sciences of the USA* **96**(2):361–365.

Chapman B.M. (1969) A review of the mechanical properties of keratin fibres. *Journal of the Textile Institute* **60**(5):181–207.

Charlet K., Jernot J.P., Gomina M., Bréard J., Morvan C. and Baley C. (2009) Influence of an Agatha flax fibre location in a stem on its mechanical, chemical and morphological properties. *Composites Science and Technology* **69**(9):1399–1403.

Charlet K., Jernot J.P., Eve S., Gomina M. and Bréard J. (2010) Multi-scale morphological characterisation of flax: from the stem to the fibrils. *Carbohydrate Polymers* **82**(1):54–61.

Chen P.-Y., McKittrick J. and Meyers M.A. (2012) Biological materials: functional adaptations and bioinspired designs. *Progress in Materials Science* **57**:1492–1704.

Chen Y.-C., Chen M., Gaffney E.A. and Brown C.P. (2017) Effect of crosslinking in cartilage-like collagen microstructures. *Journal of the Mechanical Behaviour of Biomedical Materials* **66**:138–143.

Cheng Y.T., Rodak D.E., Wong C.A. and Hayden C.A. (2006) Effects of micro- and nano-structures on the self-cleaning behaviour of lotus leaves. *Nanotechnology* **17**:1359–1362.

Clemons C. (2008) Raw materials for wood-polymer composites. In: *Wood-Polymer Composites*, K. Oksman-Niska and M. Sain (eds.) Woodhead Publishing, Cambridge, pp. 1–22.

Cousins W.J. (1976) Elastic modulus of lignin as related to moisture content. *Wood Science and Technology* **10**:9–17.

Cousins W.J. (1978) Young's modulus of hemicellulose as related to moisture content. *Wood Science and Technology* **12**:161–167.

Currey J.D. (2002) *Bones, Structure and Mechanics.* Princeton University Press, Princeton, NJ, 456 pp.

Currey J.D. and Brear K. (1990) Hardness, Young's modulus and yield stress in mammalian mineralized tissues. *Journal of Materials Science, Materials in Medecine* **1**:14–20.

Currey J.D. and Taylor J.D. (1974) The mechanical behaviour of some molluscan hard tissue. *Journal of Zoology* **173**(3):395–406.

Daniels A.U., Chang M.O., Adriano K.P. and Heller J. (1990) Mechanical properties of biodegradable polymers and composites proposed for internal fixation of bone. *Journal of Applied Biomaterials* **1**:57–78.

Das R., Kumar A., Patel A., Vijay S., Saurabh S. and Kumar N. (2017) Biomechanical characterization of spider webs. *Journal of the Mechanical Behavior of Biomedical Materials* **67**:101–109.

De With G., van Dijk H.J.A., Hattu N. and Prijs K. (1981) Preparation, microstructure and mechanical properties of dense polycrystalline hydroxyl apatite. *Journal of Materials Science* **16**(6):1592–1598.

Denny M. (1976) The physical properties of spider's silk and their role in the design of orb-webs. *Experimental Biology* **65**:483–506.

Denny M.W. and Miller I. (2006) Jet propulsion in the cold: mechanics of swimming in the Antarctic scallop, *Adamussium colbecki. Journal of Experimental Biology* **209**(22):4503–4514.

Desch H.E. and Dinwoodie J.M. (1981) *Timber: Its Structure, Properties and Utilisation.* 6th Edition. Macmillan, 424 pp.

Desoil M., Gillis P., Gossiun Y., Pankhurst Q.A. and Hautot D. (2005) Definitive identification of magnetite nanoparticles in the abdomen of the honeybee *Apis mellifera. Journal of Physics: Conference Series* **17**:45–49.

Di Landro L. and Janszen G. (2014) Composite with hemp reinforcement and bio-based epoxy matrix. *Composites Part B: Engineering* **67**:220–226.

Donaldson L.A. (2001) Lignification and lignin topochemistry—An ultrastructural view. *Phytochemistry* **57**:859–873.

Donaldson L. (2008) Microfibril angle: measurement, variation and relationships. *IAWA Journal* **29**(4):345–386.

Duc F., Bourban P.E., Plummer C.J.G. and Månson J.-A.E. (2014a) Damping of thermoset and thermoplastic flax fibre composites. *Composites Part A* **64**:115–123.

Duc F., Bourban P.E. and Månson J.-A.E. (2014b) The role of twist and crimp on the vibration behaviour of flax fibre composites. *Composites Science and Technology* **102**:94–99.

Dufresne A., Cavaillé J.-Y. and Vignon M.R. (1997) Mechanical behavior of sheets prepared from sugar beet cellulose microfibrils. *Journal of Applied Polymer Science* **64**:1185–1194.

Ebringerova A., Hromadkova Z. and Heinze T. (2005) Hemicellulose. *Advances in Polymer Science* **186**:1–67.

Eglington G. and Hamilton R.J. (1967) Leaf epicuticular waxes. *Science* **156**:1322–1335.

Eichhorn S.J., Sirichaisit J. and Young R.J. (2001) Deformation mechanisms in cellulose fibres, paper and wood. *Journal of Materials Science* **36**(13):3129–3135.

Eichhorn S.J., Young R.J. and Davies G.R. (2005) Modeling crystal and molecular deformation in regenerated cellulose fibers. *Biomacromolecules* **6**(1):507–513.

Eling B., Gogolewski S. and Pennings A.J. (1982) Biodegradable materials of poly(L-lactic acid). 1. Melt-spun and solution-spun fibres. *Polymer* **23**:1579–1593.

Elliott G.F., Huxley A.F. and Weis-Fogh T. (1965) On the structure of resilin. *Journal of Molecular Biology* **13**(3):791–795.

Elliott D.H. (1965) Structure and function of mammalian tendon. *Biological Reviews* **40**:392–421.

Elsabee M.Z. and Abdou E.S. (2013) Chitosan based edible films and coatings: a review. *Materials Science and Engineering C* **33**(4):1819–1841.

Ennos A.R. (2012) *Solid Biomechanics*. Princeton University Press, Princeton, NY, 264 pp.

Epstein E. (1994) The anomaly of silicon in plant biology. *Proceedings of the National Academy of Sciences USA* **91**:11–17.

Erikson M., Larsson S. and Miksche G.E. (1973) Gas-chromatographische Analyse von Ligninoxidationsprodukten. VII. Zur Struktur des Lignins der Fichte. *Acta Chemica Scandinavica* **27**:903–904.

Fang X., Qiu F., Yan B., Wang H., Mort A. and Start R.E. (2001) NMR studies of molecular structure in fruit cuticle polyesters. *Phytochemistry* **57**:1035–1042.

Feng Q.L., Cui F.Z., Pu G., Wang R.Z. and Li H.D. (2000) Crystal orientation, toughening mechanisms and a mimic of nacre. *Materials Science and Engineering C* **11**:19–25.

Francucci G., Rodriguez E.S. and Vázquez A. (2010) Study of saturated and unsaturated permeability in natural fiber fabrics. *Composites Part A* **41**(1):16–21.

Francucci G., Rodriguez E.S. and Vázquez A. (2011) Experimental study of the compaction response of jute fabrics in liquid composite molding processes. *Journal of Composite Materials* **46**(2):155–167.

Fratzl P., Misov K., Zizak I., Rapp G., Amenitsch H. and Bernstorff S. (1997) Fibrillar structure and mechanical properties of collagen. *Journal of Structural Biology* **122**:119–122.

Fratzl P., Elbaum R. and Burgert I. (2008) Cellulose fibrils direct plant organ movements. *Faraday Discussions* **139**:275–282.

Frazer R.D.B., MacRae T.P., Parry D.A.D. and Suzuki E. (1971) The structure of feather keratin. *Polymer* **12**(1):35–56.

Natural Materials – Composition and Combinations

Frey-Wyssling A. (1952) Deformation of plant cell walls. In: *Deformation and Flow in Biological Systems*. A. Frey-Wyssling (ed.) North-Holland Publishing Co., Amsterdam, pp. 194–254.

Frey-Wyssling A. (1955) Structure of cellulose. *Biochimica et Biophysica Acta* **18**:166–168.

Gallezot P. (2012) Conversion of biomass to selected chemical products. *Chemical Society Reviews* **41**:1538–1558.

Garrone R. (1969) Collagène, spongine et squalette mineral chez l'éponge *Haliclona rosea* (O.S.) (Dèmosponge, Haplosléride). *Journal de Microscopie* 8:581–598.

Gassan J. and Bledzki A.K. (1999) Possibilities for improving the mechanical properties of jute/epoxy composites by alkali treatment of fibres. *Composites Science and Technology* **59**(9):1303–1309.

Gaultieri A., Buehler M.J. and Redaelli A. (2009) Deformation rate controls elasticity and unfolding pathway of single tropocollagen molecules. *Journal of the Mechanical Behaviour of Biomedical Materials* **2**(2):130–137.

Gaultieri A., Pate M.I., Vesentini S., Redaelli A. and Buehler M.J. (2012) Hydration and distance dependence of intermolecular shearing between collagen molecules in a model microfibril. *Journal of Biomechanics* **45**(12):2079–2083.

Gentleman E., Lay A.N., Dickerson D.A., Nauman E.A., Livesay G.A. and Dee K.C. (2003) Mechanical characterisation of collagen fibers and scaffolds for tissue engineering. *Biomaterials* **24**(21):3805–3813.

Ghorbani M., Liebner F., van Herwjinen H.W.G., Pfungen L., Krahofer M., Budjav E. and Konnerth J. (2016) Lignin phenol formaldehyde resoles: The impact of lignin type on adhesive properties. *BioResources* **11**(3):6727–6741.

Gibson L. (2012) The hierarchical structure and mechanics of plant materials. *Journal of the Royal Society Interface* **9**(76):2749–2766. doi:10.1098/rsif.2012.0341.

Gladman A.S., Matsumoto E.A., Nuzzo R.G., Mahadevan L. and Lewis J.A. (2016) Biomimetic 4D printing. *Nature Materials* **15**:413–418.

Goetz A.J., Griesshaber E., Abel R., Fehr Th., Ruthensteiner B. and Schmahl W.W. (2014) Tailored order: the mesocrystalline nature of sea urchin teeth. *Acta Biomaterialia* **10**(9):3885–3898.

Goseline J.M., Guerette P.A., Ortlepp C.S. and Savage K.N. (1999) The mechanical design of spider silks: from fibroin sequence to mechanical function. *Journal of Experimental Biology* **202**(23):3295–3303.

Goswami L., Dunlop J.W.C., Jungnikl K., Eder M., Gierlinger N., Coutand C., Jeronimidis G., Fratzl P. and Burgert I. (2008) Stress generation in the tension wood of poplar is based on the lateral swelling power of the G-layer. *The Plant Journal* **56**:531–538.

Greenwold M.J., Bao W., Jarvis E.D., Hu H., Li C., Gilbert M.T.P., Zhang G. and Sawyer R.H. (2014) Dynamic evolution of the alpha (α) and beta (β) keratins has accompanied integument diversification and the adaptation of birds into novel lifestyles. *BMC Evolutionary Biology* **14**:249, 16 pp.

Grijpma D.W. and Pennings A.J. (1994) (Co)polymers of L-lactide. 2. Mechanical properties. *Macromolecular Chemistry and Physics* **195**:1649–1663.

Habibi Y., Lucia L.A. and Rojas O.J. (2010) Cellulose nanocrystals: chemistry, self-assembly, and applications. *Chemistry Reviews* **110**:3479–3500.

Hale M.S. and Mitchell J.G. (2001) Functional morphology of diatom frustule microstructures: hydrodynamic control of Brownian particle diffusion and advection. *Aquatic Microbial Ecology* **24**:287–295.

Hall R.H. (1951) Changes in length of stressed collagen fibers with time. *Journal of the Society of Leather Trades' Chemists* **35**:11–17.

Hamm C.E., Merkel R., Springer O., Jurkojc P., Maier C., Prechtel K. and Smetacek V. (2003) Architecture and material properties of diatom shells provide effective mechanical protection. *Nature* **421**:841–843.

Hatakeyama H. and Hatakeyama T. (2010) Lignin structure, properties and applications. *Advances in Polymer Science* **232**:1–63.

Ho M.-P., Wang H., Lee J.-H., Ho C.-K., Lau K.-T., Leng J.-S. and Hui D. (2011) Critical factors on manufacturing processes of natural fibre composites. *Composites Part B: Engineering* **43**(8):3549–3562.

Hofstetter K. and Gamstedt E.K. (2009) Hierarchical modelling of microstructural effects on mechanical properties of wood. A review. *Holzforschung* **63**:130–138.

Hofstetter K., Hellmich C. and Eberhardsteiner J. (2005) Development and experimental validation of a continuum micromechanics model for the elasticity of wood. *European Journal of Mechanics A/Solids* **24**:1030–1053.

Holloway P.J. and Baker E.A. (1970) The cuticles of some Angiosperm leaves and fruits. *Annals of Applied Biology* **66**:145–154.

Huang X. and Netravali A. (2007) Characterization of flax fiber reinforced soy protein resin based green composites modified with nano-clay particles. *Composites Science and Technology* **67**: 2005–2014.

Huang K., Liu Z., Zhang J., Li S., Li M., Xia J. and Zhou Y. (2015) A self-crosslinking thermosetting monomer with both epoxy and anhydride groups derived from tung oil fatty acids: synthesis and properties. *European Polymer Journal* **70**:45–54.

Hull D. and Clyne T.W. (1996) *An Introduction to Composite Materials*. Cambridge University Press, Cambridge.

Iguchi M., Yamanaka S. and Budhiono A. (2000) Review: Bacterial cellulose—A masterpiece of nature's arts. *Journal of Materials Science* **35**:261–270.

Iler R.K. (1979) *The Chemistry of Silica: Solubility, Polymerisation, Colloid and Surface Properties*. Wiley Interscience, New York.

Imai T. and Sugiyama J. (1998) Nanodomains of Iα and Iβ cellulose in algal microfibrils. *Macromolecules* **31**(18):6275–6279.

Ivens J., Bos H. and Verpoest I. (1997) The applicability of natural fibres as reinforcements for polymer composites. In: *Renewable Bioproducts: Industrial Outlets and Research for the 21st Century*. EC Symposium at the International Agricultural Centre (IAC), Wageningen.

Jackson A.P., Vincent J.F.V. and Turner R.M. (1988) The mechanical design of nacre. *Proceedings of the Royal Society B* **234**(1277):415–440.

Jacob D.E., Soldati A.L., Wirth R., Huth J., Wehrmeister U. and Hofmeister W. (2008) Nanostructure, composition and mechanisms of bivalve shell growth. *Geochimica et Cosmochimica Acta* **72**(22):5401–5415.

Jiao Y., Feng Q. and Li X. (2006) The co-effect of collagen and magnesium ions on calcium carbonate biomineralization. *Materials Science and Engineering: C* **26**(4): 648–652.

Joffe R., Andersons J. and Wallström L. (2003) Strength and adhesion characteristics of elementary flax fibres with different surface treatments. *Composites Part A: Applied Science and Manufacturing* **34**(7):603–612.

Kataoka Y. and Kondo T. (1996) Changing cellulose crystalline structure in forming wood cell walls. *Macromolecules* **29**(19):6356–3658.

Natural Materials – Composition and Combinations

Kelly R.E. and Rice R.V. (1967) Abductin: a rubber-like protein from the internal triangular hinge ligament. *Science* **155**:208–210.

Ker R.F. (1981) Dynamic tensile properties of the plantaris tendon of sheep (*Ovis aries*). *Journal of Experimental Biology* **93**:283–302.

Keten S. and Buehler M.J. (2010) Atomistic model of the spider silk nanostructure. *Applied Physics Letters* **96**(15):153701, 3 pp.

Khayet M. and Fernandez V. (2012) Estimation of the solubility parameters of model plant surfaces and agrochemicals: a valuable tool for understanding plant surface interactions. *Theoretical Biology and Medical Modelling* **9**(45):21.

Khot S.N., Lascala J.J., Can E., Morye S.S., Williams G.I., Palmese G.R., Kusefoglu S.H. and Wool R.P. (2001) Development and application of triglyceride-based polymers and composites. *Journal of Applied Polymer Science* **82**(3):703–723.

Kirschvink J.L. and Gould J.L. (1981) Biogenic magnetite as a basis for magnetic field detection in animals. *BioSystems* **13**(3):181–201.

Klauditz W. (1957) Zur biologisch-mechanischen Wirkung der Cellulose und Hemicellulose im Festigngsgewebe der Laubhölzer (Role of cellulose and hemicellulose in strength of wood) *Holzforschung* **11**(4):110–116.

Klemm D., Kramer F., Moritz S., Lindström T., Ankerfors M., Gray D. and Dorris A. (2011) Nanocelluloses: a new family of nature-based materials. *Angewandte Chemie International Edition* **50**:5438–5466.

Klyosov A.A. (2007) *Wood-Plastic Composites*. John Wiley & Sons, Hoboken, NJ.

Kobayashi I. and Samata T. (2006) Bivalve shell structure and organic matrix. *Materials Science and Engineering: C* **26**(4):692–698.

Kolatukkudy P.E. (1980) Biopolyester membranes of plants: cutin and suberin. *Science* **208** (4447):990–1000.

Kolattukudy P.E. (1984) Biochemistry and function of cutin and suberin. *Canadian Journal of Botany* **62**:2918–2933.

Kong K. and Eichhorn S.J. (2005) Crystalline and amorphous deformation of process-controlled cellulose-II fibres. *Polymer* **46**(17):6380–6390.

Kong K., Davies R.J., McDonald M.A., Young R.J., Wilding M.A., Ibbett R.N. and Eichhorn S.J. (2007) Influence of domain orientation on the mechanical properties of regenerated cellulose fibers. *Biomacromolecules* **8**(2):624–630.

Kooistra W.H.C.F. and Pohl G. (2015) Diatom frustule morphology and its biomimetic applications in architecture and industrial design. In: *Evolution of Lightweight Structures*, C. Hamm (ed.) Biologically Inspired Systems 6, Springer, Dortrecht, pp. 75–102.

Lam T.B.T., Kadoya K. and Iiyama K. (2001) Bonding of hydroxycinnamic acids to lignin: ferulic acid and p-coumaric acids are predominantly linked at the benzyl position of lignin, not the β-position, in grass cell walls. *Phytochemistry* **57**(6):987–992.

Landis W.J., Hodgens K.J., Song M.J., Arena J., Kiyonaga S., Marko M., Owen C. and McEwen B. (1996) Mineralization of collagen my occur on fibril surfaces: evidence from conventional and high-voltage electron microscopy and three-dimensional imaging. *Journal of Structural Biology* **117**(1):24–35.

Lee S.-H. and Wang S. (2006) Biodegradable polymers/bamboo fiber biocomposite with bio-based coupling agent. *Composites Part A: Applied Science and Manufacturing* **37**(1):80–91.

Lewin J. and Reimann B.E.F. (1969) Silicon and plant growth. *Annual Review of Plant Physiology* **20**:289–304.

Li Y., Mai Y.-W. and Ye L. (2000) Sisal fibre and its composites: a review of recent developments. *Composites Science and Technology* **60**:2037–2055.

Lichtenegger H., Reiterer A., Stanzl-Tschegg S.E. and Fratzl P. (1999) Variation of cellulose microfibril angles in softwoods and hardwoods—A possible strategy of mechanical optimization. *Journal of Structural Biology* **128**:257–269.

Liu Q. and Hughes M. (2008) The fracture behaviour and toughness of woven flax fibre reinforced epoxy composites. *Composites Part A* **39**:1644–1652.

Liu C., Han Y., Gao C., Liu C. and Chen X. (2011) Properties of biogenic magnetite nanoparticles in the radula of chiton *Acanthochiton rubrolineatus* Lischke. *Journal of Wuhan University of Technology—Materials Science Edition* **26**(3):478–482.

Lligadas G., Ronda J.C., Galià M. and Cádiz V. (2010) Plant oils as platform chemicals for polyurethane synthesis: Current state-of-the art. *Biomacromolecules* **11**(11):2825–2835.

Lowenstam H.A. (1967) Lepidocrocite, an apatite mineral, and magnetite in teeth of chitons (polyplacophora). *Science* **156**(3780):1373–1375.

Ma H.Y. and Lee I.-S. (2006) Characterization of vaterite in low quality freshwater-cultured pearls. *Materials Science and Engineering: C* **26**(4):721–723.

Ma H., Li Y., Shen Y. and Wang D. (2016) Effect of linear density and yarn structure on the mechanical properties of ramie fiber yarn reinforced composites. *Composites Part A: Applied Science and Manufacturing* **87**:98–108.

Mader A., Volkmann E., Einsiedel R. and Müssig J. (2012) Impact and flexural properties of unidirectional man-made cellulose reinforced thermoset composites. *Journal of Biobased Materials and Bioenergy* **6**(4):481–492.

Madsen B. and Lilholt H. (2003) Physical and mechanical properties of unidirectional plant fibre composites—An evaluation of the influence of porosity. *Composites Science and Technology* **63**(9):1265–1272.

Madsen B., Thygesen A. and Lilholt H. (2007) Plant fibre composites—Porosity and volumetric interaction. *Composites Science and Technology* **67**(7–8):1584–1600.

Madsen B., Thygesen A. and Lilholt H. (2009) Plant fibre composites—Porosity and stiffness. *Composites Science and Technology* **69**(7–8):1057–1069.

Maeda Y., Jayakumar R., Nagahama H., Furuike T. and Tamura H. (2008) Synthesis, characterization and bioactivity studies of novel β-chitin scaffolds for tissue-engineering applications. *International Journal of Biological Macromolecules* **42**(5):463–467.

Magdans U. and Gies H. (2004) Single crystal structure analysis of sea urchin spine calcites: systematic investigations of the Ca/Mg distribution as a function of habitat of the sea urchin and the sample location within the spine. *European Journal of Mineralogy* **16**(2):261–268.

Mark R.E. (1967) *The Cell wall Mechanics of Tracheids*, Yale University Press, New Haven, CT.

Martin J.T. and Juniper B.E. (1970) *The Cuticles of Plants*. Edward Arnold, London, 347 pp.

Marshall G.W. Jr., Balooch M., Gallagher R.R., Gansky S.A. and Marshall S.J. (2001) Mechanical properties of the dentinoenamel junction: AFM studies of nanohardness, elastic modulus, and fracture. *Journal of Biomedical Materials Research* **54**(1):87–95.

McAlpine K.J. (2016) 4D-printed structure changes shape when placed in water. http://news.harvard.edu/gazette/story/2016/01/4d-printed-structure-changes-shape-when-placed-in-water/ [Accessed 26/6/17].

Natural Materials – Composition and Combinations

McKittrick J., Chen P.-Y., Bodde S.G., Yang W., Novitskaya E.E. and Meyers M.A. (2012) The structure, functions and mechanical properties of keratin. *Journal of Materials* **64**(4):449–468.

Mercader J., Bennett T., Esselmont C., Simpson S., Walde D. (2009) Phytoliths in woody plants from the Miombo woodlands of Mozambique. *Annals of Botany* **104**(1):91–113.

Meyer K.H. and Misch L. (1937) Position des atomes dans le nouveau modèle spatial de la cellulose. Sur la constitution de la partie cristallisee de la cellulose. *Helvetica Chimica Acta* **20**:232–244.

Meyers M.A. and Chawla K.C. (2009) *Mechanical Behavior of Materials*. 2nd ed. Cambridge University Press, Cambridge.

Meyers M.A., Chen P.-Y., Lin A.Y-M. and Seki Y. (2008) Biological materials: structure and mechanical properties. *Progress in Materials Science* **53**:1–206.

Miller S.A. (2013) Sustainable polymers: opportunities for the next decade. *ACS Macro Letters* **2**(6):550–554.

Millesi H., Reihsner R., Hamilton G., Mallinger R. and Menzel E.J. (1995) Biomechanical properties of normal tendons, normal palmar apneuroses, and tissues from patients with Dupuytren's disease subjected to elastase and chondroitinase treatment. *Clinical Biomechanics* **10**(1):29–35.

Mizota M. and Maeda Y. (1986) Magnetite in the radular teeth of chitons. *Hyperfine Interactions* **29**(1–4):1423–1426.

Mohan N.H., Debnath S., Mahapatra R.K., Nayak L.K., Baruah S., Das A., Banik S. and Tamuli M.K. (2014) Tensile properties of hair fibres obtained from different breeds of pigs. *Biosystems Engineering* **119**:35–43.

Mohanty A.K., Misra M. and Drzal L.T. (2002) Sustainable bio-composites from renewable resources: opportunities and challenges in the green materials world. *Journal of Polymers and the Environment* **10**(1/2):19–26.

Mosiewicki M.A. and Aranguren M.I. (2013) A short review on novel biocomposites based on plant oil precursors. *European Polymer Journal* **49**:1243–1256.

Mulholland S.C. and Rapp G. Jr. (1992) A morphological classification of grass silica bodies. In: *Phytolith Systematics*. Springer, New York, pp. 65–89.

Mukherjee P.S. and Satyanarayana K.G. (1986) An empirical evaluation of structure-property relationships in natural fibres and their fracture behaviour. *Journal of Materials Science* **21**(12):4162–4168.

Muralhidar B.A., Giridev V.R. and Raghunathan K. (2012) Flexural and impact properties of flax woven, knitted and sequentially stacked knitted/woven preform reinforced epoxy composites. *Journal of Reinforced Plastics and Composites* **31**(6):379–388.

Nägele H., Pfitzer J., Nägele E., Inone E.R., Eisenreich N., Eckl W. and Eyerer P. (2002) Arboform®—A thermoplastic, processable material from lignin and natural fibers. In: *Chemical Modification, Properties and Uses of Lignin*, T.Q. Hu (ed.) Springer Science + Business Media, New York. pp. 101–119.

Newman R.H. (1999) Estimation of the relative proportions of cellulose I_α and I_β in wood by carbon-13 NMR spectroscopy. *Holzforschung* **53**(4):335–340.

Niedermann P., Szebényi G. and Toldy A. (2015) Characterization of high glass transition temperature sugar-based epoxy resin composites with jute and carbon fibre reinforcement. *Composites Science and Technology* **117**:62–68.

Nimz H.H. (1974) Beech lignin—Proposal of a constitutional scheme. *Angewandte Chemie International Edition* **13**:313–321.

Nishino T. (2004) Natural fibre sources. In: *Green Composites – Polymer Composites and the Environment*, Baillie C. (Ed) Woodhead Publishing, Cambridge.

Nishino T., Matsui R. and Nakamae K. (1999) Elastic modulus of the crystalline regions of chitin and chitosan. *Journal of Polymer Science B: Polymer Physics* **37**(11):1191–1196

Nishiyama Y., Langan P. and Chanzy H. (2002) Crystal structure and hydrogen bonding system in cellulose Iβ from synchrotron x-ray and neutron fiber diffraction. *Journal of the American Chemical Society* **124**:9074–9082.

Nishiyama Y., Sugiyama J., Chanzy H. and Langan P. (2003) Crystal structure and hydrogen bonding system in cellulose Iα from synchrotron x-ray and neutron fiber diffraction. *Journal of the American Chemical Society* **125**:14300–14306.

O'Donnell A., Dweib M.A. and Wool, R.P. (2004) Natural fiber composites with plant oil-based resin. *Composites Science and Technology* **64**(9):1135–1145.

Oksman K. (2001) High quality flax fibre composites manufactured by the resin transfer moulding process. *Journal of Reinforced Plastics and Composites* **20**(7):621–627.

Okuyama T., Yamamoto H., Yoshida M., Hattori Y. and Archer R.R. (1994) Growth stresses in tension wood: role of microfibrils and lignification. *Annales des Sciences Forestières* **51**:291–300.

Orgel J.P.R.O., Irving T.C., Miller A. and Wess T.J. (2006) Microfibrillar structure of type I collagen in situ. *Proceedings of the National Academy of Sciences USA* **103**(24):9001–9005.

Orgel J.P.R.O., San Antonio J.D. and Antipova O. (2011) Molecular and structural mapping of collagen fibril interactions. *Connective Tissue Research* **52**(1):2–17.

Orssengo G.J. and Pye D.C. (1999) Determination of the true intraocular pressure and modulus of elasticity of the human cornea *in vivo*. *Bulletin of Mathematical Biology* **61**(3):551–572.

O'Sullivan A.C. (1997) Cellulose: The structure slowly unravels. *Cellulose* **4**(3):173–207.

Parry D.A.D. and North A.C.T. (1998) Hard α-keratin intermediate filament chains: substructure of the N- and C-terminal domains and the predicted structure and function of the C-terminal domains of type I and type II chains. *Journal of Structural Biology* **122**(1–2):67–75.

Pasquali-Ronchetti I. and Baccarani-Contri M. (1997) Elastic fiber during development and aging. *Microscopy Research and Technique* **38**(4):428–435.

Peng X., Fan M., Hartley J. and Al-Zubaidy M. (2011) Properties of natural fiber composites made by pultrusion process. *Journal of Composite Materials* **46**(2):237–246.

Peters W. (1972) Occurrence of chitin in Mollusca. *Comparative Biochemistry and Physiology, Part B: Comparative Biochemistry* **41**(3):541–550.

Phillips S. and Lessard L. (2011) Application of natural fiber composites to musical instrument top plates. *Journal of Composite Materials* **46**(2):145–154.

Pizzi A. (2013) Bioadhesives for wood and fibres: a critical review. *Adhesion and Adhesives* **1**(1):88–113.

Preston R.D. (1952) *Molecular Architecture of Plant Cell Walls*. Chapman and Hall, London.

Prozorov T. (2015) Magnetic microbes: bacterial magnetite biomineralization. *Seminars in Cell & Developmental Biology* **46**:36–43.

Natural Materials – Composition and Combinations

Puxkandl R., Zizak I., Paris O., Keckes J., Tesch W., Bernstorff S., Purslow P. and Fratzl P. (2002) Viscoelastic properties of collagen: synchrotron radiation investigations and structural model. *Philosophical Transactions of the Royal Society of London B: Biological Sciences* **357**(1418):191–197.

Reddy N. and Yang Y. (2011) Novel green composites using zein as matrix and jute fibers as reinforcement. *Biomass and Bioenergy* **35**(8):3496–3503.

Rho J.Y., Tsui T.Y. and Pharr G.M. (1997) Elastic properties of human cortical bone and trabecular lamellar bone measured by nanoindentation. *Biomaterials* **18**:1325–1330.

Ridley B.L., O'Neill M.A. and Mohner D. (2001) Pectins: structure, biosynthesis, and oligogalacturonide-related signalling. *Phytochemistry* **57**:929–967.

Rigby B.S., Hirai N., Spikes J.D. and Eyring H. (1959) Mechanical properties of a rat tail tendon. *Journal of General Physiology* **43**:265–283.

Round F.E., Crawford R.M. and Mann D.G. (1990) *Diatoms: Biology and Morphology of the Genera*. Cambridge University Press, Cambridge, p 760.

Roy M., Rho J.Y., Tsui T.Y. and Pharr G.M. (1996) Variation of Young's modulus and hardness in human lumbar vertebrae measured by nanoindentation. In: *Advances in Bioengineering*, Vol. BED-33, S. Rastegar (ed.) ASME, Atlanta, pp. 385–386.

Sachinvala N.D., Winsor D.L., Menescal R.K., Ganjian I., Niemczura W.P. and Litt M.H. (1998) Sucrose-based epoxy monomers and their reactions with diethylenetriamine. *Journal of Polymer Science A: Polymer Chemistry* **36**(13):2397–2413.

Salmén L. (2004) Micromechanical understanding of the cell-wall structure. *Comptes Rendues Biologies* **327**:873–880.

Salmén L. and Burgert I (2009) Cell wall features with regard to mechanical performance. A review. *Holzforschung* **63**:121–129.

Sarén M.P., Andersson S., Serimaa R., Paakkari T., Saranpää P. and Pesonen E. (2001) Structural variation of tracheids in Norway spruce (*Picea abies* [L.] Karst.). *Journal of Structural Biology* **136**(2):101–109.

Sarén M.P., Serimaa R., Andersson S., Saranpää P., Keckes J. and Fratzl P. (2004) Effect of growth rate on mean microfibril angle and cross-sectional shape of tracheids in Norway spruce. *Trees* **18**:345–362.

Sarikaya M., Fong H., Sunderland N., Flinn B.D., Mayer G., Mescher A. and Gaino E. (2001) Biomimetic model of a sponge-spicule optical fiber—Mechanical properties and structure. *Journal of Materials Research* **16**(5):1420–1428.

Sasaki N. and Odajima S. (1966) Stress-strain curve and Young's modulus of a collagen molecule as determined by the x-ray diffraction technique. *Journal of Biomechanics* **29**(5):655–658.

Scalbert A., Monties B., Lallemand J.-Y., Guittet E. and Rolando C. (1985) Ether linkage between phenolic acids and lignin fractions from wheat straw. *Phytochemistry* **24**(6):1359–1362.

Schneider M.H. and Phillips J.G. (1991) Elasticity of wood and wood polymer composites in tension compression and bending. *Wood Science and Technology* **25**:361–364.

Screen H.R.C. (2008) investigating load relaxation mechanics in tendon. *Journal of the Mechanical Behaviour of Biomedical Materials* **1**(1):51–58.

Shah D. (2013) Developing plant fibre composites for structural applications by optimising composite parameters: a critical review. *Journal of Materials Science* **48**(18): 6083–6107.

Shah D.U. (2014) Natural fibre composites: Comprehensive Ashby-type materials selection charts. *Materials and Design* **62**:21–31.

Sherman V.R., Yang W. and Meyers M.A. (2015) The materials science of collagen. *Journal of the Mechanical Behaviour of Biomedical Materials* **52**:22–50.

Shimizu K., Cha J., Stucky G.D. and Morse D.E. (1998) Silicatein α: Cathepsin L-like protein in sponge biosilica. *Proceedings of the National Academy of Sciences of the USA* **95**(11):6234–6238.

Silver F.H., Ebrahimi A. and Snowhill P.B. (2002) Viscoelastic properties of self-assembled type-I collagen fibers: molecular basis of elastic and viscous behaviours. *Connective Tissue Research* **43**(4):569–580.

Simpson T.L., Langenbruch P.-F. and Scalera-Liaci L. (1985) Silica spicules and axial filaments of the marine sponge *Stelletta grubii* (Pordera, Demospongiae). *Zoomorphology* **105**:375–382.

Smith J.W. (1968) Molecular pattern in native collagen. *Nature* **219**(5150):157–158.

Snyder et al. (2007) *Silicon. Handbook of plant nutrition*, Taylor and Francis, New York.

Sobczak L., Land R.W. and Haider A. (2012) Polypropylene composites with natural fibers and wood—General mechanical property profiles. *Composites Science and Technology* **72**:550–557.

Södergård A. and Stolt M. (2002) Properties of lactic acid based polymers and their correlation with composition. *Progress in Polymer Science* **27**:1123–1163.

Sorieul M., Dickson A., Hill S.J. and Pearson H. (2016) Plant fibre: Molecular structure and biomechanical properties, of a complex living material, influencing its deconstruction towards a biobased composite. *Materials* **9**:618, 36 pp.

Spārniņš E., Nyström B. and Andersons J. (2012) Interfacial shear strength of flax fibers in thermoset resins evaluated via tensile tests of UD composites. *International Journal of Adhesion and Adhesives* **36**:39–43.

Spear M.J., Eder A. and Carus M. (2015) Wood polymer composites. In: *Wood Composites*, M.P. Ansell (ed.) Woodhead Publishing, Cambridge, pp. 195–249.

Stewart J.J., Akiyama T., Chapple C., Ralph J. and Mansfield S.D. (2009) The effects on lignin structure of overexpression of ferulate 5-hydroxylase in hybrid poplar. *Plant Physiology* **150**(2):621–635.

Strömberg C.A.E. (2004) Using phytolith assemblages to reconstruct the origin and spread of grass-dominated habitats in the great plains of North America during the late Eocene to early Miocene. *Paleogeography, Paleoclimatology, Paleoecology* **207**:239–275.

Šturcová A., Davies G.R. and Eichhorn S.J. (2005) Elastic modulus and stress-transfer properties of tunicate cellulose whiskers. *Biomacromolecules* **6**(2):1055–1061.

Subhash G., Yao S., Bellinger B. and Gretz M.R. (2005) Investigation of mechanical properties of diatom frustules using nanoindentation. *Journal of Nanoscience and Nanotechnology* **5**(1):50–56.

Summerscales J., Dissanayake N., Virk A. and Hall W. (2010a) A review of bast fibres and their composites. Part 1—Fibres as reinforcements. *Composites Part A: Applied Science and Manufacturing* **41**(10):1329–1335.

Summerscales J., Dissanayake N., Virk A. and Hall W. (2010b) A review of bast fibres and their composites. Part 2—Composites. *Composites Part A: Applied Science and Manufacturing* **41**(10):1336–1344.

Natural Materials – Composition and Combinations

Sun R.C., Sun X.F., Wang S.Q., Zhu W. and Wang X.Y. (2002) Ester and ether linkages between hydroxycinnamic acids and lignins from wheat, rice, rye, and barley straws, maize stems, and fast-growing poplar wood. *Industrial Crops and Products* **15**(3):179–188.

Szent Györgyi A.G. and Cohen C. (1957) Role of proline in polypeptide chain configuration of proteins. *Science* **126**(3276):697–698.

Thongboonkerd V., Semangoen T. and Chutipongtanate S. (2006) Factors determining types and morphologies of calcium oxalate crystals: molar concentrations, buffering, pH, stirring and temperature. *Clinica Chimica Acta* **367**(1–2):120–131.

Timell T.E. (1967) Recent progress in the chemistry of wood hemicelluloses. *Wood Science and Technology* **1**(1):45–70.

Tomas J. and Geffen A.J. (2003) Morphometry and composition of aragonite and vaterite otoliths of deformed laboratory reared juvenile herring from two populations. *Journal of Fish Biology* **63**(6):1389–1401.

Van de Velde K. and Kiekens P. (2002) Biopolymers: overview of several properties and consequences on their applications. *Polymer Testing* **21**:433–442.

Van de Weyenberg I., Ivens J. and Verpoest I. (2000) Parametric study of the relationship between the fibre and the composite properties of flax fibre reinforced epoxy. In: *Proceedings of ECCM9*, Brighton, UK, June 2000.

Van de Weyenberg I., Ivens J., De Coster A., Kino B., Baetens E. and Verpoest I. (2003) Influence of processing and chemical treatment of flax fibres on their composites. *Composites Science and Technology* **63**(9):1241–1246.

Van de Weyenberg I., Chi Truong T., Vangrimde B. and Verpoest I. (2006) Improving the properties of UD flax fibre reinforced composites by applying an alkaline fibre treatment. *Composites Part A* **37**(9):1368–1376.

Velásquez P., Skurtys O., Enrione J. and Osorio F. (2011) Evaluation of surface free energy of various fruit epicarps using acid-base and Zisman approaches. *Food Biophysics* **6**:349–358.

Vhrovski B. and Weiss A.S. (1998) Biochemistry of tropoelastin. *European Journal of Biochemistry* **258**:1–18.

Wada M., Sugiyama J. and Okano T. (1994) The monoclinic phase is dominant in wood cellulose. *Mokuzai Gakkaishi* **40**(1):50–56.

Wainwright S.A. (1970) Design in hydraulic organisms. *Naturwissenschaften* **57**:321–326.

Wainwright S.A., Biggs W.D., Currey J.D. and Gosline J.M. (1976) *Mechanical Design in Organisms*. Edward Arnold (Publishers) Ltd, London. 423 pp.

Weaver J.C., Aizenberg J., Fantner G.E., Kisailus D., Woesz A., Allen P., Fields K., Porter M.J., Zok F.W., Hansma P.K., Fratzl P. and Morse D.E. (2007) Hierarchical assembly of the siliceous skeletal lattice of the hexactinellid sponge *Euplectella aspergillum*. *Journal of Structural Biology* **158**:93–106.

Weaver J.C., Wang Q., Miserez A., Tantuccio A., Stromberg R., Bozhilov K.N., Maxwell P., Nay R., Heier S.T., DiMasi E. and Kisailus D. (2010) Analysis of an ultra hard magnetic biomaterial in chiton radular teeth. *Materials Today* **13**(1–2):42–52.

Wegst U.G.K. and Ashby M.F. (2004) The mechanical efficiency of natural materials. *Philosophical Magazine* **84**(21):2167–2181.

Weis-Fogh T. (1960) A rubber-like protein in insect cuticle. *Journal of Experimental Biology* **37**:889–907.

Weis-Fogh T. (1961a) Thermdynamic properties of resilin. *Journal of Molecular Biology* **3**:520–531.

Weis-Fogh T. (1961b) Molecular interpretation of the elasticity of resilin, a rubber-like protein. *Journal of Molecular Biology* **3**:648–667.

Woesz A., Weaver J.C., Kazanci M., Dauphin Y., Aizenberg J. and Morse D.E. et al. (2006) Micromechanical properties of biological silica in skeletons of deep-sea sponges. *Journal of Materials Research* **21**(8):2068–2078.

Woodings C.R. (1995) The development of advanced cellulosic fibres. *International Journal of Biological Macromolecules* **17**(6):305–309.

Xu Z., Yang Y., Zhao W., Wang Z, Landis W.J., Cui Q. and Sahai N. (2015) Molecular mechanisms for intrafibrillar collagen mineralization in skeletal tissues. *Biomaterials* **39**:59–66.

Yan L. and Chouw N. (2013) Crashworthiness characteristics of flax fibre reinforced epoxy tubes for energy absorption. *Materials and Design* **51**:629–640.

Yan L., Chouw N. and Jayaraman K. (2014a) Flax fibre and its composites—A review. *Composites: Part B: Engineering* **56**:296–317.

Yan L., Chouw N. and Jayaraman K. (2014b) Lateral crushing of empty and polyurethane-foam filled reinforced epoxy composite tubes. *Composites Part B: Engineering* **63**:15–26.

Yang W., Sherman V.R., Gludovatz B., Schaible E., Stewart P., Ritchie R.O. and Meyers M.A. (2015) On the tear resistance of skin. *Nature Communications* **6**:1–10. doi:10.1038/ncomms7649.

Zafeiropoulos N.E., Baillie C.A. and Matthews F.L. (2000) Modification and characterisation of the interface in flax/polypropylene composite materials. In: *Proceedings of ECCM9*, Brighton, UK, June 2000.

Zhang J.-F. and Sun X. (2005) Poly(lactic acid)–based bioplastics. In: *Biodegradable Polymers for Industrial Applications*. R. Smith (ed.) CRC, Woodhead Publishing, Cambridge, pp. 251–288.

Zini E. and Scandola M (2011) Green Composites: an overview. *Polymer Composites* **32**(12):1905–1915.

4

Designing with the Life Cycle in Mind

C. Skinner

Bangor University

CONTENTS

Environmental Drivers and the Current Design Landscape 111
Life Cycle Assessment – Overview of Methodology 113
LCA in Relation to Wood and Wood-Based Products 116
LCA and Plant Fibre Materials 119
LCA and Animal-Derived Materials 121
    Wool 122
    Leather 124
Summary 125
References 127

Environmental Drivers and the Current Design Landscape

Natural materials have been used by societies throughout history for their ready supply and renewability. The Industrial Revolution, and subsequent technological advances throughout the twentieth century, saw the advent of mass production and the introduction of many new synthetic materials. Plastics, in particular, became a default choice of material for a huge range of applications, and interest in natural materials, such as wood, plant fibres and wool, went somewhat into decline. This has been changing more recently due to the growing emphasis on the environmental issues associated with fossil-based production; issues such as resource depletion, land use and habitat degradation, the generation and disposal of wastes, and in particular the need to mitigate climate change.

In response to this shifting emphasis, many organisations have used environmental footprinting methodologies, such as life cycle assessment (LCA), to better understand the impacts associated with their products. This is driven in part by an increasingly challenging legislative landscape, and in part by the green marketing opportunities that positive environmental messages can bring. Today, many of the world's leading companies engage in some form of corporate social responsibility planning, and sustainability features

111

prominently on their websites and promotional literature. However, there is evidence that environmental performance is still not widely considered in the design-based phases of product development. Deutz et al. (2013), for example, assessed attitudes of a sample of large UK manufacturers and found that engagement with eco-design principles was generally low. They cited a range of reasons for this, including designers' level of understanding of environmental issues and 'the extent of the design space, influenced by legal requirements, economic and supply chain constraints'. Certainly, cost is often considered to be a barrier to greater uptake of eco-design principles. It is known, for example, that consumers were less willing to pay premiums for green products following the global downturn of 2007/8 (PwC, 2012). However, there is also evidence that eco-design can increase profitability when used effectively. Ploffe et al. (2011) surveyed 30 small and medium enterprises and large companies that had used eco-design principles in France and Quebec, and found that 24 out of 30 reported increased profitability as a result. Since then, the release of an international eco-design standard, ISO 140006, should facilitate further uptake. This standard provides guidance for organisations wishing to set up or improve their management of eco-design initiatives as part of a broader environmental management system (ISO, 2011).

One sector in which eco-design standards are already well defined (and which offer a significant opportunity for further use of natural materials) is that of construction. In this sector, two programmes in particular are well established. BREEAM (BRE, 2016) and LEED (USGBC, 2016) are both internationally recognised certification schemes for sustainable building design, construction and use. Each standard is based on a system of credits awarded for a wide range of sustainability metrics, spanning the full life cycle of the building. At the project design stage, BREEAM awards credits for having a clear sustainability brief in place *prior to completion of the concept design*, and further credits are awarded for appointing a sustainability champion ahead of this early stage of the process. Both schemes encourage the use of LCA for assessing the choice of materials used. BREEAM was the first scheme to launch, having been in operation since 1990, but LEED is now more widely used internationally. Collectively, they have been used to certify over 50,000 building projects worldwide (Mark, 2013).

Elsewhere, work in the furniture industry in Australia has focused on factors affecting disposal options at the end-of-life (EOL). Critchlow (2010) highlighted the large number of material components used in many modern furniture designs and the difficulty this presents in terms of their recyclability. It also noted that the designed-in inability to separate and replace worn components (in particular, upholstery) meant that whole items tend to be discarded even where the underlying structure may be undamaged. The report called for a number of changes throughout the (Australian) furniture industry. These included the wider use of replaceable upholsteries; better access to materials databases for designers (i.e. those containing environmental performance data); increased sustainability training for furniture design students; and increased sharing of international best practice.

Designing with the Life Cycle in Mind   113

In the transport sector, progressively stricter legislation around exhaust emissions has encouraged designers to use new and novel materials in order to light-weight their vehicles (as an approach to improve overall fuel efficiency). The EURO emissions standards, which define the legal limits for vehicle exhaust emissions across the continent, have been updated four times since 2000, with each revision bringing stricter limits. This highlights the increasing legislative pressure on designers in this sector to find more environmentally benign solutions to vehicle design projects.

All of this activity is increasingly underpinned by the use of analytical methods for assessing environmental performance. A number of such methods exist, but LCA is by far the most comprehensive of the quantitative approaches, and has become an established tool for the environmental assessment of products across industrial sectors.

Life Cycle Assessment – Overview of Methodology

LCA is a method for quantifying the environmental impacts of a product or service across its entire life cycle. It therefore includes all impacts associated with the extraction or cultivation of raw materials, transportation of goods, manufacturing, in-life use and EOL. Carbon footprinting is in fact one form of LCA, though it focuses on a single environmental indicator, global warming potential (GWP). Full LCA, by contrast, covers a much broader set of impacts, ranging from environmental degradation (e.g. acidification and eutrophication potentials), human health impacts (e.g. particulate matter formation; human toxicity) and resource depletion (non-renewable energy use, metals and water). The principles and methodology are outlined by two international standards, ISO 14040 & 14044 (ISO, 2006a, b), which define LCA as a four stage process.

During the first stage (Goal and Scope Definition), the unit against which the results will ultimately be reported is defined. This can be framed either in material terms (e.g. 1 kg of processed wool fibres) or in functional terms (e.g. 1 m^2 insulation with a thermal resistance of x and lifespan of y). Using *functional units* allows products of differing material make-ups to be compared on a functionally equivalent basis. This is the approach required by ISO 14040/44 when comparisons between products are to be made public. Materially based units (referred to as *declared units* in LCA) are also useful as they can provide the basis for further analyses (including those based on functionality). In practice, the choice of which type of unit to use will depend on the aim and scope of the study at hand.

A *system boundary* is then defined to further formalise the scope of the study. This indicates which stages of the life cycle, and which pathways of the production process, are to be included in the analysis. At its most comprehensive, a *cradle-to-grave* approach encompasses all stages from initial raw materials extraction or cultivation through to EOL. In practice, shorter systems are often considered, in particular when assessing product footprints from an organisational perspective. These shorter *cradle-to-gate,* or even *gate-to-gate* systems effectively model the environmental impacts over which the organisation has control, whether that be directly (through its own operations) or indirectly (through its supply chain purchasing decisions). Such studies typically draw a boundary at the point of hand-over to the customer (i.e. 'the factory gate') and often act as the basis for published eco-profiles and product declarations.

Modelling the life cycle can be relatively straightforward where a production system results in just one product. In this case, all impacts associated with the production process are attributed to the single product. However, when multiple products share a common production pathway, a method for portioning impacts across the various co-products is required. Allocation of these impacts (or avoidance of it) can be contentious in LCA since the choice of approach can significantly affect the end results for each product. Avoidance of allocation can be achieved by expanding the system boundary to include the co-products, and factoring in (as a 'credit') the avoided production of an equivalent product (known as substitution). Alternatively, the system boundary can be maintained and impacts can be allocated based on the relative masses or economic values of the products involved. ISO 14044 outlines a stepwise approach to determine allocation procedure, but different product sectors may be better suited to different approaches. With this in mind, there has been a move towards developing sector or product-specific rules (known product category rules or PCRs) to further define this (and other) aspects of the modelling process. This can be seen as part of a wider drive to increase the level of harmonisation within LCA.

Stage two of the process (Inventory Analysis) involves drawing together a balanced inventory of the material and energy flows entering and exiting the system boundary. This will include all raw materials and energy usage as inputs, and any products, wastes, and emissions sources as outputs. Once a fully balanced mass and energy inventory has been established, each datapoint is paired with a life cycle inventory (LCI) dataset containing the environmental impact information. These datasets are available through licensed databases such as Ecoinvent (Weidema et al., 2013) and Agri-footprint (Blonk Agri-footprint BV, 2015), among others, and their number is expanding over time (Ecoinvent currently hold over 11,500 unit processes). Despite this, notable gaps in the data do still exist, in particular in relation to novel additives, chemicals and emerging processes, as well as more mainstream processes (such as grid electricity generation) in regions outside of the main global economies. Expanding the regional specificity of LCI data, in particular, is an active area of research and new datasets have recently been added to databases such as Ecoinvent with this in mind.

Designing with the Life Cycle in Mind

Stage three involves the use of an Impact Assessment method to calculate the LCA results from the prepared inventory. Several such methods exist, and there is some regional variation in choice, but all present results as either *mid-point* or *end-point indicators*. Mid-point indictors are those that use units particular to the impact category in question. For example, GWP is cited in 'kg CO_2 equivalents', while fossil depletion may be cited in 'kg oil equivalents'. End-point indicators endeavour to translate this diverse information into measures of 'damage', by grouping mid-points into one of three much broader categories (namely, ecosystem, human health, and resource depletion) and providing results in units particular to these categories (see Table 4.1). This streamlines the amount of information that is presented for interpretation, but also introduces an added element of uncertainty to these results. A further step, in which end-point results are amalgamated into a 'single score' is also possible, through this is largely discredited due to the further uncertainty that it introduces (e.g. Hill et al., 2015). Finally, in the fourth stage of the LCA process, output is evaluated and interpreted ready for presentation and documentation.

As earlier indicated, there has been a drive towards greater harmonisation of approach within LCA, both in terms of the methodology and reporting

TABLE 4.1

Mid-Point and End-Point Indicators for Impact Categories Used in the ReCiPe Impact Assessment Method

Impact Category	Mid-Point Indicator	End-Point Indicator
Climate change (Human Health)	kg CO_2 eq.	
Ozone depletion	kg CFC-11 eq.	
Human toxicity	kg 1,4-DB eq.	DALY[a]
Photochemical oxidant formation	kg NMVOC	
Particulate matter formation	kg PM_{10} eq.	
Ionising radiation	kBq^{235}U eq.	
Climate change (Ecosystems)	kg CO_2 eq.	
Terrestrial acidification	kg SO_2 eq.	
Freshwater eutrophication	kg P eq.	
Terrestrial ecotoxicity	kg 1,4-DB eq.	
Freshwater ecotoxicity	kg 1,4-DB eq.	species.yr
Marine ecotoxicity	kg 1,4-DB eq.	
Agricultural land occupation	m^2a	
Urban land occupation	m^2a	
Natural land transformation	m^2	
Metal depletion	kg Fe eq.	
Fossil depletion	kg oil eq.	$

Source: Goedcoop et al. (2008).
[a] Disability adjusted life years.

TABLE 4.2

Life Cycle Modules Outlined in EN15804

Module		Sub-Stage
A	Product stage	A1 – Raw material supply
		A2 – Transport
		A3 –Manufacturing
	Construction process	A4 – Transport
		A5 – Construction/installation
B	Use	B1 – Use
		B2 – Maintenance
		B3 – Repair
		B4 – Replacement
		B5 – Refurbishment
		B6 – Operational energy use
		B7 – Operational water use
C	End-of-life	C1 – Deconstruction/demolition
		C2 – Transport
		C3 – Waste processing
		C4 – Disposal
D	Benefits and burdens beyond the system boundary	Reuse, recovery and recycling potential

Source: BSI (2013).

protocols used. One such initiative is the Environmental Product Declaration (EPD) programme, which uses PCRs to define the approach to be taken for specific product categories, often where competing co-products result from shared processes. The programme operates by reviewing and certifying product LCA results and then publishing them in a simplified, standard format (thereby facilitating product comparisons). The scheme is primarily a business-to-business tool, though the information is publically available and free to access.

EPD programmes have been particularly well used by the producers of construction materials, including wood products, and can be a good source of information for environmentally-based materials choices. A core PCR for the construction industry is EN 15804 (BSI, 2013) which outlines requirements for assessing all construction products. This presents life cycle stages and sub-stages specific to the industry and defines reporting requirements in line with this (Table 4.2).

LCA in Relation to Wood and Wood-Based Products

Wood and wood-based products have been well studied by LCA, with most of the main product categories having been assessed in one form or another.

Designing with the Life Cycle in Mind

The use of wood in the built environment has been a particular area of focus, in part because of its potential to replace large quantities of non-renewable materials (such as concrete and steel), but also due to its capacity to store carbon when used in long-lived construction products. Each tonne of dry wood contains at least 40% biogenic carbon and therefore effectively removes approximately 1.5 tonnes of CO_2 from the atmosphere for the duration of its lifespan (Hill et al., 2015). When considering wood-based construction products used in housing stock, this is estimated to be around 70 million tonnes of stored CO_2 in the UK alone (Read et al., 2009). The potential to increase this by using more wood in the built environmental is therefore of considerable interest.

When wood is used in a relatively unprocessed form (e.g. for joists, beams, sawn timber etc.), and is harvested from sustainable sources, it typically has a better environmental profile than other structural building materials. For example, Borjesson and Gustavsson (2000) and Lenzena and Treloar (2002) both assessed wooden-framed building construction relative to concrete and found the wood-frame to have lower embodied energy and greenhouse gas (GHG) emissions over its lifetime. Similarly, John et al. (2009) compared four multi-storey building designs, based on either concrete, steel, wood (laminated veneer lumber) or a 'timber plus' design. The timber plus scenario included additional wooden features such as window frames, ceilings and exterior cladding. They found that increasing the amount of timber in the building decreased the embodied energy and GHGs emissions associated with it (assuming similar in-use energy demands and identical lifespans). Ximenes and Grant (2013) looked at maximising the amount of wood used in house design. They modelled a house consisting of timber elements in all major structural aspects of the building and found that this resulted in 50% lower GHG emissions over its lifespan compared to a market-typical control.

All of the above studies focused on the construction and EOL stages of building life cycles. However, when the use phase is factored in, in-life energy consumption, especially for heating in colder climates, typically dominates the lifetime environmental profile of buildings (Adalberth, 2000). The thermal resistance of components playing a role in insulating the building should therefore be considered a critical factor in terms of sustainable design.

More processed wood products, such as wood-based panels, typically have higher process energy requirements and may include chemical additives such as resins, adhesives and/or preservatives. This increases the environmental burdens associated with these products relative to raw timber. Hill and Dibdiakova (2016) compared published EPD results for a range of panel products (fibreboard, particleboard, orientated strand board and Glulam/ laminated veneer lumber) relative to solid wood. They found considerable variation within results for each product group, but that fibreboard tended to have the highest GHGs emissions overall. Results for embodied energy were similarly variable, but again tended to be higher with increasing processing demands. In the case of fibreboard, this is primarily associated with process electricity use, which dominates the results for this product (Rivela et al., 2007; Skinner et al., 2016). Hill and Dibdiakova's work also compared EPD results for the wood-based products with those for cement, concrete, steel and brick. Drawing comparisons based on declared units (i.e. on a mass for mass basis), they showed that results for concrete and brick fell into the lower range of those associated with the wood products, while results for steel and cement fell within the upper range. However, when they factored in the atmospheric carbon stored in the wood-based products, the GHG balances for all the wood products became negative overall. This underlines the potential climate change mitigation opportunity presented by increasing the use of wood in the built environment.

It is worth noting that the role of additives in wood product LCAs is not limited to those that form an integral part of the product itself (e.g. resins in medium density fibreboard (MDF); preservatives in treated wood). Other components can enter the life cycle as secondary inputs, such as paints, varnishes or oils used in the use stage of the life cycle. Nebel et al. (2006), for example, showed that LCA results for wooden floor coverings were highly sensitive to assumptions made about glues and finishes used for their installation and maintenance. The development of more environmentally benign additives is an on-going area of research and may provide opportunities to reduce the impacts of such inputs in future.

As already indicated, one the most important aspects of using wood in long-lived products is its capacity to store biogenic carbon. In quantifying this benefit, it is necessary to understand the likely lifespan of the products involved, and in the case of very long-lived products (such as structural building components) this can be highly uncertain. In fact, service lives used in construction LCAs are typically based on generic values rather than on direct assessment (Aktas and Bilec, 2012; Rincón et al., 2013), and these can show considerable variation (see Ormondroyd et al., 2016). This therefore needs to be considered carefully when undertaking LCAs of wood-based construction products.

Following EOL, the period of carbon storage can be further extended by recycling or reprocessing wood-based products into new products. Much used wood is currently recycled into panel products (in particular particleboard) but treated or contaminated wood cannot be used in this way, and is typically incinerated or landfilled. Ormondroyd et al. (2016) have explored opportunities for the reclamation and reuse of wood from construction and demolition sites. They highlighted a need for better tracking of treated wood over its lifetime to enable better separation during demolition and EOL. They also noted potential future pathways for reusing low grade and/or contaminated wood, including the continued development of wood-plastic and wood-cement composites; advances in MDF recovery technologies; and the emerging lignocellulosic biorefinery technologies. Once wood fibres can no longer be reprocessed into useful goods, they can be incinerated in an energy recovery

Designing with the Life Cycle in Mind

facility. This allows for recovery of the embedded energy in the wood, while finally returning the carbon to the biosphere. The development of more incineration facilities capable of handling contaminated wood waste (and thereby further minimise the volumes reaching landfill) would be a good thing.

LCA and Plant Fibre Materials

There has been renewed interest in working with plant fibre materials in recent years, partly because of their renewability and partly because of the physical characteristics they can offer. Two sectors where this has been particularly pronounced are the automotive industry, where natural fibre composites (NFCs) are now present in most new models brought to market, and in construction, where natural fibre insulation and structural composites have been of interest.

In the automotive industry, NFCs have been developed primarily as alternatives to glass fibre-polypropylene (PP) composites, where the lower densities of plant fibres translate into weight-savings in the vehicle. When considered over the entire life cycle, impacts associated with vehicles are typically dominated by fuel consumption, so any reduction in overall weight of the vehicle (and the knock-on effect this has on fuel efficiency) is of great interest. Typically, plant fibres such has flax, hemp, jute and sisal have densities ranging from 1.3–1.6 g/cm^3. This compares to 1.7–2.8 g/cm^3 for more convention reinforcement materials such as glass fibre, carbon fibre and talc (Sobczak et al., 2012). This differential makes weight reductions of up to 30% possible for components containing natural fibres (Bledzki and Faruk, 2006).

More specifically, Joshi et al. (2004) reviewed three studies, each comparing a NFC with its' more standard alternative. These were a hemp fibre-epoxy composite (Wötzel and Flake, 2001), a flax fibre-PP composite (Schmidt and Beyer, 1998) and china reed fibre-PP composite (Corbière-Nicollier et al., 2001). In each case, although the overall reduction in vehicle weight was small, the associated fuel efficiency improvement was the most important factor in the vehicle's improved environmental performance over time. Other points they noted were that the natural fibres typically had lower production impacts than glass fibre; that the NFCs allowed for a higher fibre content (and therefore reduced polymer content) without loss of performance; and that (unlike some GF composites) the NFCs could be incinerated for energy recovery at EOL.

Since then, NFCs based on other plant species have also been investigated. Alves et al. (2010) compared a jute-polyester composite with a standard glass fibre mix, and also found that the lighter weight of this NFC was the key factor in its preferential LCA results. Luz et al. (2010) studied sugar bagasse fibres blended with PP, as a potential replacement for talc-PP composites. They noted that the reduction in particulate matter emissions (associated with industrial talc use) was a further benefit to this system. Conversely, Zah et al. (2007) studied a curauá fibre-PP blend finding that the heavier weight of these fibres was disadvantageous. However, they also noted that the production of this fibre still had considerable opportunities for optimisation.

Dellaert (2014) assessed sisal production in Brazil and Tanzania, and highlighted the wide scope for variation in cultivation burdens associated with these (and other) plant fibre crops. The study found that while the non-renewable energy demand of Brazilian sisal production was 93% lower than that of glass fibre production, it was also 41%–61% lower than that of Tanzanian sisal production. Similarly, GHG emissions of Tanzanian sisal production were 157% higher than that of Brazilian sisal production. This difference was explained by the widely different production systems used in these two countries, as well as by the different soils, climate and plant hybrids used. Cultivation practices are therefore an important factor when considering the environmental performance of NFCs as potential replacements to non-renewable materials (especially when the later already have highly optimised, industrial production systems place). Optimisation of these cultivation systems is therefore likely to be a key factor in further improving the environmental (and economic) profiles of the emerging NFC materials.

In terms of the construction sector, interest in plant fibres has primarily focused on natural fibre insulation (NFI). For example, Ardente et al. (2008) assessed a kenaf fibre-polyester insulation containing 85% kenaf. They found that the polyester, while only accounting for 15% of the material content, was responsible for 39% of its cradle-to-grave GHG emissions. This NFI performed less well on this measure than other insulation formats, however, when compared on the basis of equivalent thermal resistance. These other formats included a flax roll (see Schmidt et al., 2004), polyurethane (PU) foam, and mineral and glass wools. However, no account was made for energy recovery at EOL or for potential co-products from vegetable residues in this initial study.

Since then, Batouli et al. (2014) has assessed kenaf blended with PU to form structural insulation panels (SIPs). They compared SIPs containing 5%, 10% and 15% kenaf and found that although the natural fibre was associated with much lower environmental impacts than PU, this effect did not increase with increasing fibre content. Subsequent structural analysis showed that at higher concentrations the plant fibre was replacing air in the foam (rather than polymer), thereby increasing the panel's density without improving its thermal or environmental performance. They recommended further research to establish a method for better replacing the polymer content with fibres in order to realise the potential environmental improvements.

Murphy and Norton (2008) assessed a NFI containing 46% hemp fibre and 40% recycled cotton fibre. They found impacts associated with that product were marginally higher than that of stone and glass wool in several impact categories, though this was largely due to transportation (associated with importing the NFI product to the UK) and the small-scale of production. They highlighted a number of potential areas for product optimisation, suggesting that a net negative carbon balance could be achievable over the product's lifetime if these were implemented (assuming that the carbon sequestered in the biomass is taken into account).

Zampori et al. (2013) compared different allocation procedures (mass and economic) on LCA results for hemp cultivation, and subsequent NFI panel production using 85% hemp fibre and 15% polyester. Their results for cultivation showed little variation between to two allocation approaches, and by modelling biogenic CO_2 uptake, they demonstrated a negative carbon balance for the cradle-to-gate system they modelled. Comparing their results with those for rockwool, they noted a 67% reduction in non-renewable energy demand for the hemp-based product.

Apart from insulation, several challenges associated with natural fibres have limited their wider use in the construction sector. Their higher moisture absorption, lower fire resistance and weaker mechanical properties compared to synthetic composites, mean that they have typically been limited to interior, non-structural applications so far (Dittenber and GangaRao, 2012). An exception would be hemp-concrete, which has been studied in the context of load-bearing walls and relative thermal resistance (Pretot et al., 2014). This was found to have a significantly lower GHG profile than equivalent brick or concrete block walls insulated with mineral wool. The application of coatings (sand-lime for external use, and either that or hemp-lime for indoor use) over the lifetime of the wall was also modelled and found to add only marginally to the overall lifetime impacts (unless re-applications became numerous). Hemp concrete therefore appears to have good potential for further use in this area. Similarly, new developments in surface treatments, additives and coatings are likely to make other natural fibre composites viable options for a greater range of applications in the future, both within the construction sector and beyond (Dittenber and GangaRao, 2012).

LCA and Animal-Derived Materials

Animal-derived materials such as wool and leather have been used for millennia for textiles, furnishings, insulation and other uses, but they pose certain challenges when modelling LCA. Such materials are co-products of complex agricultural systems that generate significant GHG emissions, as well as other potential environmental impacts associated with land management and water quality. Allocating burdens between the various co-products of these farm systems can be achieved based on a number of methodological approaches, the choice of which can have a significant bearing on results of the study. Inevitably, this has been a source of some debate among different industry groups, which have an interest in one particular product, and one eye on the benefits of demonstrating a positive environmental profile for that product. Harmonisation of methodology is a key goal here and is in varying stages of realisation.

Wool

While wool's use in clothing has declined in recent years (Turley et al., 2009), there remains considerable interest in its performance as a renewable insulation material, both in construction applications (e.g. Mansour et al., 2016; Murphy and Norton, 2008) and temperature-controlled packaging (e.g. see www.woolcool.com). In fact, an increasing global appetite for meat means that sheep numbers are expected to increase by around 60% by 2050 (Foresight, 2011) so, at a macro level at least, there should be no shortage of supply should such uses find greater market-share in coming years.

When considering the environmental footprint of wool production, a distinction needs to be draw between fine grade Merino wool, typically bound for the clothes industry, and the courser fibres used in carpets, furnishings, and insulation etc. The economic value of each product differs widely and this has a bearing on the farm system used. Since up to 80% of the GHG emissions associated with wool production occur prior to the farm gate (Wiedemann et al., 2015), this is an important distinction to draw. Where an economic allocation is used to apportion emissions between meat and wool, this distinction becomes more pronounced. Fine-grade Merino wool may be the primary economic driver of that form of sheep rearing in Australia (Browne et al., 2011), but courser wools typically produced in the UK are considered as low value bi-products of meat production (Jones et al., 2014). In any system, inter-farm variation is also high; so, large sample sizes are important for establishing reliable on-farm data.

Using economic allocation, Jones et al. (2014), assessed GHG emissions associated with 64 sheep farms across England and Wales. They established that on average 2.3% of the economic value of sheep production was associated with wool and used this as the basis of their study. This is in line with assumptions made by other researchers looking at British wool (e.g. 2% by Edwards-Jones et al., 2009), but contrasts with figures ranging from 5%–54% for Australian systems (including Merion flocks), (Biswas et al., 2010; Eady et al., 2012). In France, where wool has little value, a 0.3% economic split to wool has been used (Gac et al., 2012).

An alternative approach is to allocate based on mass. This was also explored by Gac et al. (2012), who found that 10.4% of production burdens in their French sample would be attributed to wool using this method (compared to 0.3% using economic allocation). They found that variation between sample groups was less using this approach than economic allocation.

Exploring other potential bases for allocation, Cottle and Cowie (2016) explored an approach based on the protein content of each product and found this to be a simple and easily applied method for attributional LCA. Wiedemann et al. (2015) explored seven methods (three using biophysical allocation, plus protein mass, economic and two system expansion approaches). Results varied widely between these methods (Figure 4.1).

In view of this, efforts have been made to harmonise the way in which sheep farm systems are assessed. The International EPD System has developed a PCR for meat of mammals (Boeri, 2013) determining that allocation between co-products should be made on the basis of 'average economic value over the last 3 years' and making annual verification of updates a requirement. Another initiative, by the FAO (2015) which focused specifically on small ruminant systems, recommended a biophysical approach based on energy or protein requirements for growth and fibre production. Whichever approach is ultimately used, it is clear that burdens up to the point of slaughter are a key factor in the eco-profiles of wool, and those choices around their allocation are a critical determinant of the end results.

Post-slaughterhouse, wool production has relatively high water requirements during the scouring, dyeing and finishing processes (Turley et al., 2009), though these latter two stages are unnecessary where wool is destined for insulation. Scouring can be associated with use of bleaching agents and high organic-content effluent discharge, though this is effectively treated in well-regulated operations (EC, 2003). Murphy and Norton (2008) investigated

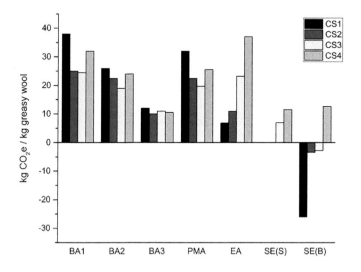

FIGURE 4.1
GHG emissions from greasy wool production for four farming case studies (CS1-4) using seven different approaches to handle wool and meat co-production. BA1-3 = biophysical allocations; PMA = protein mass-based allocation; EA = economic allocation; SE(S) & SE(B) = system expansion approaches. (Reproduced from Wiedemann et al. 2015.)

emissions associated with wool-based insulation. They estimated that the energy required for scouring was approximately 0.25 kWh electricity and 0.8 kWh gas per kg greasy wool, and that for bonding (with 15% polyester binder) the figures were 0.58 kWh electricity and 0.94 kWh gas per kg finished product (based on personal communication from a manufacturer). They found the product to be carbon-negative across the life cycle, though pre-farm gate emissions were not included in this study. Wool-based insulation also requires additives, such as fire retardant and pesticide, and the authors noted that optimising the use of these was likely to be an important factor in terms of optimising the environmental performance of this product.

Recently, there has been interest in using wool in the built environment for its ability to absorb volatile organic compounds (VOCs) such as formaldehyde. The accumulation of these chemicals in indoor spaces is of increasing concern, as buildings become more airtight in order to meet stricter energy efficiency targets. Mansour et al. (2016) have demonstrated wool's ability to absorb and store a range of these VOCs, thereby removing them from indoor air. Interestingly, the research has shown that the amount and type of VOC removed from air was dependent upon the breed of sheep and the method of wool processing. In their initial study, wool from Swaledales was better at absorbing limonene and dodecane than that from the other breeds tested, while wool from Blackfaces was better at absorbing formaldehyde. This is believed to be due to the varying polarity of the VOCs tested and suggests that wool products could be tailored to selectively absorb a range of specific VOCs from indoor spaces.

Leather

Leather poses similar challenges to wool in terms co-product allocation of on-farm burdens. Disagreement over the classification of hides and skins as either co-products, bi-products or wastes from cattle rearing, and the subsequent impact this can have on LCA results, has dominated recent discussion. For example, a report prepared for the United Nations Industrial Development Organization (UNIDO) (Brugnoli, 2012) concluded that animals were farmed for their meat and milk, but not their hides, and that assessments of leather should therefore start at the slaughterhouse (i.e. modelling hides as a waste product). This is in contrast to PCR guidance for the International EPD System (for finished bovine leather) that requires results be presented separately for three life cycle stages, one being 'upstream processes' which should contain all agricultural inputs (Pernigotti, 2011). The PCR then stipulates that allocation should be modelled on an economic basis.

Further to this, Wiedemann and Yan (2014) have investigated four methods of allocation (biophysical, economic, system expansion and a hybrid approach using both biophysical and system expansion). Their scenarios resulted in either 5.7% (economic allocation) or 13% (biophysical allocation) of cattle rearing burdens being assigned to hides. For the system expansion approaches,

Designing with the Life Cycle in Mind

they substituted leather with vinyl at the ratio of 4:1 by mass to account for the superior durability of leather, which they assumed would last for 80 years. As with wool, the debate over modelling co-production is key to establish eco-profiles of leather, since pre-slaughter processes can account for 65%–85% of leather's overall footprint (Clarke, 2014).

Post-slaughter, Notarnicola et al. (2011) have investigated leather processing and found that tanning, dye-retaining and soaking/liming processes were the environmental hotspots. Their study focused on chromium-based tanning in Europe, and noted that optimisation of these processes (with a view to reducing chemical usage) should be key aims for reducing impacts associ-ated with leather production. However, this and another study (Joseph and Nithya, 2009) also noted the large-scale shift of leather production in recent years from industrialised to developing countries, prompted by more strin-gent environmental regulation in the former. Joseph and Nithya noted that in India (the focus of their study) this had come at considerable cost to both the environment and human health. They cited discharges of untreated effluent as being a main concern, attributing this to poor management practices. This perhaps highlights a weakness in assessing the environmental performance of goods results based on areas of best-practice alone.

More recently, UNIDO have identified a number of opportunities for 'greening' tanning and leather production across the life cycle (UNIDO, 2013). These include the avoidance of damage to skins during rearing and slaughter; elimination of chemicals such as dyestuff and heavy metals; strict control and measurement of water consumption; and better recycling and treatment pro-cesses for effluent. The report also recommended the establishment of spe-cific industrial zones where clusters of leather producers can gain access to specialised treatment services (that may otherwise be prohibitively expen-sive). Finally, they noted emerging processes, such as organic aldehyde and semi-metal tanning methods that may bring further process improvements in the future.

Summary

Natural materials are currently popular choices with designers wishing to tap into a demand for beautiful yet sustainably constructed products. Wood, in particular, is a visible sign of this interest, owing perhaps to its aesthetic appeal when used in exposed areas of buildings and in quality furniture. Other natural materials are being used in less obvious ways, often in compos-ites and as interior components of construction products. The challenge now is further drive a market, and a design philosophy, that sees the greater use of natural materials becoming the long-term norm, rather than components of a cyclical trend.

In the short term, incentives may be needed to develop a market for natural insulation products which, in the UK at least, remain economically uncompetitive compared to mass-produced mineral and foam alternatives. These products enjoy greater uptake elsewhere in the EU, suggesting that they can achieve more market share when pricing makes them attractive to a wider audience.

In the automotive industry, natural fibre composites offer the benefit of vehicle weight reduction and this is driving their greater usage by designers with one eye on fuel efficiency. Many natural fibres, including flax, hemp, jute, sisal and kenaf are already incorporated into a wide range of vehicle parts, and the market for such fibres is expanding (Faruk et al., 2014). Animal-derived materials, such as wool and leather, are receiving less explicit commercial interest at present, though expectations of rapidly increasing global red meat consumption will make these materials more available and cost competitive over time.

All of this is desirable because natural materials offer renewable alternatives to situations where concrete, steel, plastics and other non-renewable materials are currently the norm. In particular, it is natural materials capacity to store biogenic carbon throughout their lifetime that is most compelling about considering their wider usage. Wood used in the built environment already represents a significant carbon sink, but the same is true for other long-lived natural material products. Hemp-concrete (for example) stores approximately 325 kg carbon per tonne throughout its life (Pervaiz and Sain, 2003). By increasing the use of these materials, while still cultivating them sustainably, they can act as a significant global carbon sink.

Challenges remain in terms of process optimisation for plant fibre products, both at farm and factory levels. Many plant fibre crops grow quickly and require few agricultural inputs, but studies show that systems vary widely from location to location (e.g. Dellaert, 2014). This can have a significant effect on the environmental profile of the cultivated fibres, and attention must be paid to potential land use issues as production of crops for fibre scales up. At the factory level, there are opportunities for technological advances and process optimisation. This can reduce energy demand, offer cleaner production methods, and drive the development of further markets. Scaling-up of production will inevitably bring economies of scale that bring further benefits in terms of LCA profiling.

In relation to environmental assessment, further development of sector-specific PCRs for use in LCA calculations will help harmonise the approach taken to some of the more complex aspects of modelling natural materials life cycles. Greater public availability of independently verified LCA results, through schemes such as EPD programmes, and the continued use of certification schemes such as BREEAM and LEED, will also help further embed life cycle thinking into the natural materials design process.

Designing with the Life Cycle in Mind

References

Adalberth, K. 2000. Energy use and environmental impact of new residential buildings, Report TVBH-1012, Department of Building Physics, Lund Institute of Technology, Lund, Sweden.

Aktas, C.B. and Bilec, M.M. 2012. Impact of lifetime on US residential building LCA results. *International Journal of Life Cycle Assessment* 17:337–349.

Alves, C., Ferrão, P.M.C., Silva, A.J., Reis, L.G., Freitas, M., Rodrigues, L.B., Alves, D.E. 2010. Ecodesign of automotive components making use of natural jute fiber composites. *Journal of Cleaner Production* 18:313–327.

Ardente, F., Beccali, M., Cellura, M., Mistretta, M. 2008. Building energy performance: a LCA case study of kenaf-fibres insulation board. *Energy and Buildings* 40:1–10.

Batouli, S.M., Zhu, Y., Nar, M., D'Souza, N.A. 2014. Environmental performance of kenaf-fiber reinforced polyurethane: a life cycle assessment approach. *Journal of Cleaner Production* 66:164–173.

Boeri, F. 2013. *Product Category Rules: Meat of Mammals. UN CPC 2111 and 2113.* Environmental Product Declaration, Stockholm.

Biswas, W.K., Graham J., Kelly, K., John, M.B. 2010. Global warming contributions from wheat, sheep meat and wool production in Victoria, Australia - a life cycle assessment. *Journal of Cleaner Production* 18:1386–1392.

Bledzki, A.K. and Faruk, O. 2006. Cars from bio-fibres. *Macromolecular Materials and Engineering* 291:449–457.

Blonk Agri-footprint BV. 2015. *Agri-Footprint 2.0- Part 1—Methodology and Basic Principles.* Gouda. http://www.agri-footprint.com/methodology/methodology-report.html (accessed 21 June 16).

Borjesson, P. and Gustavsson, L. 2000. Greenhouse gas balances in building construction: wood versus concrete from life-cycle and forest land-use perspectives. *Energy Policy* 28:575–588.

BRE, 2016. *BREAM International New Construction 2016, Technical Manual SD233 1.0.* BRE Global, Watford.

Browne, N.A., Eckard, R.J., Behrendt, R., Kingwell, R.S. 2011. A comparative analysis of on-farm greenhouse gas emissions from agricultural enterprises in south eastern Australia. *Animal Feed Science and Technology* 166–167:641–652.

Brugnoli, F. 2012. Life cycle assessment—carbon footprint in leather processing (review of methodologies and recommendations for harmonization). United Nations Industrial Development Organization. Eighteenth Session of the Leather and Leather Products Industry Panel, Shanghai, China, 01–05 September 2012.

BSI. 2013. *BS EN 15804:2012+a1:2013 Sustainability of Construction Works— Environmental Product Declarations—Core Rules for the Product Category of Construction Products.* British Standards Institute, London.

Clarke, C. 2014. System boundaries in product carbon footprints of leather. *The International Leather Maker*, Dec-Jan, 24–26.

Corbière-Nicollier, T., Gfeller Laban, B., Lundqvist, L., Leterrier, Y., Månson, J-A.E., Jolliet, O. 2001. Life cycle assessment of biofibres replacing glass fibres as reinforcement in plastics. *Resources, Conservation and Recycling* 33:267–87.

Cottle, D.J. and Cowie, A.L. 2016. Allocation of greenhouse gas production between wool and meat in the life cycle assessment of Australian sheep production. *International Journal of Life Cycle Assessment* 21(6):820–830. DOI: 10.1007/s11367-016-1054-4.

Critchlow, J. 2010. *End of Life Furniture Sustainability*. International Specialised Skills Institute, Victoria.

Dellaert, S.N.C. 2014. Sustainability assessment of the production of sisal fiber in Brazil. Master thesis, Faculty of Geosciences, Utrecht University, The Netherlands.

Deutz, P., McGuire, M., Neighbour, G. 2013. Eco-design practice in the context of a structured design process: an interdisciplinary empirical study of UK manufacturers. *Journal of Cleaner Production* 39:117–128.

Dittenber, D.B. and GangaRao, H.V.S. 2012. Critical review of recent publications on use of natural composites in infrastructure. *Composites: Part A* 43:1419–1429.

Eady, S., Carre, A., Grant, T. 2012. Life cycle assessment modelling of complex agricultural systems with multiple food and fibre co-products. *Journal of Cleaner Production* 28:143–149.

EC, 2003. Integrated Pollution Prevention and Control (IPCC) reference document on best available techniques for the textile industry, July 2003. Published by the European Commission.

Edwards-Jones, G., Plassmann, K., Harris, I.M. 2009. Carbon footprinting of lamb and beef production systems: insights from an empirical analysis of farms in Wales, UK. *Journal of Agricultural Science* 147:707–719.

FAO. 2015. Greenhouse gas emissions and fossil energy use from small ruminant supply chains. Guidelines for assessment. Food and Agriculture Organization of the United Nations. http://www.fao.org/partnerships/leap/publications/en/ (accessed 06 May 2016).

Faruk, O., Bledzki, A.K., Fink, H-P., Sain, M. 2014. Progress report on natural fiber reinforced composites. *Macromolecular Material Engineering* 299:9–26.

Foresight, 2011. The future of food and farming. Final project report. The Government Office for Science, London, UK. https://www.gov.uk/government/uploads/system/uploads/attachment_data/file/288329/11-546-future-of-food-and-farming-report.pdf (accessed 21 June 2016).

Gac, A., Ledgard, S., Lorinquer, E., Boyes, M., Le Gall, A. 2012. Carbon footprint of sheep farms in France and New Zealand and methodology analysis. In *Proceedings of the 8th International Conference on Life Cycle Assessment in Agri-food Sector, (LCA Food 2012), 1–4 October 2012, Saint Malo, France*, ed. Corson, M.S. and H.M.G. van der Werf.

Goedkoop, M. J., Heijungs, R., Huijbregts, M., De Schryver, A., Struijs, J., Van Zelm, R. 2008. ReCiPe 2008. A life cycle impact assessment method which comprises harmonised category indicators at the midpoint and the endpoint level; First edition Report I: Characterisation. http://www.lcia-recipe.net (accessed 10 May 2016).

Hill, C. and Dibdiakova, J. 2016. The environmental impact of wood compared to other building materials. *International Wood Products Journal* 7(4):215–219. DOI: 10.1080/20426445.2016.1190166.

Hill, C., Norton, A., Kutnar, A. 2015. *Environmental Impacts of Wood Composites and Legislative Obligations, in Wood Composites, (ed. M. Ansell)*. Woodhead Publishing, Cambridge, pp. 311–333.

ISO, 2011. *ISO 140006:2011. Environmental Management Systems—Guidelines for Incorporating Eco-Design*. International Organization for Standardization, Geneva.

ISO, 2006a. *ISO 14040:2006. Environmental Management—Life Cycle Assessment—Principles and Framework*. International Organization for Standardization, Geneva.

ISO, 2006b. *ISO 14044:2006. Environmental Management—Life Cycle Assessment—Requirements and Guidelines*. International Organization for Standardization, Geneva.

John, S., Nebel, B., Perez, N., Buchanan, A. 2009. Environmental impacts of multi-storey buildings using different construction materials. Research Report 2008-02. Christchurch, New Zealand. Department of Civil and Natural Resources Engineering, University of Canterbury.

Jones, A.K., Jones, D.L., Cross, P. 2014. The carbon footprint of lamb: Sources of variation and opportunities for mitigation. *Agricultural Systems* 123:97–107.

Joseph, K. and Nithya, N. 2009. Material flows in the life cycle of leather. *Journal of Cleaner Production* 17:676–682.

Joshi, S.V., Drzal, L.T., Mohanty, A.K., Arora, S. 2004. Are natural fiber composites environmentally superior to glass fiber reinforced composites?. *Composites: Part A* 35:371–376.

Lenzena, M. and Treloarb, G. 2002. Embodied energy in buildings: wood versus concretereply to B. Orjesson and Gustavsson. *Energy Policy* 30:249–255.

Luz, S.M., Caldeira-Pires, A., Ferrão, P.M.C. 2010. Environmental benefits of substituting talc by sugarcane bagasse fibers as reinforcement in polypropylene composites: ecodesign and LCA as strategy for automotive components. *Resources, Conservation and Recycling* 54:1135–1144.

Mansour, E., Curling, S., Stéphan, A., Ormondroyd, G. 2016. Absorption of volatile organic compounds by different wool types. *Green Materials* 4:1–7.

Mark, L. 2013. LEED outstrips BREEAM across the globe—including Europe. *The Architects Journal*. http://www.architectsjournal.co.uk/news/leed-outstrips-breeam-across-the-globe-including-europe/8643464.article (accessed 21 June 16).

Murphy, R.J. and Norton, A. 2008. Life cycle assessments of natural fibre insulation materials. Final report. National Non-Food Crops Centre, York, UK.

Nebel, B., Zimmer, B., Wegener, Z. 2006. Life cycle assessment of wood floor coverings. A representative study for the German flooring industry. *International Journal of Life Cycle Assessment* 11(3):172–182.

Notarnicola, B., Puig, R., Raggi, A., Fullana, P., Tassielli, G., De Camillis, C., Rius A. 2011. Life cycle assessment of Italian and Spanish bovine leather production systems. *Afinidad LXVIII* 553:167–180.

Ormondroyd, G.A., Spear, M.J., Skinner, C. 2016. The opportunities and challenges for re-use and recycling of timber and wood products within the construction sector. In: Kutnar A., Muthu S.S., (eds.) *Environmental Impacts of Traditional and Innovative Forest-Based Bioproducts*. Environmental footprints and ecodesign of products and processes, vol 7. Springer, Singapore, pp. 45–103. ISBN 9789811006531; 9789811006555.

Pernigotti, D. 2011. *Product Category Rules: Finished Bovine Leather. UN CPC 2912*. Environmental Product Declaration, Stockholm.

Pervaiz, M. and Sain, M.M. 2003. Carbon storage potential in natural fiber composites. *Resources, Conservation and Recycling* 39:325–340.

Pretot, S., Collet, F., Garnier, C. 2014. Life cycle assessment of a hemp concrete wall: impact of thickness and coating. *Building and Environment* 72:223–231.

PwC, 2012. *Sustainable Packaging: Myth or Reality.* Pricewaterhouse Coopers LLP, London. https://pwc.blogs.com/files/pwc-sustainable-packaging-report-june2012.pdf (accessed 21 June 16).

Read D.J., Freer-Smith P.H., Morrison J.I.L., Hanley N., West C.C., Snowdon P., (eds.). 2009. *Combating Climate Change—A Role for UK forests. An Assessment of the Potential of the UK's Trees and Woodlands to Mitigate and Adapt to Climate Change.* The Stationery Office, Edinburgh.

Rincón, L., Pérez, G., Cabeza, L.F. 2013. Service life of the dwelling stock in Spain. *International Journal of Life Cycle Assessment* 18:919–925.

Rivela, B., Moreira, M.T., Feijoo, G. 2007. Life cycle inventory of medium density fibreboard. *International Journal of Life Cycle Assessment* 12(3):143–150.

Schmidt, W-P. and Beyer, H.M. 1998. Life cycle study on a natural fibre reinforced component. SAE Technical Paper 982195, SAE International. DOI: 10.4271/982195.

Schmidt, A.C., Jensen, A.A., Clausen, A.U., Kamstrup, O., Postlethwaite, D. 2004. A comparative life cycle assessment of building insulation products made of stone wool, paper wool and flax. *International Journal of Life Cycle Assesment* 9(1)53–66.

Skinner, C., Stefanowski, B.K., Heathcote, D., Charlton, A., Ormondroyd, G.A. 2016. Life cycle assessment of pilot-scale wood fibre production using mechanical disc refining at different pressures. *International Wood Products Journal* 7(3):149–155. Doi: 10.1080/20426445.2016.1200825.

Sobczak, L., Lang, R.W., Haider, A. 2012. Polypropylene composites with natural fibers and wood—general mechanical property profiles. *Composites Science and Technology* 72:550–557.

Turley, D.B., Horne, M., Blackburn, R.S., Stott, E., Laybourn, S.R., Copeland, J.E., Harwood, J. 2009. The role and business case for existing and emerging fibres in sustainable clothing: final report to the Department for Environment, Food and Rural Affairs (Defra), London, UK.

UNIDO. 2013. Greening food and beverage value chains: the case of the meat processing industry. A report for the UNIDO Green Industry Initiative, United Nations Industrial Development Organization, Vienna, Austria.

USGBC, 2016. *LEED v4 for Building Design and Construction.* U.S. Green Building Council, Washington D.C.

Weidema, B.P., Bauer, Ch., Hischier, R., Mutel, Ch., Nemecek, T., Reinhard, J., Vadenbo, C.O., Wernet, G. 2013. The ecoinvent database: overview and methodology, data quality guideline for the ecoinvent database version 3, www.ecoinvent.org.

Wiedemann, S.G., Ledgard, S.F., Henry, B.K., Yan, M-J., Mao, N., Russell, S.J. 2015. Application of life cycle assessment to sheep production systems: investigating co-production of wool and meat using case studies from major global producers. *International Journal of Life Cycle Assessment* 20:463–476.

Wiedemann, S. and Yan, M. 2014. Livestock meat processing: inventory data and methods for handling co-production for major livestock species and meat products. In: Schenck, R., and Huizenga, D., (eds.) *Proceedings of the 9th International Conference on Life Cycle Assessment in the Agri-Food Sector (LCA Food 2014), 8–10 October 2014, San Francisco, USA.*

Wötzel, K. and Flake, M. 2001. Renewable materials as an alternative to plastics: a comparative life cycle analysis of a hemp fibre reinforced component and an ABS-moulded carrier. *Umweltwissenschaften und Schadstoffforschung* 13(4):237–47.

Ximenes, F.A. and Grant, T. 2013. Quantifying the greenhouse benefits of the use of wood products in two popular house designs in Sydney, Australia. *International Journal of Life Cycle Assessment* 18:891–908.

Zah, R., Hischier, R., Leão, A.L., Braun, I. 2007. Curauá fibers in the automobile industry—a sustainability assessment. *Journal of Cleaner Production* 15:1032–1040.

Zampori, L., Dotelli, G., Vernelli, V. 2013. Life cycle assessment of hemp cultivation and use of hemp-based thermal insulator materials in buildings. *Environmental Science and Technology* 47:7413–7420.

5

Restoring Credibility to Natural Materials

D. K. Spilsbury
The Wool Packaging Company Limited

CONTENTS

Introduction 133
'You Cannot Be Serious!' 134
Global Influences 135
Historic Credentials 135
The Industrial Revolution (1700–1900) and New Opportunities 139
Progress Is All 139
Pragmatism and Conservation 140
Emerging Synthetics and Natural Precursors 142
The Birth of the Plastic Age 142
The British Experience 146
Mass-Market Demand 147
Design for the Future – Utility 148
Resistance to the 'American Way' 150
Utility versus Glamour 150
Big in Japan 153
The Swinging Sixties 155
Modernism to Post-Modernism 157
The Future 158
Restoring Credibility 159
Silk 159
Wood 160
Wool – A Technical 'Smart Fibre' 160
Bibliography 163

Introduction

Historically, science has pursued a premise that Nature can be understood fully, its future predicted precisely, and its behaviour controlled at will. However, emerging knowledge indicates that the nature of Earth and biological systems transcends the limits of science, questioning the premise of knowing, prediction and control. This knowledge has led to the recognition that, for civilised human survival, technological society has to adapt to the constraints of these systems.

Simultaneously, spurred by explosive developments in the under-standing of materials (non-biological and biological), applied scientific research pursues a contrary goal of controlling the material world, with the promise of spectacular economic growth and human well-being. If adaptation to Nature is so important, why does applied research pursue a contrary course?

Adapting to Nature requires a recognition of the limitations of science, and espousal of human values. Although the concept of adapting to Nature is accepted by some, especially conservation ecologists, such an acceptance may not exist in other fields. Also, in a world dominated by democratic ideals of freedom and liberty, the discipline required for adapting to Nature may often be overridden by competition among various segments of society to exercise their respective rights. In extreme cases of catastrophic failure of Earth or biological systems, the imperative for adaptation may fall victim to instinct for survival.

In essence, although adequate scientific know-how and technological competence exists to facilitate adaptation to Nature, choosing between that and the pursuit of controlling Nature entails human judgment. What that choice may be when humans have to survive under severe environmental stress cannot be predicted.

TN Narasimhan *Limitations of Science and Adapting to Nature*[*]

Before we go any further, let's get one thing straight: natural materials are a supremely credible choice for modern purposes. The plants and animals that produce them have spent millions of years on their research and development, and have reached a degree of technical sophistication that leaves mankind's advanced materials struggling to emulate them.

If we accept this fact, and dedicate our research techniques and our cutting-edge technology to the better understanding of natural materials, we will have an opportunity not only to improve artificial products, but also to design a more sustainable world for ourselves.

'You Cannot Be Serious!'

Not the rant of an irate eighties tennis player, but the incredulous reaction of a latter-day Big Pharma buyer to the suggestion that they might consider

---

[*] Department of Materials Science and Engineering, Department of Environmental Science, Policy and Management, 210 Hearst Memorial Mining Building, University of California, Berkeley, CA 94720-1760, USA.

Restoring Credibility to Natural Materials 135

using wool fibre rather than expanded polystyrene as a packaging material. If you do make that suggestion, the buyer will look you in the eye while using their peripheral vision to check if you are wearing beads or sandals. Then they will tell you confidently that such a suggestion is too eccentric to take seriously. 'We all know that natural materials are more expensive to produce and do not perform as well as man-made'.

This belief is so deeply entrenched that it has become one of those things that everyone just knows. It's axiomatic. But it's also simply and straightforwardly wrong – and in some contexts, we know it's wrong. Who would prefer a suit made from polyester to one made of wool? So, how did this state of affairs come about for packaging? How did natural materials lose their credibility? This is the question that this chapter will seek to answer.

Global Influences

Over the past 200 years, a series of cultural and scientific events have changed the popular perception of natural materials, and this change has resulted in a situation where consumers understand the need for sustainability but are poorly informed about what it means in practice. Terms such as 'organic', 'fair trade' and 'environmentally friendly' have been exploited by marketeers, and as a result, their original meanings have been sacrificed on the altar of market share. People are torn between the desire for an ever-improving lifestyle that necessitates the use of man-made products created from finite resources and the more sensible use of natural materials as a credible alternative, but often have too little information to make an informed choice as to where to strike the balance.

If we step back to the beginnings of the Industrial Revolution and run the tape forward to the present day, we can chart these cultural events and identify the stages of growth that have brought us to modern environmentalism. At the same time, we can highlight how the development of man-made polymers has played its part in our prejudicial view of natural materials.

Historic Credentials

To begin with, let us take a look at our original understanding of natural materials. The use of natural fibres to make textiles was the largest global industry for most of recorded history. And for nearly 5,000 years, up until the late 19th century, textiles were made primarily from flax, hemp, silk, wool and cotton.

The civilisation of Egypt, China and Peru used cotton for cloth and as long ago as 1500 BC. India established a cotton weaving industry that produced fine cloths with complex and sophisticated designs. When Alexander the Great invaded that country, his troops started wearing light cotton clothes in preference to wool. Cotton manufacture was introduced to Europe by the Umayyad conquest of the Iberian Peninsula and Sicily in the 8th and 9th centuries. The knowledge of cotton weaving spread to northern Italy in the 12th century, when Sicily was conquered by the Normans, and from there to the rest of Europe. Spinning wheels, and the dangers of being pricked by them, entered popular mythology.

A distracted spinner: detail from the Smithfield Decretals (Illuminated manuscript of about 1350).

Flax, too, played a major role in ancient economies. The earliest evidence of humans using wild flax as a textile comes from the Caucasus, where spun, dyed and knotted fibres were found in the Dzudzuana Cave in the modern state of Georgia and dated to the Upper Palaeolithic, some 30,000 years ago. Flax spread through Mesopotamia, Syria, Lebanon, Jordan, Palestine and northern Egypt, where it was used to mummify corpses. Samples tested by the Polish Institute of Natural Fibres have shown that flax fabrics were also present in Turkey in 6500 BC and in Sudan in 2000 BC. There was a parallel development in China and India, where domesticated flax was cultivated at least 5,000 years ago. In Europe, textile production began in Switzerland and Germany by 3000 BC.

Restoring Credibility to Natural Materials 137

Egyptians harvesting flax.

Silk is a natural protein fibre. According to legend, Empress Hsi Ling Shi, wife of Emperor Huang Ti (the Yellow Emperor), discovered its potential as a fabric, while sipping tea under a mulberry, when a cocoon fell into her cup and began to unravel. The empress became fascinated by the fine, shiny threads and followed their source to the *Bombyx mori* silkworm in the leaves above her. From this discovery, the Chinese derived sericulture, the cultivation of silkworms, and the subsequent invention of the silk reel and loom. For nearly 3,000 years, the Chinese maintained a global monopoly on silk production. So valuable was this material that it was traded at the price of gold, and the penalty for industrial espionage was death.

The first silk road.

The story of wool begins in Asia Minor during the last days of the Upper Palaeolithic, about 10,000 years ago. Primitive man living in the Mesopotamian Plain used sheep and goats for three basic needs: food, clothing and shelter. Later, humans learned to spin and weave certain specialised fibres taken from their coats: woollen textiles originated from the Akkadian Empire in around 2300 BC. The warmth of woollen clothing and the mobility of the herds then allowed civilisations to spread far beyond the warm climate of Mesopotamia.

Ancient sheep.

Between 3000 BC and 1000 BC, the Persians, Greeks and Romans distributed sheep and wool throughout Europe and worked to improve their breeds. The Romans took sheep everywhere they went; on the British Isles, they established a wool processing operation in what is now Winchester as early as 50 AD.

The wool trade in English medieval history.

The Industrial Revolution (1700–1900) and New Opportunities

The Industrial Revolution took shape in Britain in the second half of the 18th century. By the early 19th century, Britain was producing more than three-quarters of the mined coal, half the cotton and iron goods and a majority of steam engines in Europe. The industrial techniques and processes that originated in England and Scotland then spread rapidly through northern Europe and the USA.

Richard Trevithick's 'Catch-Me-Who-Can'.

This period also marked the early stages of the modern population explosion. In the mid-1700s, the world population of humans stood at 700 million; around 1800, the billionth baby was born.

The revolution increased material wealth, extended life and was a powerful force for social change. It also undermined the centuries-old class structure in Europe and helped to confirm the 'enlightenment' view that the proper use of science and technology could bring humanity a more fruitful quality of life.[*]

Progress Is All

The Industrial Revolution heralded the beginning of an era of progress and excess. Alongside the material gains brought by new energy sources, urban

---

[*] http://www.victorianweb.org/technology/ir/irchron.html – The Industrial Revolution – A Timeline. The Industrial Revolution: A Timeline George P. Landow, Shaw Professor of English and Digital Culture, National University of Singapore.

expansion, deforestation and the widespread use of chemicals began to place pressures on the environment, resulting in the beginning of pollution, the depletion of natural resources and reductions in biodiversity in habitats exploited by the new industries.

The idea that 'progress' through growth and improvement was the destiny of human and natural life was consolidated during the 19th century. Charles Darwin's theory of evolution, first expressed in *Origin of Species* in 1859 was taken as a nature's confirmation of this.

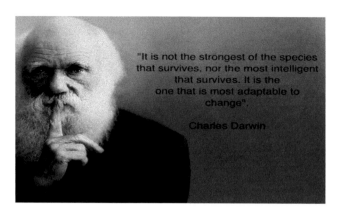

Charles Darwin.

Herbert Spencer extended Darwin's discovery into the social realm, and David Ricardo and John Stuart Mill argued that growth and improvement in the economic sphere was best delivered by unregulated competition. Little did they anticipate that in less than 200 years, this philosophy would result in the global depletion of natural resources and the alteration of the globe's climate to an extent that casts a shadow over the future possibilities of life on Earth.

Pragmatism and Conservation

The USA plays a major role in the global environmentalism story. This was partly through the Transcendentalist movement of the 1820s, associated with Ralph Waldo Emerson and Henry David Thoreau. Some 50 years later, this current of thought, with its reverencing of nature, flowed into the Pragmatism movement that began with Charles Sanders Peirce. Together they led to a distinctively American 'can do' mentality that inspired new measures to help conserve nature.

Beginning in the 1860s, the US government began to create parks and preserve wild territories for the environment and public benefit. Yosemite

Restoring Credibility to Natural Materials 141

became the first National Park in 1872. The Audubon Society was founded in 1872 and Sequoia and General Grant parks were established. Twenty years after that, John Muir was elected president of the Sierra Club, one of the first large-scale environmental preservation organisations in the world, and one that continues to lobby for green policies.[*]

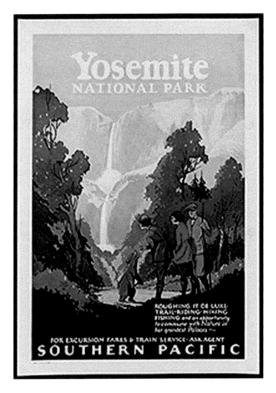

Yosemite was California's most popular tourist destination in the 1920s (California State Library).

Although the federal government had begun taking actions to preserve lands, it was Muir and President Theodore Roosevelt who publicised and popularised conservation. By 1916, the National Park Service had been established and preservationism became part of a mass movement. Led by such figures as Roosevelt and Gifford Pinchot, the first chief of the US Forestry Service, these early conservationists focused on the wise and efficient use of natural resources.

In the meantime, the development of synthetics was coming up on the rails.

---

[*] www.webecoist.momtastic.com/2008/08/.../a-brief-history-of-the-modern-green-movement.

Emerging Synthetics and Natural Precursors

'Plastic' existed as a concept long before it became an industrial product. As early as the 1600s, the term was used to denote something that could be easily shaped, and was derived from the Latin word *plasticus* and the Greek word πλαστικός both meaning 'able to be moulded, pertaining to moulding'. Most likely, Greeks used 'plastikos' to describe clay before it was baked.

It is not until the 1800s that we begin to see the emergence of the first synthetics, and once we did, their development started to overlap and overtake natural materials.

The Birth of the Plastic Age

As Bismarck predicted, some damned silly thing in the Balkans triggered the global tragedy of two World Wars and the Great Depression of the 1930s. The conflicts created a demand for war materiel that natural materials could not meet. Manufacturers therefore turned to the burgeoning chemicals industry to supply the deficit; cellulosic, acrylic, nylon, and especially phenolic and polyethylene became valuable materials. Production was increased. Material manufacturers, machine builders, mould makers and processors all prospered.

Teflon from DuPont chemicals.

Restoring Credibility to Natural Materials 143

The groundwork for the explosion of chemical products in the Second World War was laid in the 1920s. Cellulose acetate was marketed as a substitute for celluloid. Vynilite was also brought onto the market as a medium for gramophone records.

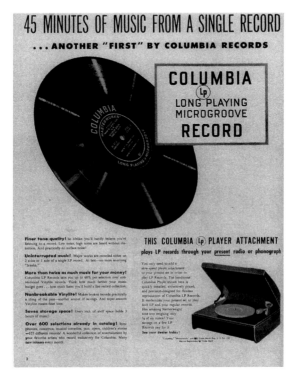

A shellac '78.

Probably the greatest discovery of that era was urea formaldehyde resin which, unlike Bakelite, offered a full colour range with no loss of strength including, for the first time, white. The first of these resins was known as 'beetle' and was available in the USA under licence from the UK.

Other materials available before the Second World War included Lucite, the glass-like acrylic manufactured by Du Pont, and its rival Plexiglas®, made by Rohm & Haas. There was also the best-known consumer plastic of all, i.e. nylon. This emerged in 1938 as a sleek and inexpensive alternative to silk for women's stockings. But, it soon became just as important for the military, supplying parachutes, tents, ponchos and ropes for the Allied war effort.

The end of the Second World War marked the beginning of a seemingly endless boom for the US economy. Increasingly, affluent Americans began to expect that their quality of life would continuously improve, and they increasingly saw a clean, safe and beautiful environment as an essential component of their rising living standards. This, rather than a concern for the

144 *Designing with Natural Materials*

A beetle clock on offer in 1945.

management of natural material resources, was the driver behind modern environmentalism. The introduction of plastics meant that luxury materials such as ebony, alabaster, onyx and amber could be imitated and made available to all. Consumerism was revolutionised.

Vending machine for Nylons in the US 1950's – Getty Images.

Restoring Credibility to Natural Materials 145

The aggressive marketing of plastics resulted in hundreds of consumer products being redesigned to make better use of these new materials.[*] An executive of General Plastics summed up the marketing campaigns by urging colleagues to 'make decoration symbolic of our modern age, using simple machine-cut forms to get that verve and dash that is so expressive of contemporary life'.[†] Together with the good reputation that plastics had acquired during the war, the campaign helped plastic to become a reputable material, shaking off its 'substitute' image. The look and feel of the 1950s' USA – a 'modern' world of moulded plywood furniture, fibreglass, plastics, and polyester – had its roots in the materials innovations of the war. The glamour of Plexiglas® cockpit covers on Allied bombers naturally transferred to the bubble domes of futuristic cars in peacetime.

Plastics products had the added advantage of being relatively easy and cost-effective to make. Products that had been moulded had no need of labour-intensive hand finishing. A radio case, for example, 'popped' out of its mould and was ready for assembly. This was a boon to the manufacturing industry as it had products that could be sold at a low price and high margin. For their part, American consumers had well-designed, affordable products to choose from.

The bubble car – Getty Images.

---

[*] *Plastics, Social Attitudes & Domestic Product Design*, Plastics Historical Society 6 Dec 2016 p.2.
[†] *The Plastic Age from Modernity to Post-modernity*, Meikle JL. Victoria & Albert Museum, 1990.

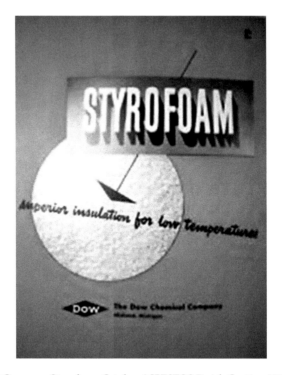

(DOW Chemical Company Styrofoam Catalog ASBESTOS Finish Coating 1954).

The British Experience

The USA's wholehearted enthusiasm for plastic was not replicated in Britain. Before the war, plastic had been seen as suspect, an ersatz substitute for 'real' materials. Towards the end of the 1930s, there was resistance to 'machine age' style, and the popularity of plastics had begun to lessen. British propaganda, led by the Arts and Crafts movement, and motivated to some degree by envy of the advances made by the Germany chemical industry, began accusing plastics of being 'dishonest'.

When the war ended, the British people expected to return to the genuine materials denied to them by wartime privations. Natural materials seemed to be linked with a return to 'normality', to the values and traditions the war had disrupted. At the *Britain Can Make It* exhibition of 1946 there was a section called 'Designers Look Ahead', which showed how advances made during the war could be developed into domestic markets. The 'War to Peace' section showed '… designs springing from wartime innovations in materials'.[*]

---

[*] Woodham JM 1983.

Restoring Credibility to Natural Materials 147

The following year saw the first *Ideal Home Exhibition* since the war, featuring a section entitled 'Science Comes Home' that also demonstrated how wartime research could be applied to domestic situations. It must have seemed as though the potential for plastics in peacetime was unlimited.

(Ideal Homes Exhibition 1947).

The British consumer, however, did not have the reaction to plastics the industry had hoped for. A number of factors became apparent that had been overlooked by the enthusiastic optimism of an industry convinced of its post-war success.

Mass-Market Demand

There were also indications that a 'mass-culture' was emerging. In the decade that followed the war, mass media had a greater impact on society than ever before. These, together with newer forms of communication were fast becoming part of daily life: as one commentator put it, 'More than ever before, the effects of the press, the radio, television and the cinema entered the lives of practically every individual in the industrialised world'.[*]

This resulted in a dramatic cultural change. New values, ambitions and expectations were emerging, greatly influenced by the media, which was freely available to most people. Although this 'move forward' was welcomed

---

[*] Sparke P (Ed), *The Plastic Age—From Modernity to Post-Modernity* Victoria & Albert Museum, 1990.

by some, it was criticised by others, notably the Council for Industrial Design (COID) that had been set up to guide consumers and designers, and to educate the tastes of the masses. The council felt that 'many post-war developments represented a fundamental threat to all that was quintessentially British about Britain, championing the English virtues of tradition and craftsmanship as bulwark against American manufacturing and selling techniques'.[*]

Design for the Future – Utility

The Board of Trade Utility Scheme was introduced in 1942 in response to the declining availability of natural materials, such as timber, during the Second World War. Its remit was to supply consumers with well-designed products made using a minimum of materials. Its ruling was that only those products that conformed to utility standards could be produced after 31 January 1943.

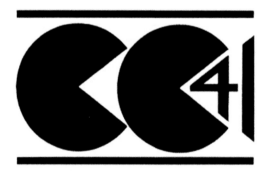

Utility mark. Source: Board of Trade, *Utility Furniture Trader's Leaflet* UFD/8, 1942, Utility Furniture Distribution Committee (1942–1944), The National Archive, BT 64/1749.

The adopted 'utility' aesthetic followed the spirit of the Council for Art and Industry's 'working class home report' of 1937, which had stressed the need for well-designed furniture and domestic equipment to be marketed at an affordable price, with emphasis on good proportions, pleasant colours, surfaces and textures.[†] The Utility scheme was generally seen as being a golden opportunity to put into practice the good design doctrine with regards to the use of plastics developed in the 1930s.

---

[*] Catteriall C, 1990.
[†] Woodham JM, 1983.

Utility dining room and living room furniture, 1945. Source: Design Council Archive, University of Brighton Design Archives.

The Board of Trade was not alone in its desire to promote good design. Organisations such as the Council for Encouragement of Music and the Arts (CEMA) were also working towards public design awareness. It promoted exhibitions such as 'Homes to Live In', which sought to demonstrate the ties between social problems, town planning and interior design.[*] CEMA also produced a travelling exhibition aimed at 'improving taste' in everyday life. This travelled around a circuit of educational establishments complete with a permanent lecturer. The main emphasis of the display was placed on the rooms in a house, and the domestic products that could be used to furnish them.

A great number of these organisations, such as the Council for Art and Industry and the Central Institute for Art and Design, initiated contracts with the plastics industry to promote research that would help to boost post-war exports. In 1945, CEMA staged a 'Design at Home' exhibition, held at the National Gallery in London, prior to touring the country. Its intention was to stimulate interest in how a post-war home might be furnished. Noel Carrington, in his exhibition catalogue introduction, put forward a case for the acceptance of mass production and the synthetic materials that were on display. Milner Gray, who, with other members of the Ministry of Information, was responsible for organising the exhibition, displayed a Perspex washbasin.[†]

---
[*] Ibid.
[†] Ibid.

The Council for Industrial Design (COID) was formed in 1944 as an independent body concerned with the production and marketing of British industrial goods. The aim was to improve Britain's ability to compete for market share once the war had ended, and good design appeared to be the key. The problem was that, although British design had a good reputation for workmanship, it lacked the appeal of American, German, Swiss, Swedish and Italian products, and although there was a certain amount reluctance to change it became clear that in order to succeed in peacetime, British design would have to adopt American industrial techniques: good mass produced design was seen as essential to economic prosperity.

The Festival of Britain, held throughout the United Kingdom in the summer of 1951, was another opportunity for the COID to convince the public that good design would substantially improve the quality of their lives. The Homes and Gardens Pavilion was set up as the showcase for the best domestic products. However, British reluctance to embrace 'mass culture' continued to distort the message, and the country's industrial designers struggled to compete against the high quality plastic products being produced by their counterparts in Italy and Germany.

Resistance to the 'American Way'

Another complicating factor was the ambiguous attitude of British industry towards their American cousins. The United States' influence on British industrial design had been evident before the Second World War, and there was a desire to reproduce its commercial effectiveness. Raymond Loewy, the father of modern American design, opened an office in London in the thirties and was made an honorary member of the Faculty of Royal Designers in 1939, and in 1949 the Anglo-American council on productivity sent a study group to the United States to examine the possible adoption of American design styles, resulting in the recommendation of expanding their use in Britain.

Utility versus Glamour

Even in the early 1940s, reservations had been expressed about the 'superfluous' aerodynamic streamlining of household objects such as refrigerators, and this hardened during the 1950s. As design historian Jonathan Woodham put it, 'People were afraid that Britain was going to follow the United States

Restoring Credibility to Natural Materials 151

in the submersion of people's individualism in large-scale, monopolistic commercial and industrial networks'.[*]

This distaste for American 'brashness' was offset by a fascination by what was to be found in American kitchens and garages. *Design Magazine*, the mouthpiece of the COID, regularly featured American goods, and arguments for and against Americanisation were a standard theme. To some extent, the question was academic: Britain's economic position after the war was too precarious to allow mass consumerism, and economic controls, rationing and the Utility Scheme persisted into the 1950s; when they were over, it was clear that the kind of design control represented by CEMA and COID, which took as its starting point 'the greatest good for the greatest number', was also over.

By comparison with the consumer goods on display at the 1946 'Britain Can Make It' exhibition, utility goods seemed to represent all that was dull, uniform and uninspiring. The British, starved of such novelties throughout the war and bored with the sensibleness of utility items, yearned for colour, decoration and expressive forms. What was more, money was not necessarily a problem. Visitors to the exhibition were advised: 'Some people believe that what is mass-produced and low-priced cannot be well designed. They think that 'good' must mean 'expensive'. On the contrary, design is not a matter of price. Limited sums spent carefully can buy good design'.[†]

---

[*] Ibid.
[†] Sparke P, 1986.

A dining room in a suburban home, as imagined by the 'Britain Can Make It' Exhibition of 1946 (Image courtesy of Flickr Design Archives).

Designed by David Booth, ARIBA, NRD 'A young curate, who is a keen naturalist and a great reader, has a wife who likes to collect modern pottery, three children, who do their homework in the dining room – and not much money. The dining room designed for them has the whole wall adjoining the kitchen designed for service – the oak sideboard is in a recess between glass-fronted shelves painted white, on which china can be placed from the kitchen and taken out in the dining room. The chairs are upholstered in red'.

Design for the modern home saw a new use of materials, particularly plastics and aluminium. Colours and patterned designs were coming to the fore. Manufacturers such as Heals, Ernest Race and Dunns, capitalised on the new optimism by commissioning items from the post-war generation of designers who could see that the way ahead was to create a modern aesthetic to herald the 'new age'.

Sadly, Britain made few advances in this area. Germany quickly improved on the quality of mass manufactured plastic products, whereas Italian manufacturers began to exploit the aesthetic potential of plastics. The poor quality of plastic goods available in Britain, in both material and design, did little to persuade the consumer of the material's potential. As one history of the period put it 'the widespread application during post-war austerity reinforced the disgraced reputation plastics already had and which the industry is still in the process of setting right'.[*]

---

[*] Sparke P (Ed), *The Plastic Age—From Modernity to Post-Modernity*. Victoria & Albert Museum. 1990.

Restoring Credibility to Natural Materials 153

Big in Japan

The publicity of the war years must have been ringing in the ears of the consumer when peace brought, not the bubble car that had become an icon of techno-utopia, but an influx of badly designed, poor quality plastic items. These were generally produced by the Japanese plastics industry, financed by the USA, which had swamped the international market with cheap injection moulded products manufactured from war surplus scrap. 'Bargain prices bought shoddy goods – faded colours, crazed mugs, cracked boxes and designs that came off in the wash'.*

In the 1950s, companies such as Sony began to understand the importance of good design and the Japanese plastics industry tasked its product designers with the creation of high quality goods. Italy, meanwhile, had also made great headway in the creative design of plastics and increased the pressure on Japan in the international market place.

In Britain, British Industrial Plastics (BIP) saw itself as the standard-bearer for the image and reputation of this material, and employed a designer, a design assistant and four draughtsmen. This group set out to achieve three aims: to rework existing designs to make them more suitable for plastic manufacture; to design moulds for companies that did not employ their own designers; and to suggest suitable applications for BIP's technical developments. As the company put it, 'It is hoped that by introducing well designed products to a discerning public, they will force manufacturers to maintain a reasonable standard of modern design'.†

BIP was one of the largest plastics makers in Britain at the time and directed the BIP design team to 'contact buyers of chain stores, large public corporations, wholesale houses and the bazaar trade, and offer to interpret their ideas for new products in terms of plastics'.‡

Their aim was to help and encourage mainstream retailers to supply a range of domestic plastic goods to a high standard of both material and design. The impact of this was felt most strongly in the lives of women. Women had worked with industrial machinery in munitions factories during the war and were aware of advances in modern technology. In post-war Britain, this knowledge and the changing role of the housewife meant that women now had consumer power and consequently became the target for advertisers encouraging people to buy plastics.

Housewives in the years immediately following the war were under a great deal of pressure to be successful in their role of homemaker, not least from women's magazines that had increased in popularity. An untidy or unclean

---

* Farr M, *Design and British Industry—A Mid-Century Survey.* Cambridge University Press. 1955.
† Ibid.
‡ Attfield J and Kirkham P, *A View from the Interior—Feminism, Women and Design.* The Women's Press. London 1989.

home indicated failure. Plastics, in the form of wipe-clean furniture and kitchen appliances, were a vital weapon in their daily battle against the dirt and grime. Magazines that featured articles on how to keep your home spotlessly clean were also full of advertisements for domestic products in plastics.

The rise of linoleum-enabled Domestic Goddess, as imagined by Vintage Life Magazine in the 1950s.

Magazines such as *Good Housekeeping*, *Woman's Journal*, *Every Woman* and *Ideal Home* popularised contemporary domestic design. *Ideal Home* not only featured modern design, but also commissioned new designs for a wide range of domestic items, whereas *Woman's Journal* included features on labour-saving devices.

Women's organisations, many of which were nationwide, began to be consulted in the 1950s on the quality of consumer products. Such groups

Restoring Credibility to Natural Materials 155

were encouraged to put forward their opinions on the strengths or weaknesses of domestic products. In 1951, the British Standards Institute set up the Women's Advisory Committee, which gave women an influence on the innovation of domestic products, although this tended to be more practical than aesthetic in nature.

The 1950s also saw an acceleration in the rate at which British women acquired consumer skills, partly because there was a flood of new commodities, and partly because the government's reconstruction programme encouraged them to reassume the role of homemaker with a new fervour 'as part of their patriotic duty in the battle for peace'.[*] This role was to be different from the way it was before the war. The consumption of new goods and services became part of the housewife's expanded job description. Women were not only expected to consume but to consume in a certain way. Women's magazines played an important part in educating women to do so in a disciplined and 'responsible' way, exercising restraint and 'good taste' by choosing well-designed, useful and efficient goods.[†]

The Swinging Sixties

The 1960s saw Europe competing with the US' cultural leadership of 'the West'. One example was the British pop art revolution, which was predominantly based in London. Its influence was worldwide, and lasted most of the decade. Mass culture was quick to take its influences from the avant-garde subculture of the time and the lasting lesson of the 1960s was that 'plastics could represent both mainstream and alternative culture without a hint of contradiction'.[‡]

During the 1960s, attitudes towards design changed radically. The elitist design attitudes of the COID began to be rejected by a newly affluent new youth culture that embraced pop art and design. The Kings Road and Carnaby Street represented a liberated 'anything goes' design aesthetic, and young designers energetically explored their new freedom of expression. Expendability and a rapid turnover of styles came into vogue and were seen as a healthy aspect of social behaviour. Plastics seemed to epitomise this mass-produced, throwaway culture. Cheap, useful, products with no intrinsic value, such as ballpoint pens, razors and inflatable chairs embodied the contemporary love of disposability.

---

[*] Ibid.

[†] Holdsworth I, *Social Attitudes towards the Use of Plastics in Domestic Product Design, 1920-1960.* The Plastics Historical Society 2015.

[‡] Reilly P, 'The Challenge of Pop' *Architectural Review*. October 1967; Slosson EA, 'Chemistry in Everyday Life'. *The Mentor*. April 1992.

156   *Designing with Natural Materials*

One value of plastic was that it had no value.

Restoring Credibility to Natural Materials 157

As a result of this, many of the traditional prerequisites of industrial design came into question. Paul Rielly of the COID, later the Design Council, spoke out about the narrowness of orthodox style. He observed that 'to reject the ephemeral per se is to ignore a fact of life'.*

Design theorists such as the Independent Group began to analyse mass culture in the early years of post-war Britain. This group based its analysis on the public acceptance of expendability and styling in products and began to reject the ideas of 'good' and 'bad' taste put forward by the COID. While this was going on, certain British manufacturing companies were, by the beginning of the 1960s, responding directly to the needs of the public, and were producing consumer goods that reflected mass values. They took their lead from pop culture and exploited the qualities of expendability, fun and symbolism in designed objects. Consumption rather than functionalism began to appear as the new design criteria.

As a result of all this, the 1960s were a great time for plastics. The plastics industry invested in developing new polymers, and designers across the board worked on ways to translate the characteristics of the material into the exciting design that expressed the ideals of the new era. Expandable polyurethane 'fun-foams' meant that furniture designers were no longer tied to the traditional construction methods, but could develop a freer style of construction.

'Pop' designers were feted by the plastics industry, which was well aware of the benefits that good design could bring to the reputation of the material. The *Prospex 67* exhibition sponsored by Imperial Chemicals Industries and held at the Royal College of Art was an important link between the plastics industry and young designers. It not only helped promote the aesthetic qualities of Imperial Chemicals Industries Perspex, but also gave students the opportunity to experiment with acrylic in adventurous ways. Dunlop also used this opportunity to promote its soft foam by instigating an annual design competition.

Modernism to Post-Modernism

The 1960s saw a clearly defined move from modernism to post-modernism, the latter representing a culture that subverted the dictates of taste and encouraged a playful attitude to form. The boundaries of 'high' and 'low' culture became more plastic, in the Latin sense of the term, and consequently the outputs of the petrochemical industry came to be seen as the modern material *par excellence* – something that was 'capable of incorporating all the meanings that are possible within a post-modern epoch'. As Roland Barthes

---

* Ibid.

wrote, 'plastic is the very idea of its infinite transformation ... it is ubiquity made visible'.[*]

Within this new movement, the emphasis lay with the meaning of an object rather than its method of manufacture. There was no longer a need to find an 'authentic' aesthetic for plastics, as the designers of the 1950s had believed. Plastics had taken on a fundamental importance symbolising better than any other material the ideology of 'pop' culture. Young people responded by grabbing the idea of pop with both hands, despite this often being in direct opposition to values held dear by their parents. Durability and permanence had been exchanged for flexibility, change and expendability. The functionalist ideal of the modern movement, as much ethic as aesthetic, gave way to the idea that an object's form and its expressive relationship with the ideals and aspirations of the society into which it was destined to play a part were more important'.[†]

The newly emergent plastics of the 1960s were the ideal materials with which to represent the key themes of pop culture. They represented a commitment to the future and technology, were flexible and expendable, and were highly suited to bright colours and the application of surface patterns.

The years following the Second World War had seen plastics struggle to achieve an acceptable status as a useful, aesthetic material. The public of the 1960s, however, whole-heartedly embraced plastics, with designers, industrial bodies and the consumer all in agreement about the material's potential. Indeed, this remained the general attitude until today.

The Future

Society's attitude to plastic materials and products no longer views them as life-enhancing, exotic and glamorous. Since the 1960s, we have come to see the method of plastic manufacture before use and their disposability after use as threats on a planetary scale. In less than 200 years – the blink of an eye in the *longue durée* of human history – we have discovered, developed and overexploited the chemical possibilities of oil. The wheel has now come almost full circle, and this natural resource is likely to deplete within the next few generations, but not before it has fundamentally altered the world's climate systems. The environmental, social and cultural inheritance of this behaviour has created a carbon-conscious world whose values are diametrically opposed to those of the sixties.

---

[*] Sparke P. Barthes, Roland *Mythologies,* Farrar, Straus and Giroux, 1972. Page 90.
[†] Ibid.

The principle theoretical response to this situation is to move towards an economy based on renewable energy and circular manufacturing. 'Our industrial economy has hardly moved beyond one fundamental characteristic established in the early days of industrialisation: a linear model of resource consumption that follows a 'take-make-dispose' pattern. Companies harvest and extract materials, use them to manufacture a product, and sell the product to a consumer – who then discards it when it no longer serves its purpose. In 2020, approximately 82 billion tonnes of raw materials are expected to enter the economic system'.[*]

The call for a new economic model is getting louder. Substantial improvement in resource performance is needed across the economy and businesses must explore ways to reuse products or their components to recycle their materials, energy and labour.

Restoring Credibility

It is as part of this renewable ethos that natural materials can recover the credibility that they enjoyed before the dawn of the plastic age, and it would be wrong to think that nothing is being done to reintroduce natural materials to modern day industrial applications. Indeed, there are many examples that show this can be accomplished using state-of-the-art manufacturing technology, combined with new performance data produced by academic research using the latest analytical equipment and methodologies.

Let us take a look at just a few examples.

Silk

In pharmaceutical applications today, silk is used to fight oedema, cystitis, impotence, adenosine augmentation therapy, epididymitis and cancer (Dandin et al. 2007). Silk protein derivative, Serratio peptidase, is used as anti-inflammatory, anti-tumefacient for treating acute sinusitis, tonsiloctomy, oral surgery, tooth filling, cleaning and extractions (Ramesh et al. 2005).

The derivatives of silk fibre are also used as non-steroidal anti-inflammatory agents for treating rheumatoid arthritis.

---

[*] Ellen MacArthur Foundation 2013—Towards the Circular Economy—Economic and Business Rationale for an Accelerated Transition.

Wood

Accoya® wood represents a major development in wood technology and enables the consistent supply of durable, dimensionally stable, non-toxic solid wood. An ideal material for manufacture of high performance exterior timber products. Accoya® is based on the Acetylation of plantation-grown softwood, the gold standard in the academic field of wood science for over 80 years. A break-through in closed-loop engineering, culminating in 2007, led to the first commercial scale production of the product in the world. Critically, this high performance product is competitively priced to traditional hardwood options and widely available.

The physical properties of any material are determined by its chemical structure. Wood contains an abundance of chemical groups called 'free hydroxyls'. Free hydroxyl groups absorb and release water according to changes in the climatic conditions to which the wood is exposed. This is the main reason why wood swells, shrinks and decays.

Acetylation effectively changes the free hydroxyls within the wood into acetyl groups. Accsys Group's proprietary acetylation process involves reacting the wood with acetic anhydride, which comes from acetic acid (commonly known as vinegar when in its diluted form). When the free hydroxyl group is transformed to an acetyl group, the ability of the wood to absorb water is greatly reduced, rendering the wood more dimensionally stable and extremely durable.

Accoya® wood – an ideal material for high performance, exterior solutions. (Image: Accoya® wood Moses Bridge Netherlands).

Wool – A Technical 'Smart Fibre'

Wool is waterproof, elastic, fire-resistant, insulative, antibacterial and absorbent.

Today, there are many man-made fibres, but as yet, science has been unable to produce another fibre having all of the complex properties of wool.

Restoring Credibility to Natural Materials 161

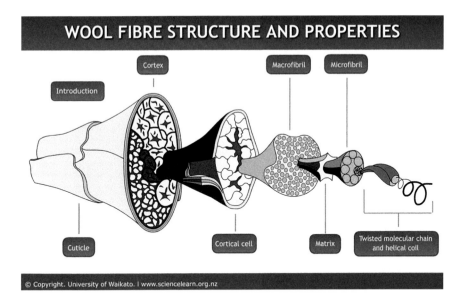

The secret of wool lies in the structure of its fibres, which absorb moisture, insulate against heat and cold, resist flame, and maintain their resilience. Unlike cotton, linen, silk or polyester, wool fibres are composed of a central protein core that is covered with tiny scales.

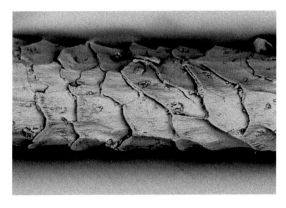

(Wool fibre 1800 magnification – Photo courtesy of Bangor University).

With the application of new scientific research, the amazing properties of wool as a sustainable, technical fibre are being increasingly developed for demanding industrial applications. These include high-tech composite materials for civil engineering, aerospace, automotive, military, food and most recently insulated pharmaceutical packaging for vaccines – helping to save human lives.

162   *Designing with Natural Materials*

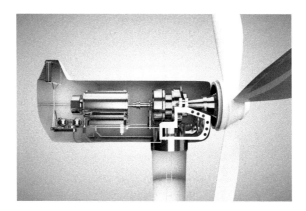

(Wool felt lubricating seals in wind turbines – reducing wear, tear and service frequency for remote locations).

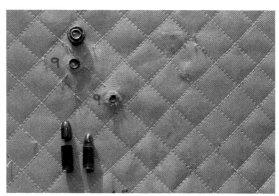

(Wool combined with Kevlar to deliver improved shock resilience and protection).

Since 2005, The World Health Organization continues to report over 50% vaccine wastage around the world. A substantial amount of the responsibility for this lies at the door of cold chain distribution and in particular, conventional man-made insulated packaging materials.

Avoidable Factors associated with Vaccine Wastage

•	**Cold Chain Failure**
	Loss of efficacy or potency (exposure to high temperature)
	Inactivation of the vaccine (exposure to freezing)
	Damage caused by temperature excursions during transit (packaging failure)
•	**Poor Stock Management**
	Vaccine reaches expiry date before use
	Lost, broken, stolen vials
	Over supply
•	**Poor Vaccination Technique**
	Administration of more than the recommended 0.5ml per injection

Source: World Health Organisation – IPV vaccine administration-Module 4 – February 2015

Restoring Credibility to Natural Materials 163

Today, new scientific research into the amazing properties of wool, a natural, genuinely sustainable, technical fibre has enabled the development of superior insulated packaging materials that substantially exceed the temperature performance criteria required by the 2013 EU and FDA GDP (Good Distribution Practice) regulations for the distribution of medicines for human and veterinary use.

The potential for using a natural 'smart' fibre to ensure the safe delivery of temperature sensitive vaccines and medicines globally is enormous - ultimately helping to save human lives.

(Wool insulated packaging for the safe delivery of temperature sensitive vaccines and medicines).

These applications of natural materials should not be confused with biomimetics, or bio-mimicry, as it is otherwise known.

There are hundreds of definitions, but we could define biomimetics as the study of the structure and function of biological systems as models for the design and engineering of materials and machines (*Wikipedia*), or as the attempt to learn from nature; it deals with the development of innovations on the basis of investigation of natural, evolutionarily optimised biological structures, functions, processes, and systems (Gletch et al. 2010).

The time is right to argue the case for the greater use of sustainable natural materials in product design in order to take the concept of a 'circular economy' one step further and to promote its potential for businesses and economies worldwide.

The challenge for a new generation of product designers is to stand up and demand change now.

Bibliography

Attfield J. and Kirkham P. *A View From the Interior—Feminism, Women and Design*. London. The Women's Press. 1989.

164   *Designing with Natural Materials*

Catteriall C. *The Plastics Age From Modernity to Post-Modernity edited by Penny Sparke.* London. Victoria & Albert Museum. 1990.

Dandin S.B. and Kumar S.N. Bio-medical uses of silk and its derivatives. *Indian Silk.* **2007**, *45*(9), 5–8.

Farr M. *Design and British Industry-A Mid Century Survey.* Cambridge. Cambridge University Press. 1955.

Katz S. *Plastics/Design and Materials.* London. Macmillan Publishers Ltd. 1978.

Meikle J.L. *The Plastic Age From Modernity to Post-Modernity edited by Penny Sparke.* London. Victoria & Albert Museum. 1990.

Ramesh S., Kumar C.S., Seshagiri S.V., Basha K.I., Lakshmi H., Rao C.G.P., and Chandrashekaraiah. Silk filament its pharmaceutical applications. *Indian Silk.* **2005**, *44*(2), 15–19

Reilly P. The Challenge of Pop. Architectural review. October 1967 Slosson, E. A. Chemistry in Everyday Life. The Mentor. April 1992.

Sparke P. (ed.). *The Plastic Age—From Modernity to Post-Modernity.* London. Victoria and Albert Museum. 1990.

http://www.victorianweb.org/technology/ir/irchron.html – The Industrial Revolution –A Timeline. The Industrial Revolution: A Timeline George P. Landow, Shaw Professor of English and Digital Culture, National University of Singapore.

The next 50 years – European Environment Agency.

Ellen MacArthur Foundation 2013 – Towards the Circular Economy - Economic and business rationale for an accelerated transition.

Von Gletch, A. Pade, C. Petschow, U. Pissarskoi, E. (2010) *Potentials and Trends in Biomimetics*, Springer.

6

Natural Materials in Automotive Design

Kerry Kerwan and Stewart Coles
Warwick University

CONTENTS

Introduction 165
    Historical Material Usage in Automotive... 165
    Reasons for Change.. 166
        Legislation 167
Benefits of Natural Materials.. 168
    Lightweighting 168
    Insulation... 168
    Aesthetics... 169
    Biodegradation Opportunities 169
Existing Barriers to Implementation 170
    Properties... 171
    Cost... 171
    Availability 171
Current and Historic Products.. 172
What Is the Real Potential of Natural Materials? 174
References 177

Introduction

Historical Material Usage in Automotive

Many of the basic materials used in the first gas-powered automobile, the Benz Patent Motor Car, are still used in today's automotive industry albeit in a different form. The Benz car utilised a tubular steel frame, wire wheels, rubber tyres and a wooden seat [1]. Of course, the design has changed substantially since that time with comfort, safety and performance all driving changes behind the look of a commercial passenger vehicle.

Early car designs were based on having a separate body and frame (known as body-in-frame), an idea taken from horse-drawn carriages. The idea was that the rigid frame would support the engine with the body effectively being draped on top. These original frames were made out of wood, although this was quickly surpassed by the use of steel ladder frames in the early 1930s. This effectively two-part system was easy to manufacture and modify, which is how it gained prominence.

Alternative methods were investigated, with a single piece construction which incorporated both the body and the frame was made out of 2-mm-thick steel by Lancia in 1922 and followed up by Citroen in 1934 [2]. This 'unibody' approach did not take hold in mass production vehicles until the 1960s, when issues such as performance, fuel consumption and safety started to take hold of which the unibody design showed clear advantages in all three areas. Again, steel was the dominant material in terms of construction. Today, nearly all small passenger vehicles are made by this method with only trucks and larger sports utility vehicles still being made using the body-in-frame method.

There has been more significant change in the materials used in the interior of the vehicle. This can be expected since most early vehicles were based on an open design and the use of compartments was not common until the early part of the 20th century, which naturally came after the development of using glass as a windscreen. Wooden seats gained padded cushions, often covered by leather. Similarly, arm rests were added and even fabric rugs included in the rear. With polymeric materials emerging from the 1940s, interior design shifted towards using these materials, allowing more easily mass-produced parts with greater reproducibility.

Nowadays, steel, aluminium and petrochemically derived polymers are well established in the automotive materials industry, having been utilised for decades in successful high-volume environments. Given the understandably risk-averse nature of the sector, moving away from these materials causes problems for manufacturers, but they are being driven to change by societal pressures and government legislation.

Reasons for Change

The automotive industry is moving towards sustainable mobility, e.g. through electric powertrains, biofuels and even a shift in business models around car ownership [3]. Europe is the world's largest producer of vehicles, with one third of the estimated 50 million vehicles produced worldwide being manufactured within the European Union (EU) [4]. The growth of vehicles on the road leads to significant increases in fuel demand, material requirements and air emissions [5]. Because of this, sustainability advancements are a pressing concern for the automotive industry.

Improvements need to be made, namely to:

- Reduce the large environmental impact of the sector – the transport sector is a significant contributor to greenhouse gas (GHG) emissions and local air and noise pollution.

Natural Materials in Automotive Design　　　　　　　　　　　　　　167

- Increase energy security and independence from foreign suppliers – growing demand for vehicles puts pressure on local governments to secure reliable sources of oil and this had led to a high dependency on foreign sources. Using alternative fuels and technologies could help to reduce this dependency.
- Search for new sources of economic growth and competitiveness – demands for sustainability improvements can induce innovation both in terms of new technologies and business models [6].

Legislation

Alongside more general environmental concerns that surround the daily use of automobiles, the industry itself, like many large industries, is subjected to meeting highly stringent legislative requirements and expectations across the whole of its operations. The single biggest piece of legislation driving the uptake of natural materials is the EU's End of Life Vehicle (ELV Directive), which effectively limits the choice and nature of the materials that can be used in a vehicle and requires innovative approaches to design.

A breakdown of a typical car is shown in Figure 6.1 [7], and it is widely accepted that approximately 75% of a vehicle was already recycled before the ELV directive came into force [8], simply driven by the scrap metal value inherent within the vehicle. The challenge to the automotive industry, to both manufacturers and disposers, is how to deal with the remaining 25% of the vehicle that remains stubbornly problematic, and this is where natural materials may have a greater role to play. The classification areas of fluids,

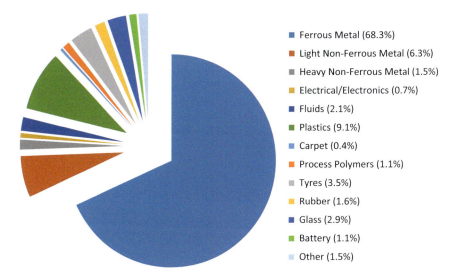

FIGURE 6.1
Breakdown of materials found in typical automotive vehicle [7].

plastics, carpet, process polymers, tyres, rubber and others all have potential application opportunities for natural materials and, although it is unlikely that all 19.3% of the car that these areas represent will become natural, major benefits could accrue [9–11].

As of 2015, 85% of a vehicle must be able to (theoretically) be 'reused' and 'recycled', with a further 10% 'recovered', with energy reclamation being a valid option – a route where natural materials offer several advantages over more traditional materials (e.g. they are often easier to burn and have lower toxicity) – or even biodegradation, again a route only available to certain material types.

Benefits of Natural Materials

Lightweighting

As so much emphasis is placed on reducing greenhouse gas emissions and improving fuel efficiency in the automotive industry, manufacturers (including suppliers, assemblers and component producers) are investing in lightweight technologies and solutions to meet these targets. Weight reduction is known to be a highly cost-effective means of reducing greenhouse gas emissions and fuel consumption – it is estimated that for every 10% of weight eliminated from the combined weight of a vehicle, fuel economy is improved by 3%–7%. Additionally, for every kilogram of weight reduced from a vehicle, there is about 20 kg of carbon dioxide reduction [12]. As well as improving fuel efficiency, weight reduction can also lead to improved braking, acceleration and improved performance handling dynamics [5]. To achieve these reductions in vehicle weight, manufacturers have been replacing the use of steel with aluminium, magnesium, composites and foams. Researchers have suggested that by replacing steel with aluminium or a composite material, it increases fuel economy and also the overall life cycle emissions and energy consumption [12]. The specific properties of natural fibres are often better, for example, in energy absorption (54.3 J/g hemp/epoxy vs. 19 J/g for metals) [9], which therefore offers the greatest potential for weight reduction. However, due to natural variations and incompatibilities, observed properties are often lower than theoretical, so the overall benefits are not being realised.

Insulation

Natural materials have been shown to be effective in providing both acoustic and thermal insulation in the automotive industry. One of the simplest ways is to use bio-based replacements of existing materials; SABIC produce polyethylene foam for the automotive industry and it is possible to produce the base polymers from renewable sources [13]. Natural fibres also

Natural Materials in Automotive Design   169

have significant potential in this area through the application of composite structures [14]. An example of this was the use of non-woven natural fibres (kenaf, jute, cotton and flax) in acoustic insulation showing a significant reduction in noise as part of a flooring system [15]. The properties of this system could be further improved by using a bio-based polyethylene terephthalate (PET) resin (of which a 30% bio-based resin is commercially available) to add thermal insulation as well as the acoustic.

Aesthetics

Some natural materials can also be aesthetically appealing, which leads them to be included as premium materials as part of an interior trim package. Leather is a very commonly used natural material within the automobile cabin and wood veneers are becoming increasingly common within this sector as customers choose to have it added in as an optional extra.

However, there does remain some concern over the customer acceptance and general aesthetic appeal and quality impression of other natural based materials such as biopolymers and composites and research is ongoing to develop a greater understanding of the complex issues surrounding the customer opinion [16].

Literature that has explored customer acceptance of renewable and recycled materials in vehicle interiors have found an overall positive attitude towards this. Hetterich et al. found that car buyers favoured the social image associated with purchasing a car interior made with natural or renewable materials [17]; they also perceived these materials to be innovative, and it was found that the majority of participants would pay a price premium for an ecologically sustainable car interior if given the choice. However, there were some reservations articulated by customers, such as quality issues (optic, haptic and wear), a reduction in comfort and an unpleasant odour from natural fibres. Ultimately, this highlights the need for sustainable materials to meet the expectations of customers in order to provide satisfaction for a given product [18].

Automotive OEMs have also been vocal about the changing consumer market. BMW have predicted that the growth of the sustainable luxury car market will increase considerably by 2025 [19]. Additionally, IBM in their report entitled 'Automotive 2020' have found key issues influencing the automotive industry [20]. Within this, it was determined that consumers are increasingly demanding economy, environmental responsibility and sustainability from automotive companies. Overall, car buyers by 2020 will be more informed, more demanding and more environmentally conscious.

Biodegradation Opportunities

As stated previously, biodegradation is a valid route for 'recovery' within the ELV directive. where bio-based materials are actively digested in the environment (in a manner similar to composting). Some synthetic materials are also structurally similar enough to natural materials to be broken down in this way [21,22] as the organisms recognise them as food sources. The main benefit of biodegradable materials is the potential reduction of waste compared to e.g. synthetic polymers, which can remain in the environment for hundreds or thousands of years if put to landfill; however, if digestion is uncontrolled, then it can result in methane generation which is an undesired greenhouse gas. For many materials, biodegradation does not occur when left in the general environment, instead requiring specialised composting facilities with a very controlled environment that optimises the process. One issue is that there are not many of these facilities in the United Kingdom [23].

If more biodegradable materials were utilised in the manufacture of cars, then this process would be of greater interest to automotive manufacturers as it requires little or no process energy at end of life and producing little waste. The lack of facilities to carry out effective biodegradation of materials suitable for automotive use is however hampering their uptake in the industry despite their obvious potential benefits.

Existing Barriers to Implementation

As with most things, there are both advantages and disadvantages to increase the use of natural materials in industrial applications.

In addition to the previously identified opportunities, natural materials (especially ones grown locally) reduce our dependency on synthetic equivalents and ultimately fossil oil which is both volatile in price and ease of availability. They also offer extra income strands for agricultural or bio-based communities and often exhibit unique properties such as biodegradability, passive humidity control [24], enhanced vibration [25] and/or sound proofing [26]. As more of these materials are being adopted, more favourable economies of scale are appearing and costs are reducing.

More negatively, some materials have a limited life span and very often natural materials do exhibit less favourable mechanical/chemical properties to their synthetic counterparts, although considered design and material selection can minimise these effects. There are other common concerns about the wider adoption of natural materials including their variability in performance and supply, the use of arable land for growing non-food crop feedstocks, knock on effects of intensive farming (e.g. the use of pesticides, nitrate run off etc.) and their interactions with existing processing/waste streams e.g. polymer recycling or composting facilities. Some of these areas are looked at over the following pages.

Properties

Although the theoretical properties of natural materials are generally superior to synthetic counterparts, their real world performance is generally inferior due to a number of reasons [27,28]. The first is that natural materials generally contain a larger number of imperfections or flaws, simply because of their growing processes. Whilst 'post harvesting' processes (e.g. fibre treatment or polymerisation) can overcome some of these flaws, they cannot overcome all of them.

The second key (related) reason is that they also remain comparatively poorly understood materials compared to synthetic alternatives, especially in terms of chemical and mechanical characteristics, which can again change due to natural variability (e.g. type of crop, weather, harvesting, storage etc.) [29,30]. These variations in mechanical and chemical properties also present significant challenges when utilising natural materials in production processes that have been designed for synthetic counterparts who have much better understood and tightly controlled properties and subsequent processing requirements. Overall, this generally results in products being over engineered to ensure performance levels and removes many of the lightweighting or environmental benefits that were desired in the first place.

Cost

The cost of natural materials remains a significant problem within the automotive industry. As mentioned above, climatic conditions can affect the overall properties but it can also affect the amount of material available. A variable supply with even relatively consistent demand will give rise to fluctuating prices, which is difficult to be tolerated within the business model of the automotive (or manufacturing in general).

It is often also possible to source materials that have superior properties for the same cost. Certainly, petrochemically derived polymers and glass fibre have been shown to have superior tensile strengths for the same unit cost as their bio-based counterparts [31], which makes it difficult to break into the automotive market, particularly at the high volume end of the market. Oil prices have also dropped significantly; the price in July 2016 was around $43/barrel compared with nearly $110/barrel in July 2013 [32]. This has made it even more difficult for natural materials to break into the automotive industry in large volumes.

Availability

The availability of useful biomass, whilst apparently plentiful, is in fact problematic for the growth of the bioeconomy. It has been noted by the European Bioeconomy Panel [33] that in order to meet the European Commission's 2012

172   *Designing with Natural Materials*

Strategy and Action Plan [34] for the implementation of bio-based materials, a sustainable supply of biomass must be in place to support an emerging bioeconomy. The most likely scenario for industry to adopt has been suggested to be focusing on locally available feedstocks to facilitate their adoption [35].

Current and Historic Products

Table 6.1 presents a number of natural fibres grouped into their relevant classifications. Although there are a multitude of plant fibres available, only a few are suitable for automotive applications. Flax, kenaf and hemp tend to be the most prominent fibres due to their strength properties. They also have a high CO_2 assimilation rate and are therefore able to consume large quantities of CO_2 when used [36].

Table 6.2 presents examples of different automotive components that can be produced using these natural fibres.

TABLE 6.1

A List of Vegetable and Cellulose Fibre Classifications [36]

Bast	Leaf	Seed	Fruit	Stalk	Wood Fibres
Flax	Sisal	Cotton	Coconut	Bamboo	Hardwood
Hemp	Manila	Kapok	Coir	Wheat	Softwood
Jute	Curauna			Rice	
Kenaf	Banana			Grass	
Ramie	Palm			Barley	
Banana				Corn	
Rattan					

TABLE 6.2

Examples of Interior and Exterior Automotive Parts Produced from Natural Materials [36]

Vehicle Part	Material Used
Interior glove box	Wood/cotton fibres moulded, flax/sisal
Door panels	Flax/sisal with thermoset resin
Seat coverings	Leather/wool backing
Seat surface/backrest	Coconut fibre/natural rubber
Trunk panel	Cotton fibre
Trunk floor	Cotton with PP/PET fibres
Insulation	Cotton fibre
Exterior floor panels	Flax mat with polypropylene

Natural Materials in Automotive Design 173

The use of these materials is not new. From as early as 1996, Mercedes-Benz used an epoxy matrix with jute – a vegetable fibre – within the door panels in its E-class vehicles. Also, in 2000, the door trim panels in Audi's A2 were made of polyurethane reinforced with a mixed flax/sisal material [37]. Table 6.3 provides a list of manufacturers and the models that incorporate natural fibre composites. Since this, BMW have launched their i3 and i8 models, both of which are designed to promote a stance on sustainability. Specifically, the i3 uses materials such as wool blend fabric, kenaf, eucalyptus and olive leaf tanned leather [19].

TABLE 6.3

Automotive Applications of Natural Fibre Composites [38]

Manufacturer	Model	Components
Audi	A2, A3, A4, A4 Avant, A6, A8, Roadster	Seat back, side and back door panel, boot lining, hat rack, spare tire lining
Citroen	C5	Interior door panelling
BMW	3, 5, 7 series	Door panels, headliner panel, boot-lining, seat back, noise insulation panels, moulded foot well linings
Lotus	Eco Elise	Body panels, spoiler, seats, interior carpets
Fiat	Punto, Brava, Marea, Alfa Romeo 146, 156	Door panel
Opel	Astra, Vectra, Zafira	Instrumental panel, headliner panel, door panels, pillar cover panel
Peugeot	406	Front and rear door panels
Rover	2000 and others	Insulation, rear storage shelf/panel
Toyota	Raum, Bevis, Harrier, Celsior	Door panels, seat backs, floor mats, spare tire covers
Volkswagen	Golf A4, Passat Variant, Bora	Door panel, seat back, boot lid finish panel, boot liner
Mitsubishi	Space star, Colt	Cargo area floor, door panels, instrumental panels
Renault	Clio, Twingo	Rear parcel shelf
Daimler-Benz	Mercedes A, C, E, S class, trucks, EvoBus (Exterior)	Door panels, windshield/dashboard, business table, pillar cover panel, glove box, instrumental panel support, insulation, moulding rod/apertures, seat backrest panel, trunk panel, seat surface/backrest, internal engine cover, engine insulation, sun visor, bumper, wheel box, roof cover
Honda	Pilot	Cargo area
Volvo	C70, V70	Seat padding, natural foams, cargo floor tray
General Motors	Cadillac Deville, Chevrolet, Trailblazer	Seat backs, cargo floor area
Saturn	L3000	Package trays and door panel
Ford	Modeo CD 162, Focus, Freestar	Floor trays, door panels, B-pillar, boot liner

What Is the Real Potential of Natural Materials?

In 2009, engineers from WMG at the University of Warwick built the world's most sustainable racing car – the WorldF3rst Formula 3. The aim of this programme was to demonstrate the true potential of natural and sustainable materials in an extreme environment. The car itself is in Figure 6.2.

Of the natural or bio-based materials used, highlight included:

- The fuel that was bio-based from several different oils and waste fat products [11].
- The lubricants that were plant oil based and readily biodegraded in the natural environment – particularly important when there is an oil spill.
- The steering wheel (Figure 6.3) was nanocellulose-based (with the cellulose being sourced from carrots) – their mechanical performance was between that of glass and carbon fibre equivalents.
- The seat (Figure 6.4) was produced via an autoclaving process and consisted of a flax shell, padded with a bio-based polyurethane foam and covered in recycled carpet.
- Wing Mirror cases were produced from potato starch foam that was subsequently waterproofed.

FIGURE 6.2
WorldF3rst F3 car tackling the hill climb at the Goodwood Festival of Speed 2009.

Natural Materials in Automotive Design

FIGURE 6.3
Carrot fibre steering wheel.

Two key structural aerodynamic parts were also produced from natural composites, namely the 'Bib' from flax (Figure 6.5) and the 'Barge Board' from 3D woven hemp (Figure 6.6). These materials were also examined as alternatives to carbon fibre within crash structures and found to perform very favourably [9,39].

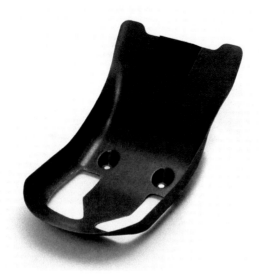

FIGURE 6.4
Flax seat shell and finished product.

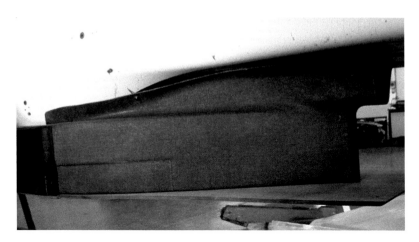

FIGURE 6.5
The flax bib.

FIGURE 6.6
The hemp barge board.

Natural Materials in Automotive Design 177

At the time of writing, all of these technologies are still fully functional and operating at the same performance level as they did during early trials and subsequent demonstration of the vehicle.

References

1. Daimler. Benz Patent Motor Car, the First Automobile (1885–1886) https://www.daimler.com/company/tradition/company-history/1885-1886.html (accessed Aug 5, 2016).
2. Genta, G.; Morello, L.; Cavallino, F.; Filtri, L. *The Motor Car; Mechanical Engineering Series*; Springer: Dordrecht, Netherlands, 2014.
3. Firnkorn, J.; Müller, M. Selling mobility instead of cars: New business strategies of automakers and the impact on private vehicle holding. *Bus. Strateg. Environ.* **2012**, *21*(4), 264–280.
4. Zorpas, A.A.; Inglezakis, V.J. Automotive industry challenges in meeting EU 2015 environmental standard. *Technol. Soc.* **2012**, *34*(1), 55–83.
5. McAuley, J.W. Global sustainability and key needs in future automotive design. *Environ. Sci. Technol.* **2003**, *37*(23), 5414–5416.
6. Beltramello, A. *Market Development for Green Cars*; OECD Publishing: Paris, 2012.
7. DTI. End of Life Vehicle (ELV) Waste Arisings and Recycling Rates, 2002.
8. ACORD. Automotive Consortium on Recycling and Disposal, Annual Report; London, UK, 2001.
9. Meredith, J.O.; Ebsworth, R.; Coles, S.R.; Wood, B.; Kirwan, K. Natural fibre composite energy absorption structures. *Compos. Sci. Technol.* **2012**, *72*(2), 211–217.
10. Wood, B.M.; Coles, S.R.; Kirwan, K.; Maggs, S. Biocomposites: Evaluating the potential compatibility of natural fibers and resins for new applications. *J. Adv. Mater.* **2010**, *42*(2), 5–16.
11. Wood, B.M.; Kirwan, K.; Maggs, S.; Meredith, J.; Coles, S.R. Study of combustion performance of biodiesel for potential application in motorsport. *J. Clean. Prod.* **2015**, *93*, 167–173.
12. Ghassemieh, E. Materials in automotive applications, state of the art and prospects. In *New Trends and Developments in Automotive Industry*; Chiaberge, M., Ed.; InTech, London, 2011; pp. 365–394.
13. SABIC. SABIC Launches New Renewable Polyolefins Portfolio https://www.sabic.com/europe/en/news-and-media/news/2014/20140423--SABIC-launches-new-renewable-polyolefins-portfolio (accessed Aug 5, 2016).
14. Dunne, R.; Desai, D.; Sadiku, R.; Jayaramudu, J. A review of natural fibres, their sustainability and automotive applications. *J. Reinf. Plast. Compos.* **2016**, *35*(13), 1041–1050.
15. Parikh, D.V.; Chen, Y.; Sun, L. Reducing automotive interior noise with natural fiber nonwoven floor covering systems. *Text. Res. J.* **2006**, *76*(11), 813–820.
16. Karana, E. Characterization of "natural" and "high-quality" materials to improve perception of bio-plastics. *J. Clean. Prod.* **2012**, *37*, 316–325.

17. Hetterich, J.; Bonnemeier, S.; Pritzke, M.; Georgiadis, A. Ecological sustainability—A customer requirement? Evidence from the automotive industry. *J. Environ. Plan. Manag.* **2012**, *55*(9), 1111–1133.
18. Ljungberg, L.Y. Materials selection and design for development of sustainable products. *Mater. Des.* **2007**, *28*(2), 466–479.
19. BMW. Future Relevance of Sustainable Luxury Goods the Customers' Perspective http://www.ifb.unisg.ch/~/media/internet/content/dateien/instituteundcenters/ifb/sim/3bmwsustainableluxury.pdf (accessed Jul 29, 2016).
20. IBM Institute for Business Value. Automotive 2020: Clarity Beyond The Chaos http://www-07.ibm.com/shared_downloads/6/IBM_Automotive_2020_Study_Clarity_beyond_the_Chaos.pdf (accessed Jul 29, 2016).
21. Stevens, E.S. *Green Plastics: An Introduction to the New Science of Biodegradable Plastics*; Princeton University Press: Princeton, NJ, 2001.
22. Plackett, D. Biodegradable polymer composites from natural fibres. In *Biodegradable Polymers for Industrial Applications*; Smith, R., Ed.; Woodhead Publishing Limited: Cambridge, 2005.
23. Slater, R.; Frederickson, J. Composting municipal waste in the UK: Some lessons from Europe. *Resour. Conserv. Recycl.* **2001**, *32*(3), 359–374.
24. Palumbo, M.; McGregor, F.; Heath, A.; Walker, P. The influence of two crop by-products on the hygrothermal properties of earth plasters. *Build. Environ.* **2016**, *105*, 245–252.
25. Senthil Kumar, K.; Siva, I.; Jeyaraj, P.; Winowlin Jappes, J.T.; Amico, S.C.; Rajini, N. Synergy of fiber length and content on free vibration and damping behavior of natural fiber reinforced polyester composite beams; *Mat. Des.* **2014**, *56*, 379–386.
26. Prabhakaran, S.; Krishnaraj, V.; Kumar, M.S.; Zitoune, R. Sound and vibration damping properties of flax fiber reinforced composites. *Procedia Eng.* **2014**, *97*, 573–581.
27. Dittenber, D.B.; GangaRao, H.V.S. Critical review of recent publications on use of natural composites in infrastructure. *Compos. Part A Appl. Sci. Manuf.* **2012**, *43*(8), 1419–1429.
28. Koronis, G.; Silva, A.; Fontul, M. Green composites: A review of adequate materials for automotive applications. *Compos. Part B Eng.* **2013**, *44*(1), 120–127.
29. Sherman, L.M. Natural Fibers: The New Fashion in Automotive Plastics http://www.ptonline.com/articles/natural-fibers-the-new-fashion-in-automotive-plastics (accessed Aug 10, 2016).
30. Bismarck, A.; Mishra, S.; Lampke, T. Plant fibers as reinforcement for green composites. In *Natural Fibers, Biopolymers, and Biocomposites*; Mohanty, A.K., Misra, M., Drzal, L.T., Eds.; CRC Press: Boca Raton, FL, 2005; pp. 37–108.
31. Ashby, M.F. *Materials and the Environment—Eco Informed Material Choice*, 2nd ed.; Butterworth-Heinemann: Oxford, 2013.
32. Nasdaq. Crude Oil Prices http://www.nasdaq.com/markets/crude-oil.aspx?timeframe=3y (accessed Aug 10, 2016).
33. European Commission. European Bioeconomy Panel -2nd Plenary Meeting https://ec.europa.eu/research/bioeconomy/pdf/bioeconomy-panel-summary-2nd-meeting_en.pdf (accessed Aug 10, 2016).
34. European Commission. Innovating for Sustainable Growth: A Bioeconomy for Europe http://ec.europa.eu/research/bioeconomy/pdf/official-strategy_en.pdf (accessed Aug 10, 2016).

35. Hennig, C.; Brosowski, A.; Majer, S. Sustainable feedstock potential—A limitation for the bio-based economy?. *J. Clean. Prod.* **2016**, *123*, 200–202.
36. Holbery, J.; Houston, D. Natural-fiber-reinforced polymer composites in automotive applications. *JOM* **2006**, *58*(11), 80–86.
37. Mohanty, A.K.; Misra, M.; Drzal, L.T. *Natural Fibers, Biopolymers and Biocomposites*; CRC Press: Boca Raton, FL, 2005.
38. Faruk, O.; Bledzki, A.K.; Fink, H.P.; Sain, M. Progress report on natural fiber reinforced composites. *Macromol. Mater. Eng.* **2014**, *299*(1), 9–26.
39. Meredith, J.; Coles, S.R.; Powe, R.; Collings, E.; Cozien-Cazuc, S.; Weager, B.; Müssig, J.; Kirwan, K. On the static and dynamic properties of flax and Cordenka epoxy composites. *Compos. Sci. Technol.* **2013**, *80*, 31–38.

7

Rediscovering Natural Materials in Packaging

Angela F. Morris
The Wool Packaging Company Limited

CONTENTS

Packaging 181
Glass.. 183
Metals.. 184
Paper and Cardboard 185
Wood 186
Natural Fibres 188
Wool 189
Case Study: Wool Packaging for Superior Insulation.................. 191
Conclusion 195
References 198

Packaging

Packaging is not just a product of the modern world. The first hunter-gatherers carried tools, artefacts and food with them, and to do so they fashioned containers from leaves, grasses, reeds, nuts and gourds, hollowed tree trunks, animal skin and fleeces (Figure 7.1).

When man developed pastoral and agricultural societies during the Neolithic revolution – about 10,200 BC – they began to stay longer in one place, growing crops, keeping animals and manufacturing goods. This created the need for larger storage and transportation containers, such as barrels, boxes and crates made from wood. It was also found that wicker could be used to make baskets, cellulose fibres could be converted into paper, other natural fibres could be taken from vegetable and animal sources, and metal containers could be made from ores and compounds.

In the period between the Stone Age and the plastic age, humankind evolved an ever-increasing number of ways of applying these basic materials to a multitude of uses. It is this collective knowledge that we should retain if we wish to maximise our ability to adapt to the conditions of the twenty-first century.

 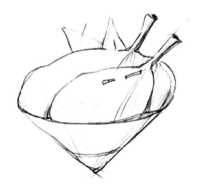

FIGURE 7.1
Food 'packaging' made from Banana leaves [1].

The principal force behind a renaissance in our use of natural materials is the pressure to use environmental and sustainable materials, rather than those derived from petrochemicals. This is coming from governments, international organisations and public opinion, which is increasingly sensitive to the need to use sustainable and environmentally friendly materials. Companies are under pressure to acknowledge and address the environmental impact of their products. Tighter laws and regulations are forcing businesses to take responsibility for their packaging – the resources that are used to make it and what happens to it after it has been used. It is important that the packaging industry views this as an opportunity to be socially responsible and support developments that help to conserve finite resources.

The result is that many industries are looking again at the enormous wealth of natural materials available for use in everyday applications. Combining the innate properties of natural materials with the latest technological advances in materials science will enable the development of new ways of using these ancient products.

That said, there are many questions to answer. What can we use natural materials for? Are we able to increase their supply by the extent that we may need to? Are there natural materials with properties that we have not yet explored? The packaging industry should encourage designers and manufacturers to revisit the technology of natural materials and the solutions they can offer.

Before going any further, I should declare an interest in this question. I am a packaging professional and a board member of the Natural Materials Association (NMA). In 2009, I founded the Wool Packaging Company, which produces wool-based packaging products on a commercial scale and also conducts research into the insulation and protective properties of wool fibres with the aim of finding novel uses of natural materials and their commercialisation. To that end, we provide resources, support and education to

Rediscovering Natural Materials in Packaging

industries wishing to benefit from natural alternatives to artificial materials. Education, I believe, is essential if we are to change perceptions, develop credibility, create awareness, build knowledge and ensure a sustainable impetus for the use of natural materials.

The packaging industry as a whole should embrace the opportunities that the sustainability issues offer by researching, developing and commercialising natural materials for packaging applications. Their rediscovery will give packaging technologists wider scope for innovative solutions that challenge the position of existing man-made alternatives.

In what follows, I will give brief descriptions of the principal natural materials used for packaging today, their history and natural properties, and I will present a case study of how wool is used for one application that required a material with superior insulation properties.

Glass

It is thought that glass-making began around 7000 BC as an offshoot of firing ceramics, and became industrialised in Egypt in 1500 BC. The method of creating the glass was simply to mix and melt the ingredients in a kiln, then mould them while they were still hot. All that was required was sand for silica dioxide and ashes for sodium carbonate, which was required to lower the melting point of the silica. Calcium carbonate, in the form of limestone, was also needed, as the sodium made the glass soluble in water. All of these materials were in plentiful supply.

The main difficulty was to form the material into the desired shape. At first, ropes of molten glass were arranged in coils and fused. By 1200 BC, it was pressed into moulds to make cups and bowls. Glass blowing was discovered by the Phoenicians in 300 BC and proved to be a critical breakthrough in the technology: not only was productivity improved, but it was possible to create round containers. Colours were available from the beginning owing to impurities in the ingredients, and clear 'aqua' glass was developed and became widely used across Europe over the next 1000 years (Figure 7.2).

The split mould was developed in the seventeenth and eighteenth centuries, and made it possible to design irregular shapes and raised decorations. The name of the maker and the product could also be moulded into the glass container as it was made. As techniques were further refined in the eighteenth and nineteenth centuries, the price of glass containers decreased. One major development that enhanced the process was the first automatic rotary bottle-making machine, patented in 1889. This method is still employed today, and modern equipment can produce in excess 20,000 bottles a day [3].

FIGURE 7.2
Egyptian glass vessel [2].

Metals

Ancient boxes and cups, made from silver and gold, were much too valuable for common use. Metal did not become a common packaging material until other metals, stronger alloys, thinner gauges and coatings were eventually developed.

One of the 'new metals' used in packaging was tin. Tin is a corrosion-resistant metal, and ounce-for-ounce, its value is comparable to silver. However, tin can be 'plated' in very thin layers over cheaper metals, and this process made it economical for containers. The process of tin plating was discovered in Bohemia in 1200 AD, and cans of iron coated with tin were known in Bavaria as early as the fourteenth century. However, the plating process was a closely guarded secret until the 1600s.

Thanks to the Duke of Saxony, who stole the technique, it progressed across Europe to France and the United Kingdom by the early nineteenth century. After William Underwood transferred the process to the United States via Boston, steel replaced iron, which improved both output and quality. The term 'tin can' referred to a tin-plated iron or steel can and was considered a cheap item. Tin foil also was made long before aluminium foil. Today, many still refer to metal cans as 'tin cans' and aluminium foil as 'tin foil', carryover from times well past.

Rediscovering Natural Materials in Packaging 185

In 1764, London tobacconists began selling snuff in metal canisters, another type of today's 'rigid packaging'. But no one was willing to use metal for food since it was considered poisonous.

The safe preservation of foods in metal containers was finally realised in France in the early 1800s. In 1809, General Napoleon Bonaparte offered 12,000 francs to anyone who could preserve food for his army. Nicholas Appert, a Parisian chef and confectioner, found that food sealed in tin containers and sterilised by boiling could be preserved for long periods. A year later (1810), Peter Durand of Britain received a patent for tinplate after devising the sealed cylindrical can.

Since food was safe within metal packaging, other products were made available in metal boxes. In the 1830s, cookies and matches were sold in tins, and by 1866, the first printed metal boxes were made in the United States for cakes of Dr. Lyon's tooth powder. The invention of cans also required the invention of the can opener in 1858! Initially, a hammer and chisel was the only method of opening cans.

The first cans produced were lead-soldered by hand, leaving a 1 ½-inch hole in the top to force in the food. A patch was then soldered in place but a small air hole remained during the cooking process. Another small drop of solder then closed the air hole. At this rate, only 60 cans per day could be manufactured.[*] Over 200 years later, in excess of one billion aluminium cans are produced globally every day and over half are recycled, with this figure increasing as waste collection and processes develop in a response to environmental concerns.

Paper and Cardboard

Paper may be the oldest form of 'flexible packaging'. Sheets of treated mulberry bark were used by the Chinese to wrap foodstuffs as early as the first or second centuries BC. During the following 1500 years, paper-making techniques were refined and travelled westwards to the Middle East and Europe, reaching the United Kingdom in 1310 and the USA in 1690.

These papers were different from those used today. The base material was usually flax fibres and later linen rags. It was not until 1867 that paper originating from wood pulp was developed. Paper bags were first manufactured in Bristol in 1844; however, it wasn't until eight years later that Francis Wolle invented the bag-making machine in Savannah, Georgia. Further advances during the 1870s included glued paper sacks and the gusset design. After the turn of the century, in 1905, machinery was invented to produce in-line printed paper bags. The development of the glued paper sack provided

---

[*] Archival copy: http://edis.ifas.ufl.edu.

a cheap alternative to cotton flour bags, especially after 1925, when a sturdy multi-walled version was invented, with ends that could be sewn shut.

Paper can also be used as 'semi-flexible' packaging – the ubiquitous cardboard boxes and paperboard cartons that fill supermarket shelves today. The box came along first. After being invented by the Chinese in the seventeenth century, a commercial product was made in England in 1817. Corrugated paper appeared in the 1850s and about 1900, shipping cartons of faced corrugated paperboard began to replace wooden crates to store trade goods.

As with many innovations, the development of mass-produced cartons was accidental. In 1879, a printer and paper-bag maker called Robert Gair was printing an order of seed bags at his Brooklyn factory when the metal rule used to create creases shifted in position and cut through thousands of bags. Gair realised that cutting and creasing in one operation would enable him to make a carton more efficiently than had been possible before.

As the means of production developed, so did the demand for flaked cereals advance the use of paperboard cartons. The Kellogg brothers were first to use cereal cartons at their sanatorium in Battle Creek, Michigan. When this 'health food' for the select few was marketed to the masses, the flaked maize was placed in a plain box and wrapped in a heat-sealed bag of Waxtite paper printed with the brand name and advertising copy. Today, of course, the plastic liner protects cereals and other products within the printed carton.

Paper and paperboard packaging increased in popularity well into the twentieth century. Then with the advent of plastics as a significant player in packaging – this occurred in the late 1970s and early 1980s – paper and its derivatives tended to fade from use. Lately that trend has halted as designers try to respond to environmental concerns by using more biodegradable materials [3].

Wood

While wooden crates and boxes were also widely used for trade, shipping cartons made from corrugated fibreboard started to replace these.

Wood packaging, ranking among the oldest packaging materials, has been used for around 5000 years, is used widely in the packaging of heavy and large fragile cargo due to its durability and hardness and in the packaging of fresh fruit and vegetables due to its ventilation properties. Besides these, it is also used in the packaging of extremely large machinery and motorised vehicles as well as various other products.

As wood has natural antibacterial properties, it has traditionally been used for centuries in the preparation, packaging, storage and transportation of food. However, the hygiene credentials of wood have been disputed based on the fact wood is an absorbent and porous material (Figure 7.3).

Rediscovering Natural Materials in Packaging 187

FIGURE 7.3
Ancient wooden container [4].

Results from R&D projects tell a different story, indicating that wood has excellent hygienic properties. Good manufacturing quality, good handling practice and proper sanitation treatments makes wood a highly suitable material for most applications in the food industries. The results necessitate a review of the existing guidelines and regulations for the use of wood in the food industry.

Wood is a complex natural material and therefore can interact with food, like any other material for packaging. However, history (and now modern research) shows us that this does not cause problems. Wood in contact with food is traditionally used not only in single-use packages or reusable packaging but also in cutting boards and countertops, utensils and kitchen utensils, kebab skewers, toothpicks, ice pops, wine barrels and more. If we use all these items without trouble, we can do it in packaging, too.

Wood can be perceived as a less suitable material for single-use containers compared with those made of smooth materials. However, comparative studies refute this fact. Wood is even considered more difficult to clean when used for refillable and reusable containers; but effective sanitation protocols have been developed and make it extremely viable.

There is no perfect material for every situation. It is important to know the intrinsic qualities of each material and its suitability for use and other factors including the type of food, surface, contact time and operating conditions such as temperature and humidity.

Wood packaging is used to pack, transport, handle, preserve, present and give added value to fruit and vegetables, fish and seafood, wines and spirits,

188   *Designing with Natural Materials*

oils, cheese and dairy, meat and conditioned meat, bread and bakery and dried fruits.*

Further developments include new lignin glue applications, recyclable food packaging, rubber tyre replacement applications, anti-bacterial medical agents, and high strength fabrics or composites. As scientists and engineers further learn and develop new techniques to extract various components from wood, or alternatively to modify wood, for example by adding components to wood, new more advanced products will appear on the marketplace. Moisture content electronic monitoring can also enhance next generation wood protection.

Interestingly, although wood still has widespread use for packaging and transportation of products, there is little information relating to its specific use for packaging. Searches will always lead you to information for its use in construction, furniture making and artwork.

Finally, it must not be forgotten that paper and cardboard are made from wood pulp!

Natural Fibres

Natural fibres are generally classified as plant or animal (See Table 7.1). They have been vitally important to human societies for the last 9000 years. They

TABLE 7.1

Natural Fibres

NATURAL FIBRES								
Vegetable - Cellulose or Lignocellulose							Animal Protein	
Seed	**Fruit**	**Bast (or Stem)**	**Leaf (or Hard)**	**Wood**	**Stalk**	**Cane, Grass & Reed**	**Wool / Hair**	**Silk**
Cotton	Coir	Flax	Pineapple		Wheat	Bamboo	Sheep's Wool	Tussah Silk
Kapok		Hemp	Manila-hemp		Maize	Sugar Cane	Goat Hair	Mulberry Silk
Milkweed		Jute	Henequen		Barley	Esparto	Angora Wool	Spider Silk
		Ramie	Sisal		Rye	Sabai	Cashmere	
		Kenaf			Oat	Phragmites	Yak	
					Rice		Horsehair	

Source: The Wool Packaging Company Limited © 2017.

---

* http://www.fefpeb.org/wooden-packaging/properties-of-wood.

Rediscovering Natural Materials in Packaging

are biodegradable and their production doesn't harm the environment. Growing natural fibres actually recycles carbon dioxide and cleans soil that has been polluted by heavy metals. They have a wide range of desirable qualities that make them useful in a multitude of applications. They can be used as a raw material for packaging applications, insulation, they absorb dyes without the use of addition of chemicals or processes, they resist mildew, have antibacterial properties and block UV radiation [5].

Wool

Wool has been the basis of human textile production for the past 7000 years, and was the mainstay of the UK economy up to the industrial revolution. The reason that it has played such a leading role in our past is to do with the complex physical structure of wool fibre (see Figure 7.4). Sheep and goats evolved wool as the ultimate natural packaging for their bodies, which have to continue to function despite extremes of hot and cold, regardless of humidity. Millions of years of evolutionary refinements have created a naturally 'smart' fibre that offers a range of remarkable benefits to the user.

First among these is insulation. This is also what makes it good for clothing, of course, but it does not stop there. It is believed that during its Alpine crossing, Hannibal's army used fleeces from sheep to collect ice and snow from the mountains brought down to the camps to store perishable foods.

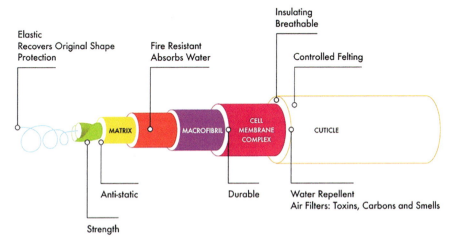

FIGURE 7.4
The structure of a wool fibre.

Source: **wwwchem.uwimona.edu.jm** Diagram The Wool Packaging Company Limited © 2016.

Second, the material 'breathes' – that is, it absorbs and releases humidity from the atmosphere. This means that garments are able to adapt to different climates and situations.

Third, it is a safe material: its high water and nitrogen content are a natural flame retardant, and therefore allow material made from wool to meet international regulations without the need for chemical treatments. This led to its use in containers containing ammunition, where its innate anti-inflammable, anti-static and hygroscopic properties are particularly desirable. Wool can also absorb carbon gases and particles in the atmosphere, helping to provide a cleaner environment.

As a result of these characteristics, wool has a vast and expanding global market that covers fashion, active wear, flooring and interior decor, aviation, architecture, manufacturing, healthcare and protective wear. Furthermore, it has recently been rediscovered as a packaging material and is used in the transportation of temperature-sensitive products such as foodstuffs, medicines and vaccines. The marrying of this ancient material with state-of-the-art technologies has resulted in new packaging solutions, and has highlighted the potential of natural materials and fibres as a viable option in other packaging applications [6] (Figure 7.5).

To give a more detailed explanation of these general points, I want to refer to a case study where the insulating properties of wool made it preferable to man-made alternatives.

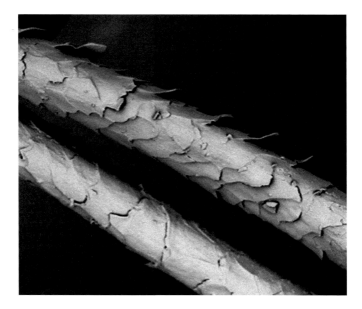

FIGURE 7.5
Wool fibre magnification × 1640.

Source: **The Wool Packaging Company Limited © 2013.**

Case Study: Wool Packaging for Superior Insulation

The brief for the Wool Packing Company was from a client whose environmental policy ruled out the use of artificial packaging. What they needed was a sustainable material with low environmental impact that could be used to replace polymer-based materials for the transport of temperature-sensitive products.

Research was undertaken into packaging material that could keep food chilled below 5°C for at least 24 hours, while it was in transit from supplier to consumer. Bio-plastics and starch-based packaging were considered initially, but after testing were found to be unsuitable for this application.

Various insulation materials were considered and trialled to the standards required for this packaging application. Pure sheep's wool that had been washed and scoured to Pollution, Prevention and Control Regulations 2000 was traditionally needle felted to a thickness of 35 mm and a weight of 800 GSM, using a mechanical needling process that did not involve applying heat. The resulting felt was then covered with a recyclable medium-density polyethylene (MDPE) sleeve to create internal liners for the insulation of a corrugated cardboard box (Figure 7.6).

Using a controlled protocol, a timed temperature trial was then undertaken. This compared the performance of a moulded expanded polystyrene (EPS) box, a corrugated cardboard box lined with polyethylene foam and

FIGURE 7.6
100% Pure wool felted for packaging insulation.

Source: **The Wool Packaging Company Limited © 2013.**

the wool-lined corrugated cardboard box. All had the same product volume, same type and weight of phase-change material (ice-packs) and the same weight and type of food product. The ambient temperature was recorded, and industry standard temperature data loggers were used within the boxes. Wool quite clearly outperformed the man-made insulation materials, keeping the food products cooler for longer (See the dark blue line Figure 7.7).

In recent history, there is little reference to wool being used within packaging as an insulator; therefore, information on its performance for this application is limited. Intellectually, it made sense that lining a box with wool should provide a good insulated package. However, the results of the trial well exceeded expectations and highlighted the amazing potential wool insulated packaging could have in a variety of industry sectors.

Further trials using various climatic profiles for both food and pharmaceutical applications proved without doubt that wool had superior insulative qualities to current man-made insulated packaging solutions.

Generally, insulation whether for packaging or construction applications is selected by its thermal conductivity value. The scientific unit used to measure a material's thermal conductivity is its K-value (sometimes also referred to as its Lambda value λ). All commercial insulating material is tested to determine its K-value, expressed as watts per square metre of surface area for a temperature gradient of 1 Kelvin per metre thickness, simplified to W/mK. The lower the value, the better the thermal efficiency of the material.

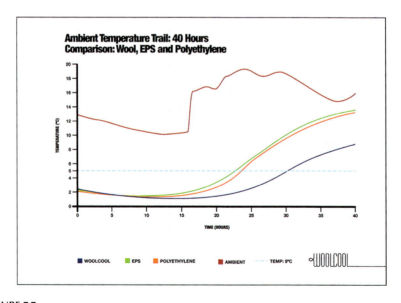

FIGURE 7.7
Temperature performance comparison graph wool (Blue), EPS/Polystyrene (Green) and Polyethylene (Orange).

Source: **The Wool Packaging Company Limited © 2016.**

Rediscovering Natural Materials in Packaging

Although wool performs in a different manner to man-made insulation materials, it was important to our potential markets that a Lambda value λ was established for the wool insulation. Depending on the weight of wool per square metre and the felting process, when wool was tested independently, values were recorded as low as 0.027 W/mK, bearing in mind that still air, which is the most effective natural insulator has a value of 0.026 W/mK. See Table 7.2 for the insulative values of common packaging materials.

It was also found that the optimisation of wool fibre properties, as listed in Table 7.2, would enable vaccines and medicines to be delivered according to the pharmaceutical industry's stringent time and temperature requirements, thus providing a superior alternative to the oil-based packaging that is usually used.

As far back as 2005, the World Health Organization (WHO) reported that up to half of vaccines transported globally were ruined due to poor distribution procedures [7]. The source of most of this damage is the failure of cold-chain logistics, and the culprit is often the artificial packaging materials that are supposed to control the temperature (2°C–8°C) of the vaccine. It appears that polymer-based packaging such as polystyrene is not sufficiently effective at maintaining temperature control (nor are they sustainable or environment friendly).

Independent trials of wool-based packaging have proven that due to its superior insulative properties, substantially smaller packs and significantly less phase-change material are required to maintain sensitive pharmaceutical product temperatures within the packaging to between 2°C and 8°C for up to 144 hours (Figure 7.8).

Designers should examine and research the chosen natural material thoroughly, as for example in the case of wool, where the fibre has properties which are not generally known and that can both enhance the performance of the product and also deliver environmental benefits (See Table 7.3).

TABLE 7.2

K-Values of Packaging Materials

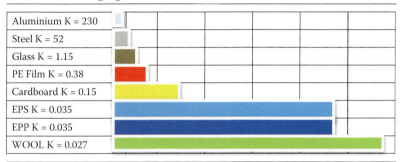

Source: The Wool Packaging Company Limited © 2016.

FIGURE 7.8
Wool insulated packaging.

Source: **The Wool Packaging Company Limited © 2016.**

TABLE 7.3

Known and Researched Properties of Wool

Thermal	A natural thermostat. Stable climate control, warm and cold	[8]
Hygroscopic	Absorbs and releases moisture up to 35% of its own weight	[9]
Elasticity	Naturally returns to its original size and shape	[10]
Carbon sink	Wool absorbs carbon from the atmosphere as sheep graze	
Absorbs toxins	Proven to absorb and lock in toxins such as formaldehyde	[11] [12]
Absorbs bacteria	Natural antimicrobial properties	[13]
Eco-friendly	Wool is compostable and releases nitrates back into the soil	

Source: Morris, A. and Spilsbury, K. (2016) Commercialisation of a natural material – wool: a bio-based PCM, *Green Materials*, 4(GMAT2), pp. 89–97. ICE Publishing holds the copyright.

Conclusion

This innovative use of wool-based felt has given pharmaceutical companies a way to ameliorate the serious issue of vaccine wastage using an ancient material in a way that out-performs any equivalent material yet devised by the world's chemical industry. Further research into the use of natural fibres will be crucial to the further development of the sustainable materials we will rely on in the future. These materials and products must be socially and environmentally responsible as well as economically effective.

Replacing oil-based packaging with a natural smart fibre that is available globally, will not only save lives by ensuring the safe and effective delivery of medicines, vaccines and blood products, but will also have more diffuse benefits to communities around the world by protecting their environments.

Clearly, efficient and effective packaging has a part to play in the wider goal of developing sustainable product lifecycles, yet it appears that the packaging industry is primarily focused on the challenge of minimising product waste, which allows it to continue with the expedient use of convenient, but unsustainable packaging materials.

As a packaging technologist, I am all in favour of using a holistic, closed-loop approach to identify genuinely sustainable supply-chain solutions, but the packaging itself still has to stand up to scrutiny in its own right. As more and more factors are lumped into some kind of 'efficiency' calculation, my worry is that it will actually be easier for FMCG (fast-moving consumer goods) companies, retailers and even packaging manufacturers to create and hide behind a sustainability smokescreen. For example, polystyrene producers can argue that their product is recyclable and great in landfill as it does not break down and release methane into the atmosphere. Polylactic acid (PLA) manufacturers can use the 'bio' prefix confidently – although those on the other side of the fence could wonder how sustainable is the use of agricultural land for crops to turn into 'plastic' bags. There will always be arguments for and against.

There is a common historical misconception that anything to do with protection of the environment means higher cost and more expense for the consumer. Industry has abused this by paying lip service to the principles of sustainability for profit gain and delivering this to the consumer at a deliberately higher price point.

'Environmentally Friendly', 'Organic' and 'Fair Trade' have all been tarnished in this way. Unfortunately, it appears that 'Sustainability' is suffering the same fate. Retail policy that drives down consumer demand for environmentally friendly products takes the pressure off FMCG companies and allows them to maintain the status quo using finite raw materials for packaging. Thus, 'packaging efficiency' improvement alone is simply a less challenging, more expedient option.

Packaging technologists, designers and manufacturers should not be content with merely 'reducing' film weights or 'minimising' waste. We must also acknowledge the reality that using recycled materials ultimately has a cost to the environment, since the final waste product is still destined to end in landfill [14].

That said, the reality is that true and complete sustainability will not be achievable in the short term. Re-educating industry and consumers will take time. However, this cannot be used as an excuse not to embark on the journey. To achieve any level of sustainability, we need to engage creative packaging designers and technologists who can turn current perceptions on their head by combining radical thinking with modern scientific research tools and thereby give due consideration to alternative, natural and sustainable materials.

Until recently, the education of product designers did not include end-of-life considerations. The general assumption being that it was destined for a one-way trip to landfill. This left creative designers free to choose from a vast array of increasingly 'high-tech' materials, most of them polymer-based, to create their products. What happened to those products when they were discarded was viewed as a matter for the waste collection authorities, continuing the 'not my problem' syndrome.

That situation is changing dramatically. Packaging responsibility is falling increasingly upon the shoulders of manufacturers. Emphasis on sustainability is becoming ever more pronounced and the damage that we have been doing to the environment is clearly and dramatically evidenced by global news media reports. Packaging designers most take some responsibility as an extremely high percentage of the plastic in our oceans is from packaging.

> Estimates suggest that there are more than 5 trillion pieces of plastic floating on the surface of the world's oceans. It has been claimed that there is now enough plastic to form a permanent layer in the fossil record. Dr Ceri Lewis, scientific adviser to the explorer Pen Hadow's Arctic Mission Expedition, based at the University of Exeter, has previously warned that people produce around 300 million tons of plastic a year, roughly the same weight as all the humans on the planet. Around half of all plastic produced is used once and then thrown away. [*]
>
> Sir David Attenborough has called for the world to cut back on its use of plastic in order to protect oceans. His new BBC TV series, Blue Planet II, is to demonstrate the damage the material is causing to marine life.
>
> Speaking at the launch of Blue Planet II, which will be broadcast 16 years after the original series, the broadcaster and naturalist said action on plastics should be taken immediately and that humanity held the future of the planet in the palm of its hands. [†]

---

[*] The Guardian *How did that get there? Plastic chunks on Arctic ice show how far pollution has spread.* 23rd September 2017.

[†] The Guardian – David Attenborough urges action on plastics after filming of Blue Planet, 15th October 2017.

Rediscovering Natural Materials in Packaging

John Ryley, Head of Sky News introduced the Sky Ocean Project with some 'Horrifying Ocean Facts'.

> Between 4.8 million and 12.7 million tons of plastic end up in the world's oceans every year and can be found everywhere from the Poles to the Equator, on coastlines, on the sea surface, and on the seafloor (Figure 7.9).
>
> Plastic makes up 95% of the rubbish in our oceans, mainly in the form of bags, food and drink containers, and fishing equipment; from past studies, it is estimated that as many as 90% of the world's seabirds have plastic in their stomachs.
>
> It will be a major challenge to put the legacy of pollution we are passing on into reverse, but we owe it to our children and future generations to acknowledge the problem and change our behaviour.*

Packaging designers must now operate with a clearer idea of what their responsibilities are, and what the future consequences of their material choices will be. The advantage of this greater awareness and charge of

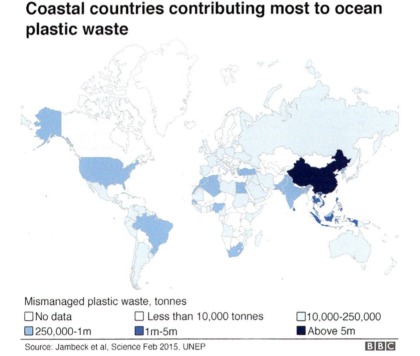

FIGURE 7.9
Ocean waste world map.

Source: **Jambeck et al, Science Feb 2015, UNEP.**

---

* Sky Ocean Project launched 24th Janauary 2017.

environmental responsibility is that sustainable packaging using natural materials presents a long-term challenge and genuine opportunity for astute and visionary companies to differentiate themselves in the market place and thereby create a sustainable legacy for their future stakeholders and the planet.

References

1. Laistrooglai, A., P. Mosikarat, and M. Wigran. (2009). Packaging Design with Natural Materials: A Study for Conservation.
2. A Warne & Co Ltd. (2013). The History of Food Packaging Part 1: A Brief Glimpse, http://www.awarne.com/, 5th Apr 2013.
3. Hook, P. and J.E. Heimlich. A History of Packaging, CDFS-133.
4. Ahmad, M. (2011). History of Food Packaging, Foodpackage.blogspot.co.uk.
5. Kozlowski, R.M. (2012). *Handbook of Natural Fibres, Volume 1.* Woodhead Publishing, Cambridge.
6. Handbook of Natural fibres Vol 1 Page 171.
7. WHO Department of Immunisation (2005). *Monitoring Vaccine Wastage at Country Level: Guidelines for Programme Managers.* World Health Organization, Geneva, WHO/V&B/03.18.Rev1.
8. Morris, A. and K. Spilsbury. (2016). Commercialisation of a natural material—wool: A bio-based PCM, *Green Materials*, 4(GMAT2), pp. 89–97.
9. Cassie, A.B.D. (1958). The physics of fibres with special reference to wool. *British Journal of Applied Physics*, 9, 341.
10. Harris, M., L.R. Mizell and L. Fourt. (1942). Elasticity of wool as related to its chemical structure. *Journal of Research of the National Bureau of Standards*, 29, 73–86.
11. Curling, S.F., C. Loxton and G.A. Ormondroyd. (2011). *A Rapid Method for Investigating the Absorption of Formaldehyde from Air by Wool.* Springer Science and Business Media, LLC.
12. Wortmann, G., S. Thome, A. Sweredjuk, G. Zwiener and F.J. Wortmann. (2009). *Chemisorption of Protein Reactive Indoor Air Pollutants by Wool.* University of Manchester, Manchester.
13. Ammayappan, L. and J. Jeyakodi Moses. (2009). Study of antimicrobial activity of aloevera, chitosan and curcumin on cotton, wool and rabbit hair. *Fibers and Polymers*, 10(2), 161–166.
14. Morris, A. (2012). Sustainable packaging is a reality not a myth. *Packaging News*, 27th June 2012.

8

Designing Tall Buildings with Natural Materials

Richard Harris
University of Bath

Wen-Shao Chang
The University of Sheffield

CONTENTS

What Is Tall? 199
Why Use Timber for Tall Buildings? 201
Types of Construction for Tall Timber Buildings 202
Cross-Laminated Timber 204
Glued Laminated Timber Frame Structures 206
Fire and Tall Timber Buildings 207
Dynamics and Tall Timber Buildings 208
Stability System and Tall Timber Buildings 211
The Future of Natural Materials in Tall Buildings 213
References 214

What Is Tall?

Historically, the desire to live and work close to one another for economic or security reasons has driven technology to enable city dwellers to construct tall buildings. In Bologna, in the twelfth and the thirteenth centuries, stone towers of 60 metres (equivalent to 20 storeys) were built; in Shibam, Yemen, mud-brick buildings of between 5 and 11 storeys still stand. Wood was widely used to construct buildings of considerable size and height. The Sakyamuni Pagoda of Fogong Temple in China was built in 1056 and is 67 metres tall. The wooden spire of Salisbury Cathedral, built 1320, was constructed (59 metres high) on top of a 64-metre high stone tower (total height: 123 metres). Japan's Toji Temple has a 55-metre high pagoda, originally dating from 826 and rebuilt in 1644.

199

The development of steel released the opportunity to build to greater heights than ever before, and in the twentieth century, it was the material of choice for tall building construction. In the twenty-first century, with the need to build more sustainably and with the rise in energy costs, wood is being developed as a natural material for use in tall building construction.

In discussing tall buildings, there is the question of 'what is tall?' The tallest building in the world is the Burj Khalifa in Dubai, standing at an extraordinary height of 829.8 metres. For buildings over 300 metres tall, the Council on Tall Buildings and Urban Habitat (CTBUH) uses the term 'supertall'. There are fewer than one hundred supertall buildings in the world and they clearly belong in a category of their own. The CTBUH recognises that there is no absolute definition of what constitutes a 'tall building' and so defines tall in terms of height relative to context, proportion and technology. In the centre of Chicago or Hong Kong, a 14-storey building would be considered as of modest height, but in the suburbs of London, it would certainly be considered as tall. A building that is very slender, with a small cross-section are relative to its height would be considered tall. In terms of absolute height, the CTBUH proposes that a building of more than 50 metres high (approximately 14-storeys) be considered as the threshold for 'tall'.

The point at which buildings made from natural materials become 'tall' brings in consideration of the available and developing technology. The mud-brick buildings of Shibam are certainly 30 metre tall and there is no evolving technology to build higher in this material. Although there were taller timber buildings constructed in the past, twentieth century timber buildings were unlikely to exceed 10 metres in height. Thus, the recently developed technology taking timber building construction to 30 metres and higher are tall in the context of timber construction.

The reason for the domination of steel in the twentieth century construction of buildings over ten storeys in height was its high ultimate strength and excellent strength-to-weight ratio. Steel construction could take advantage of prefabrication through the use of stiff and strong welded or bolted connections. But steel is a high embodied energy material, and building developers have started to seek out other, more sustainable, materials to use.

The world's tallest tree is a coastal redwood in Northern California, which is 115 metres tall. It is clear that, with ingenuity and good engineering, buildings of this height are structurally feasible and there is pressure on the engineering community to realise them. Thus, this chapter will explore the technology of tall timber construction. It will address the construction types available for wood construction of tall buildings, and it will look in more detail at the issue of dynamic response to loads.

Designing Tall Buildings with Natural Materials

Why Use Timber for Tall Buildings?

As for steel, the strength-to-weight and stiffness-to-weight ratios of timber offer good structural opportunities (Figure 8.1), particularly for tall timber buildings.

As the number of stories increases the weight of the structure (known as the 'self-weight') becomes the predominant load. At 20 stories, the self-weight is very significant and good strength-to-weight and stiffness-to-weight ratios are necessary for efficient and economic design.

On some sites, the saving in weight can be the driver for using timber in the construction. At the ten-storey Dalston Lane in Hackney, London, the designers were able to reduce the weight of the structure by 30% as compared with an equivalent concrete frame building. This was an important consideration as the building is located over the running tunnels of the high-speed rail route linking London with France. Reduced weight also means a significant reduction in the number of construction deliveries to site.

The experience of disastrous fires in tightly packed medieval cities had led many countries to use prescriptive regulations to limit the height of timber buildings. According to Östman B and Källsner B [1], as recently as 1990, all countries of Europe restricted timber buildings to two stories. The need to trade more efficiently and to innovate has brought about a move from prescriptive regulation to performance regulation, which has removed blanket restrictions and opened the opportunity for timber in higher rise building construction.

The use of performance criteria to regulate construction has enabled fire in timber buildings to be addressed by the use of fire engineering. In this way, risk-based assessments of fire load on buildings enable safe predictions of structural performance for standard fire durations to be made. Fire resistance is achieved by a combination of consideration of the natural charring

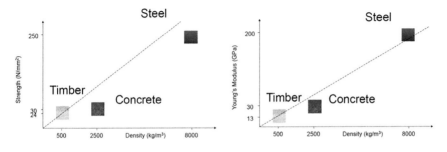

FIGURE 8.1
Strength-to-weight and stiffness-to-weight for timber, steel and concrete.

FIGURE 8.2
Typical low-rise stud wall construction in New Zealand.

of timber, which insulates and protects the inner core of a structural timber element and the application fire protection boards such as plasterboard.

There are clearly further technical challenges to be overcome to address the indisputable fact that wood is a natural product that is generally available in modest size and length. It shrinks and creeps, particularly perpendicular to the grain. It is relatively weak and flexible, particularly perpendicular to the grain and in its connections (Figure 8.2).

Types of Construction for Tall Timber Buildings

The most widely used form of construction for timber house uses load-bearing walls made of timber joists nailed to sheathing of plywood, oriented strand board (OSB) or plasterboard. For low-rise buildings, in the past, the studs have been brought to site and fixed in place, with the sheathing being applied on site. Figure 8.3 shows this form of construction, which is used throughout the world, being termed '2 by 4' construction in the United States.

In the United Kingdom, this form of construction has become factory-based, enabling prefabricated panels to be made. Wall panels incorporate window and door openings and, in the more sophisticated factories, the panels can be turned to allow both sides of sheathing to be fixed in place. With this 'closed-panel' form of fabrication, it is possible to install insulation, services, windows and cladding in the factory, before delivery to site, which leads to very fast construction and high quality buildings.

Designing Tall Buildings with Natural Materials 203

FIGURE 8.3
Prefabricated wall panels being manufactured in a UK factory.

Initial developments, in taking timber buildings taller than they had been in modern construction, were led by a UK research project, Timber Frame 2000 (TF2000). This was one of the most significant timber research projects undertaken in recent years (Figure 8.4). It was a collaborative project between the UK government, TRADA Technology, the Building Research

FIGURE 8.4
Timber frame 2000. Platform timber frame incorporating engineered wood products. Image copyright TRADA.

Establishment (BRE) and the UK timber industry; it involved the construction of a 6-storey, timber frame building – the tallest of its type in the world at that time.

The TF2000 research established multi-storey timber frame construction in the United Kingdom. The work was published as design guidance, addressing, structural stability and robustness, fire safety and differential movement as well as construction process and procedures [2].

The form of construction used for TF2000 is usually termed 'platform timber frame'. The name is derived from the construction process. A single storey of load-bearing walls is erected and then the floor is built on top of them, creating a 'platform' to erect the next set of walls, which in turn supports the next level of floor.

Having completed TF2000, the market was opened to the use of timber frame. Now, in terms of construction volume, Platform Timber Frame is the most established sector of the timber industry in the United Kingdom. It is used extensively in private and social housing, student accommodation, hotels, medical facilities and care homes. Multi-storey construction, of six floors, is common.

Cross-Laminated Timber

Cross-laminated timber (CLT) is a laminated solid wood product for structural load-bearing use consisting of layers of wood. Each layer is laid at right angles to the adjacent layer and the layers are bonded together with a structural adhesive (Figure 8.5).

In the United States, cross-laminated timber is promoted as 'Jumbo-Ply', and the name gives a good indication of its make-up. Just as with plywood, it uses timber laid up in alternating grain direction; but instead of veneers, it uses sawn structural size elements.

Single layers can be replaced by laminar engineered timber products (e.g. laminated veneer lumber, OSB, plywood or single- or multi-layer solid wood panels), which consist of pieces of timber glued together on their edges and, if multi-layer, on their faces. The introduction of layers of engineered wood products can give improved performance (e.g. air tightness, rolling shear resistance, acoustic response, aesthetic or tactile finish).

Just as the work of TF2000 transformed the UK platform timber frame industry, the introduction of cross-laminated timber has released a series of projects that benefit from its special advantages. It is also uses off-site prefabrication; in this case, precision-engineered panels using computer aided design/computer aided manufacturing technology and computer numerical control machinery. CLT has provided the impetus for increasing the height of timber buildings. In 2013, Šušteršič [3] showed a strong trend

Designing Tall Buildings with Natural Materials 205

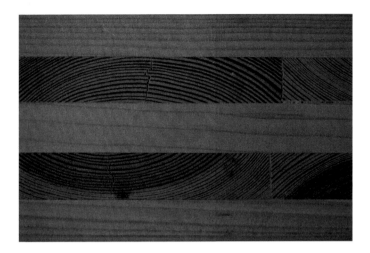

FIGURE 8.5
Cross-laminated timber (CLT).

towards taller buildings using CLT as their primary construction material (Figure 8.6), and in subsequent years, his prediction has proved to be correct.

The groundbreaking building was the eight-storey Stadthaus building at Murray Grove in Hackney, London, designed by Waugh Thistleton Architects with structural engineers, Techniker. The building uses the platform method of construction but with prefabricated CLT walls.

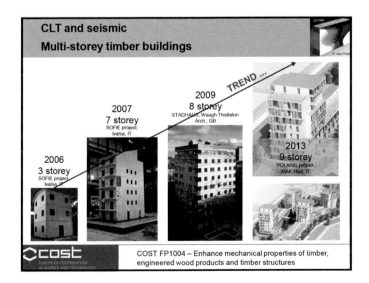

FIGURE 8.6
The increasing height of timber buildings. Iztok Šušteršič R&D Project Manager, CBD & University of Ljubljana Slovenia, 'Use of CLT in Slovenia in seismically active areas'.

Cross-lamination in CLT creates a strong, dimensionally stable panel. Dimensional stability is derived from the parallel-to-grain fibres reinforcing the perpendicular-to-grain fibres in the adjacent laminations to which they are bonded. Dimensional stability provides the capability of accurate pre-fabrication, which gives the opportunity for very fast high-quality construction. The combination of speed of construction and quality is driving rapid growth in the use of CLT for buildings of 10–12 storeys. Several new CLT manufacturing facilities have come into operation in North America to meet the growing demand.

Glued Laminated Timber Frame Structures

Economic growth and increasing land value, towards the end of the nineteenth century, led to the development of tall buildings in New York and Chicago.

In his book, on the history of building design and construction, Addis [4] plots the development of tall buildings in steel.

The need for fireproof construction generated buildings in iron and masonry. The use of cast and wrought iron buildings developed to enable longer spans and increased efficiency resulting from the smaller size of vertical structure. Early American 'skyscrapers', built between 1884 and 1939 were enabled by the technological development of fireproof steel frames, improved foundations, lift design and improved electrical systems.

In 1884, the ten-storey (42 metre high) Home Insurance Building was constructed in Chicago and became the first tall building to use structural steel in its frame. This building used a non-load bearing envelope, using iron framing to carry the walls rather than load-bearing partition walls to support the floors. Part of the success was in simplification, rationalisation and opening up of floor plans. By 1902, the Flatiron Building (originally called the Fuller Building) a 22-storey steel-framed building was completed in Manhattan, New York City. These groundbreaking structures give an indication of how technology releases new forms of construction and how long it takes to develop a new form of building.

The development of tall timber buildings parallels the development of tall steel buildings, which took place 120 years earlier. We have seen the use of glued-laminated timber release the opportunity for longer spans and of cross-laminated timber to create the opportunity for safe tall buildings. It is likely that we will see the development of timber frame construction, which releases the need to align load-bearing ones one above the other.

The tallest modern timber building is the Treet (Norwegian for Tree) Building in Bergan in Norway. The technological development came from timber bridge design. Nordic countries received a boost to their technology

Designing Tall Buildings with Natural Materials

through the development of large timber structures for the 1994 Winter Olympics in Lillehammer. This showed that that the technology existed to build timber structures that could rival steel. Since 1993, several hundred modern timber bridges have been built in Norway. These bridges typically use large timber trusses as their main structural elements. The concept for the Treet Building was to use triangulated trusses, of a similar scale to those used in bridge structures, but aligned vertically instead of horizontally.

The vertically aligned trusses of the Treet Building rise 14 storeys from a concrete structure containing basement parking. The frame uses large glulam sections, which are block glued into typical column sizes of 405×650 and 495×495 mm with typical diagonals of 405×405 mm. The glulam carries all vertical load but the building arrangement is unusual. The whole building is made using prefabricated timber modules, manufactured in Estonia and brought to site by ship. The construction sequence consisted of installing five storeys of modules on the concrete base and then erecting the glulam truss structure around them. At the fifth floor, a concrete slab was installed, and the process repeated for a further five storeys of modules, with another concrete slab at the tenth floor and then repeated again to take the building to its full height. The modules were completely fitted out in Estonia, leaving the minimum of finishing work to be completed on site. Balconies were added, and the whole structure was enclosed in a weatherproof envelope, completely protecting the timber in the building from rain.

This building demonstrates a number of key aspects of tall timber building construction. They are as follows:

- The opportunity to innovate
- A new benchmark for timber building height
- The suitability for the use of timber in pre-fabrication, providing highly accurate, fast construction in the harsh Norwegian environment
- Fire safety can be addressed in tall timber buildings
- For buildings of this height, the key structural issue is building stiffness and dynamic response to wind loads

Fire and Tall Timber Buildings

Timber is a material that has been used for centuries in building construction and, throughout history, there have been 'Great Fires' in cities of the world (e.g. of Rome, Constantinople, Hangzhou, Utrecht, Amsterdam, London, Baltimore, Tokyo), and this has led to strict restrictions on the use of timber in construction within conurbations, with a preference for non-combustible structure [5]. Steel and concrete present challenges in fire design and the technology to address these has been developed over the past 100 years. The advantages and sustainability credentials are obvious but timber is combustible. The special challenge of timber is that the combustible structure remains, even when the fuel within the building has been exhausted.

Fire in timber buildings of all types is being addressed by research. This includes building greater understanding of fire safety issues relating to the structural fire resistance for both solid timber and engineered timber elements in glulam or CLT, which are the key structural materials for tall timber buildings. The fire dynamics of compartmentation using timber-separating elements is also being addressed. In the context of current practice, which has relied for many years on non-combustible materials for this purpose, current research is leading the development of a new approach for the safe use of combustible materials in fire resistance.

Dynamics and Tall Timber Buildings

As it is linked to vibration acceptability of occupants in the building, dynamic performance is often one of the major factors that govern the success of design in tall timber buildings. Proper understanding of dynamic behaviour of tall timber buildings is essential for the design of a serviceable tall timber building and this often requires specific dynamics analysis.

Sources of vibration in a structure may include earthquakes, human and machinery movement, wind and traffic. It is important that the occupants of tall timber buildings should not feel excessive vibration from the building during its operation. Wind-induced vibration of tall timber buildings is not easily predicted and so it is a major challenge for designers.

Figure 8.7 shows the acceptance curves for wind-induced horizontal vibrations for offices (Line 1) and residential units (Line 2) from ISO 10137 [6]. The vertical axis is the peak acceleration for one-year return period and the horizontal axis is the natural frequency of the structure in Hz. These acceptance criteria are often used to judge the performance of a completed building from measurements made of vibration due to wind excitation.

Eurocode 1 [7], which provides a method for predicting the characteristic peak acceleration of a structure, requires the natural frequency and damping ratio as well as the geometry of the building and the geographic characteristics of the site.

There are several ways of estimating the natural frequency of tall timber buildings, the easiest, in the design stage, being the use of computer models. Accurate estimation of the dynamic properties of tall timber buildings relies

Designing Tall Buildings with Natural Materials

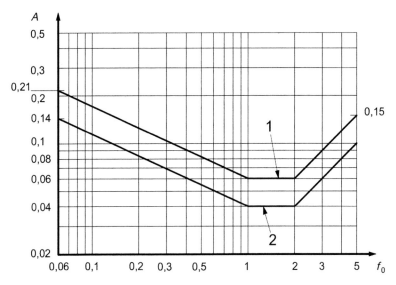

FIGURE 8.7
Acceptability curves for offices and residential units – (ISO Figure D.1 – Evaluation curves for wind-induced vibrations in buildings in a horizontal (x, y) direction for a one-year return period.

on correct assumptions of the parameters used in the model, which include accurate stiffness of the structural components and reasonable values for the damping ratio of the structure. Nonlinear behaviour of timber connections, in the in-service condition (i.e. within the Serviceability Limit State of Eurocode), makes it difficult to make accurate estimates for the parameters. Eurocode 5 gives damping ratios for timber floors as 1% (2% from UK National Annex), whereas ISO 10137 provides a range of 1.5%–4%. Eurocode 1 provides a damping ratio of 6%–12% for timber bridges. None of these standards provides damping information for tall timber structures and this lack of damping information, and the phenomenon of non-linear damping, makes accurate prediction of dynamic performance in tall timber buildings structures particularly challenging.

On completion of construction, the dynamic properties of a tall timber building can be obtained from measurement. Vibration measurement of tall timber buildings can be separated into the two categories of ambient and forced vibration. Ambient vibration tests of tall timber buildings mainly rely on wind, sometimes with some input from road traffic, as a source of excitation. Forced vibration tests can be carried out by exciting the structure with a shaker or by use of an impulse hammer, the input force of the former being more controllable then the latter.

Using ambient vibration measurement to extract the dynamic properties of tall timber buildings has proved to be a simple but effective

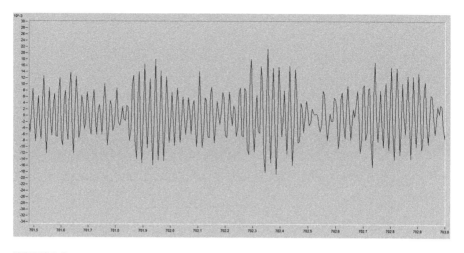

FIGURE 8.8
Ambient vibration measurement data for a 6-storey timber office building (Vertical axis – acceleration. Horizontal axis – time).

method [8], which is used when it is not possible to actively excite the structure. In this method, if it is assumed that the input generated by wind (and traffic) is a zero-mean white noise, the excitation source need not be considered. Figure 8.8 shows an example of ambient vibration measurement data.

As an alternative to measuring ambient vibration, forced vibration of tall timber buildings can be generated by a shaker or an impulse hammer. A shaker can be controlled to generate sinusoidal or random excitation to the structure so that the input force or acceleration is known. An impulse hammer produces input that is less controllable than that generated by a shaker but being equipped with an accelerometer or a load sensor to record the input, engineers can analyse the input data.

Analysis of measured data can fall into two main categories, frequency domain analysis and time domain analyses. Frequency domain analysis usually involves the use of a fast Fourier transform (FFT) of the data and then peak picking to determine the natural frequency of the structure. This is a fast and simple way to determine both the natural frequency and damping ratio. However, when the measured data involves significant amount of noise, it is less reliable. There are several ways to deal with data in time domain; the one that often used to extract the dynamic properties of tall timber buildings is the use of curve fitting techniques on the random decrement signature [8]. This method is particularly useful when only the natural frequency and the damping ratio are needed. Figures 8.9 and 8.10 illustrate the procedure and equipment set-up for measuring dynamic properties of tall timber buildings after they are completed.

Designing Tall Buildings with Natural Materials   211

FIGURE 8.9
Procedure for measuring dynamic properties of tall timber buildings.

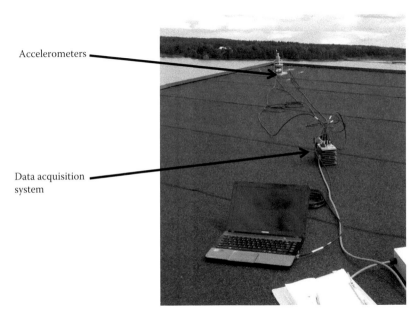

FIGURE 8.10
Typical set up for measurement of vibration on a tall timber building using accelerometers.

Stability System and Tall Timber Buildings

All structures are designed to resist lateral loads such as those due to wind and earthquake. The two types of stability systems used in tall timber buildings are the use of a core structure or the use of bracing. Unlike steel and reinforced concrete buildings, as joints in timber structures are likely to have too much slip, it is not usual in timber structures to use the strength and stiffness of moment-resisting connections. Because the framing of timber columns and beams cannot easily be stiffened, this is unlikely change in the future, even if stiff joints are developed.

FIGURE 8.11
Lifecycle tower one project in Dornbirn, Austria.

The core of tall timber buildings structures can be constructed in reinforced concrete or in timber. Reinforced concrete cores are often used as they provide good stiffness to reduce horizontal movement and to resist the lateral force. One example of using concrete core is Lifecycle Tower One project in Dornbirn, Austria (Figure 8.11). Here, the engineers chose a concrete core instead of timber cores to satisfy fire regulations and to meet cost constraints. Care is needed to account for an eccentric location of concrete core, which will induce torsional behaviour of the structure in vibration.

Replacing the concrete core with timber has been attempted in several projects (Figures 8.12 and 8.13) and vibration performance has been proven

FIGURE 8.12
Kampa Bauinnovationszentrum in Aalen-Waldhausen. Eight storeys entirely in wood.

Designing Tall Buildings with Natural Materials 213

FIGURE 8.13
Woodtek headquarters. Five storey building – the first CLT building in Taiwan.

to be acceptable [9]. The lower stiffness of a timber core produces a tall timber building with lower natural frequency and larger displacement as compared with a concrete core. As larger deformation results in larger energy dissipation within the structure, it is the experience of the authors that tall timber buildings with a timber core is likely to have higher damping than a tall timber building with a reinforced concrete core. Research is continuing on this behaviour.

Bracing is also an effective system to provide stability of tall timber structures and is well utilised by engineers. Several projects have used bracing to help tall timber buildings to resist lateral force (Figure 8.14).

The Future of Natural Materials in Tall Buildings

Natural materials in general and timber in particular offer advantages in strength/stiffness to weight ratio, which leads to opportunities for prefabrication leading to fast, high quality construction. The tall timber buildings constructed to date demonstrate these advantages and are prototypes for a new form of construction.

The key challenges to address are fire safety and dynamic response. The technology is developing, and as for the development of taller buildings in steel and concrete in the late eighteenth and early nineteenth

FIGURE 8.14
Timber bracing to provide stability in a tall timber building at the Earth Sciences Building at the University of British Columbia.

centuries, increase in height will progress as the technology develops to equip engineering designers with techniques to enable good predictions of performance.

References

1. National building regulations in relation to multi-storey wooden buildings in Europe. Östman B and Källsner B. Wooden Eco-buildings Report 5 (Building regulations) by Växjö University within the EU project Concerto-SESAC.
2. Multi-storey timber frame buildings: a design guide. Grantham R, Enjily V, Milner M, Bullock M, Pitts G (2003). BRE Report 454, BRE Press, London, UK.
3. Dujic B. Innovative timber composites—Improving wood with other materials. COST Action Conference Nicosia. October 2013. ISBN 1 85790 178 9.
4. Addis B. (2007). *Building: 3000 Years of Design, Engineering and ~Construction*. Phaidon Press Ltd, London, pp. 388–408. ISBN 978 0 7148 4146 5.
5. Bisby L. and Frangi A. (2015). Editorial for special edition on Special Issue on Timber in Fire. *Fire Technology*, 51, 1275–1277. Doi: 10.1007/s10694-015-0539-1.
6. ISO 10137:2007(E). Bases for design of structure—Serviceability of buildings and walkways against vibrations.

Designing Tall Buildings with Natural Materials

7. BS EN 1991-1-4:2005. Eurocode 1: Actions on structures—Part 1–4: General actions—Wind actions. British Standard Institution.
8. Reynolds T, Harris R, Chang W-S, Bregulla J, Bawcombe J. (2015). Ambient vibration tests of a cross-laminated timber building. *Proceedings of ICE—Construction Materials*. 168: 121–131.
9. Hein C, Kaluzni L, Twohig B, Chang W-S. Global stability of high-rise timber buildings. World Conference on Timber Engineering 2016. Vienna, Austria.

9

Emerging Nature-Based Materials and Their Use in New Products

Morwenna Spear

CONTENTS

Functional Properties: Biomimicry at Work 218
Structure and Protection 220
    Structure in Mineral Composites 221
        Tensile and Compressive Strength 221
        Protection from Crushing and Impact: Mollusc Shells 226
        Strength and Hierarchy in Nacre 230
        Toughness in Nacre 231
        Hierarchy in Bone 233
        Toughening Mechanisms in Bone 236
        Antler 240
    Microstructure of Organic Composites 241
        Hierarchy in Collagen 242
        Plant Cell Walls as Composites 245
        Strength in Compression and Tension: Wood and Plant Stems 247
        Hierarchy in Wood 250
        Natural Fibre Composites to Mimic Plant Fibres 252
Surfaces and Textures 253
    Texture and Superhydrophobicity 253
        Lotus Leaf Effect 255
        Texture and Adhesion 258
        Texture for Other Applications 259
    Chemistry and Adhesion 260
Form and Structure 261
    Shape and Form 261
    Cellular Materials 269
        Honeycombs and Foams 271
        Cellular Materials for Shape Modulation and Motion 277
    Fluid Dynamics 280
        Temperature and Humidity Control 280
        Conduction, Convection and Radiation 282
        Small-Scale Fluid Dynamics and Diffusion 283

Moisture Harvesting 285
Hydrostatic Support 285
    Cells: Containing and Using Hydrostatic Pressure 285
    Turgor Pressure and Motion 288
  Responsive Forms – Actuators 292
Biological Systems: Harnessing Natural Processes.....................................298
  Enzymes for Manufacturing...298
  Self-Assembly and Biotemplating 300
Conclusions and Outlook 303
Acknowledgements 305
References 305

Look deep into nature and you will understand everything better

Albert Einstein

Functional Properties: Biomimicry at Work

In Chapter 3, a wide range of natural materials were introduced and their constituent components described. These are the organic polymers and biominerals that provide the structural elements of our bodies and the organisms around us. In many cases, these are composite materials, and it is remarkable what a range of properties is possible, simply by changes in the way that they are structured. The process of material deposition within or by cells is controlled to construct both soft and hard materials, which can be strong, flexible or resilient, making up everything from cell walls to adhesives, and bones and exoskeletons to tendons or cartilage. An enormous amount is now known about many biosynthesis processes, and the often complex chemical composition of many of these materials. It is possible to not only explain their mechanical properties and performance in the living organism but also exploit these materials when collected, harvested and applied in new structures. Much can also be learnt by considering the form, and design of the materials, giving inspiration for development of new materials or structures. As Jeronimidis and Atkins (1995) point out, engineers extract maximum benefit from materials, whereas Nature extracts maximum benefit from structural shapes made up of indifferent materials.

Many bio-based materials have properties that can be well harnessed within biomimicry, such as hydrophilicity and related shape change or hydrophobicity and self-cleaning surfaces. We will also see that specific crystalline structures can give unique mechanical properties, while the incorporation of amorphous material within crystalline systems can dramatically increase toughness. In their review of the mechanical efficiency of natural materials, Wegst and Ashby (2004) commented that virtually all materials obtained from plant and animal sources are composites. They consist of a relatively few polymeric and ceramic components, proteins, polysaccharides and minerals, as have been introduced in Chapter 3. It is the combination of these elements in different microstructural forms that gives the great diversity of properties. This chapter aims to highlight the potential new materials that can be achieved by combining the individual components. Drawing on the structures and textures seen in natural materials, some modern analogues are given to demonstrate the potential of recombining nature-based materials using bio-inspired design to achieve new or desirable effects.

Countless examples of bio-inspired design, and of biomimetics, exist. The term biomimetics was first used in the late 1950s by Otto Schmidt to point out the overlooked potential for biophysics to inspire technology, as opposed to its more common use applying physics and technology to understand biological systems. It covers many disciplines in which solutions are derived by emulating the strategies, mechanisms and principles found in nature (Badarnah 2017). While some products are widely cited as examples of biomimetics or bio-inspiration, for example, Velcro as a product inspired by the burrs that clung to the coat of George de Mestral's dog after a walk (Velcro 1955, cited by Vincent et al. 2006), and the hydrophobic nature of the lotus leaf as inspiration for self-cleaning surfaces coated with Lotusan paint (Barthlott and Neinhuis 1997). Many others are listed by Vincent et al. (2006), alongside others of less certain, or more apocryphal, derivation, such as the structure of the Eiffel Tower being based on the structure of trabecular struts within the human femur. Yet, as we shall see later, scientists even now are returning to the mechanics and design of near-hollow bird bones or beaks to develop not only lightweight structures but also shape recovering lightweight materials and applying these new materials within emerging manufacturing technologies such as additive printing (Schaedler and Carter 2016). It is increasingly possible to test bio-inspired designs and principles in previously unconsidered technological applications.

Biomimicry, or biomimetics, is the art or science of harnessing nature-based design within materials. This spans many topics, whether in engineering ventilation systems based on principles observed in nature or designing materials using a honeycomb structure to reduce weight while retaining load transfer between the tensile and compressive faces of a component in bending. It may also be applied to systems design, for example, adjusting the urban environment related to behavioural biology, and to colour morphing paint inspired by the textural colour of feathers and beetle carapace. Many biomimetic products may not use bio-based materials – there is no bio-prerequisite. However, many of the developments in biomimetics do centre on utilising natural polymers or inorganic components with a biomimetic approach. This short review offers some examples of the application of biomimetic concepts to the development of new materials based on bio-derived components.

Bio-inspiration can draw on the composition, but more importantly the construction of the material. In fact, six main themes have been identified within biological systems that demonstrate the elegance and efficiency of biological materials (Chen et al. 2012):

- self-assembly,
- multi-functionality,
- hierarchy within the structure,
- hydration within the structure,
- mildness of synthesis conditions and
- their development within environmental constraints.

Biomimetics goes beyond copying the form or function of a biological system, to many combinations of these traits, leading to materials or systems that are efficient mechanically, or in manufacture, or in their suitability for purpose. Schmidt himself (the scientist who first used the term biomimetics) applied the term biomimetics to the development of an electronic system that harnessed the concept underlying nerve firing in squid for electronic engineering, the Schmidt trigger. Often the bio-inspiration gives rise to viable materials or systems that reflect the original inspiration source, but also apply this inspiration in an adapted way; yet, this does not diminish the original inspiration.

This is a rapidly expanding field, and it is impossible to exhaustively cover all the frontiers of current research, even when limiting the topic to bio-inspiration within bio-based materials. It is frequently commented that designers and materials scientists only utilise a small proportion of the ideas and designs that nature may provide us. It is therefore intended that this chapter will introduce some key concepts and highlight the multitude of options for microstructure or form of the bio-based material to deliver superior properties for its intended purpose.

Structure and Protection

The majority of the materials introduced in Chapter 3 were structural in some capacity. The strength, rigidity and resistance to impact of bones, shells and exoskeletons is what has led scientists and engineers to consider them for inspiration in materials or buildings. This structural role, and the mechanisms by which strong materials can also be made tough, will be the first aspect we consider. The incredible efficiency of bio-based materials is another aspect, and the way in which these materials may also be made lightweight will be considered in 'Form and Structure' section where we look at the way in which nature has harnessed the shape and form of the material.

Emerging Nature-Based Materials 221

For the organism, the most relevant properties of these materials are often the structural support – strength and stiffness, and the resistance to impacts incurred due to predation, or generated by walking, running or fighting. It should be noted that in many biological materials the mechanical and dynamic properties are highly related to the orientation in which they are tested. Some natural materials have obvious lamellar or fibrous structures, influencing properties. In plant stems, bones or shells, this anisotropy generally favours the orientation in which the material is of greatest evolutionary benefit to the organism during its life. It also relates to the process of biosynthesis or deposition of the tissue. The biopolymer and presence or absence of mineral components may vary, yet each has developed preferential alignment of these polymeric chains, or deposition of mineral or other inherent structural elements, which serve to increase strength or impact resistance, depending on function.

Structure in Mineral Composites

Tensile and Compressive Strength

Many biological materials are composites, in which some form of reinforcing element is held together by a more compliant matrix. The rule of mixtures has already been introduced as a convenient relationship between the properties of the reinforcing fibre, the matrix and the composite as a whole, based on the volume fraction. Similar relationships can be proposed for short fibre composites, based on the shear lag model (Cox 1952; Hull and Clyne 1996), or using the Eschelby method, the effect of spherical particles, elliptical particles and plates as reinforcing elements can also be modelled (Eschelby 1957, 1959; Hull and Clyne 1996). Thus engineers can predict the properties of short fibre composites, or of composites reinforced with glass particles or mineral crystals of a wide range of shapes. Within both models, the relationship between stress distribution along the fibre or particle and the shear stress at the particle–matrix interface are the important criteria. If shear stress at the interface is higher than the interfacial shear strength, fibre pullout will occur. If the stress within the fibre exceeds the fibre's ultimate tensile stress, fibre breakage will occur. Failure mode, and crack propagation, can be related to these two inherent properties of the composite.

Figure 9.1 gives an example of the displacement of a single fibre in a matrix under tension. The fibre is assumed to have a higher elastic modulus (MOE) than the matrix, and deformation in the matrix is restricted by the presence of the stiff fibre. When each segment is considered individually, it is possible to define the strain field at each location and distance from the fibre axis by considering the shear strain and the shear modulus of the matrix. Strain at the interface between the fibre and matrix is highest near the fibre ends, whereas stress within the fibre is highest in the centre and lowest at the ends. For fibres of different lengths, the profile of tensile stress is different,

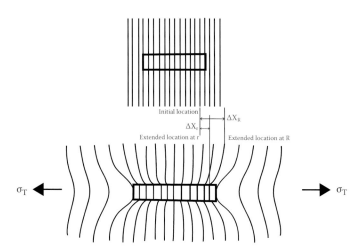

FIGURE 9.1
Schematic illustration demonstrating the development of tensile strain in the fibre and the matrix for a single fibre loaded longitudinally. Upper diagram indicates the relaxed state, and lower diagram under stress. Red labels indicate the relative displacement at two locations: within the fibre (at radius r from fibre axis), and at a distance R into the matrix.

as indicated in Figure 9.2a. It is then possible to define a critical fibre length to achieve maximum load within the fibre, based on the fibre aspect ratio (length to diameter ratio). Likewise the fibre length alters to profile of the shear stress at the interface (Figure 9.2b). If the MOE of the fibre and the matrix, and the interfacial shear strength of the two materials are known, it is possible to predict whether fibre pullout or fibre breakage will occur (Hull and Clyne 1996). In many of the biological composite materials, and the bio-based composites, the properties of the interface play a large part in determining behaviour.

Nacre is a simple composite, based on approximately hexagonal plates of aragonite within a protein-based matrix. It is found on the interior of many mollusc shells; frequently, the nacre of the abalone shell (*Haliotis rufescens* and other species) is studied, due to ease of sampling from the thick nacreous layer of 'mother of pearl'. Nacre consists of approximately 95%–98% aragonite, arranged in layers of hexagonal tablets, separated by 2%–5% organic matter (Chen et al. 2012; Olson et al. 2012). The tablets in abalone nacre are 8–12 µm in length, and 0.4 µm in thickness (Meyers et al. 2010), while other species of bivalve, cephalopod or gastropod have thicknesses of 0.30 to 1.10 µm, and a range of plate lengths 2.7 to 8.4 µm (Currey 1977; Olson et al. 2012). The thickness of these tablets, or mosaic single crystals, is of the same order of magnitude as the wavelength of light, contributing the iridescent surface (Chen et al. 2012). The layers of aragonite hexagonal tiles form a 'bricks and mortar' structure, or in some cases a 'stack of coins' structure.

The tablets of aragonite form sheets of tessellating tiles between very thin layers of the matrix substance (which can be present at as low as 2%). If these

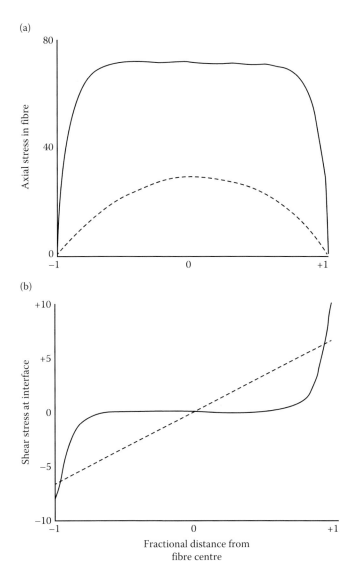

FIGURE 9.2
(a) Stress distribution within two fibres of differing aspect ratios, solid line aspect ratio = 50, dashed line, aspect ratio = 5. (b) Shear stress distribution at the fibre–matrix interface for the same two aspect ratios.

layers of aragonite were continuous (not made up of tiles), the rule of mixtures would apply, and the modulus could be predicted to be very near to 98% of the Young's modulus of aragonite itself (81–100 GPa). But the tablets mean that load is not continuously transferred within the plane by aragonite, there are dislocations between adjacent tiles, at which the load must be transferred through the matrix to the neighbouring tile. The shear lag mechanism must be considered. In fact, the Young's modulus of nacre is approximately 70 GPa, which is very high for a particulate reinforced composite, and the tensile strength is 70–110 MPa (Currey 1977). Shear lag-based models are required as each aragonite tablet is able to slip individually within the layer, and stress concentrations will occur along the tablet–matrix interface.

In the diagram of nacre (Figure 9.3), it is possible to see how, if tested in tensions along the plane, the bricks and mortar structure allows shear within the thin matrix layer. If we assume that any shear within the rigid aragonite is negligible, we can approximate the shear stress within the organic component based on experimental results. As this layer is very thin relative to the tablets, for example, 0.02 μm and 0.4 μm, respectively, the volume fraction of the organic layer is 0.05; thus, for a shear strain of 0.44 in nacre as a whole, the shear strain of the organic layer is 8.8, and shear stress of 5.96 MPa, similar to experimentally reported values of up to 10 from Sarikaya et al. (1990). Specially designed shear test apparatus with a shear gap designed to force shear within the organic layer of nacre was used by Lin and Meyers (2009), and revealed a shear strength of 36.9 ± 15.8 MPa. Testing in tension frequently results in pullout of individual tablets, leaving alternate plates and voids on each fracture face (Figure 9.4). Cracking of the tablets themselves is very rare (Menig et al. 2000).

Nacre is a strongly anisotropic material, and strength values vary greatly with the orientation of applied force. Tensile strength within the plane of the platelets is approximately 170 MPa, compared to only 5 MPa perpendicular to the plane (Figure 9.5, Meyers et al. 2008). In this mode of loading, the relatively weak protein matrix between layers is unreinforced, and the matrix, or the interface, exceeds its ultimate tensile strength and delamination tends to occur between the hydroxyapatite layers.

When nacre is tested in compression, the strength perpendicular to the plane is over 100 times higher than tension, resulting from the orientation of the platelets, and the minimal organic content. Data for tensile and compressive strength from different researchers is compiled in Table 9.1, showing the range

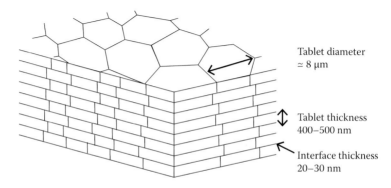

FIGURE 9.3
Simplified schematic of the structure of nacre, showing aragonite tablets embedded in a thin organic matrix.

Emerging Nature-Based Materials

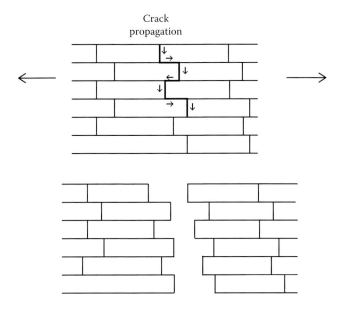

FIGURE 9.4
Schematic showing the influence of the tiled structure on the route of crack propagation through nacre when tested in tension. The same mechanism is seen within the tensile surface of a bending specimen.

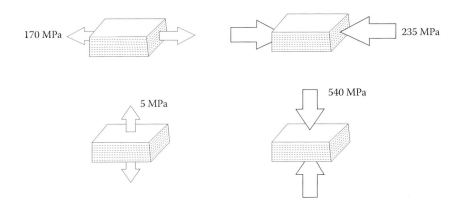

FIGURE 9.5
Schematic indicating the relative strength values for nacre when tested parallel to the plane in tension and compression, and for the same tests perpendicular to the plane.

of properties reported within the plane, or perpendicular to it. In some cases the magnitude of modulus or strength varies due to differences in test method, for example, many recent studies use nano-indentation to determine modulus of elasticity, reporting near perfect data for individual tablets of hydroxyapatite (e.g. Barthelat et al. 2006), whereas the early studies using tensile or

226   *Designing with Natural Materials*

TABLE 9.1

Selected Mechanical Properties of Nacre and Calcite or Aragonite Tested within the Plane and Perpendicular to the Plane

Property		Orthogonal	Perpendicular	Reference
Young's modulus (GPa)	Nacre	70		Jackson et al. (1988)
	Nacre tablet	81	79	Barthelat et al. (2006)
	Nacre tablet		92	Bruet et al. (2006)
Young's modulus	Calcite	137		Currey and Taylor (1974)
Young's modulus	Aragonite	144	82	Barthelat et al. (2006)
Modulus of rupture (MPa)	Nacre	116	56	Currey (1977)
		185–220		Sarikaya and Aksay (1992)
		197	177	Menig et al. (2000)
		233	194	Wang et al. (2001)
		200		Currey et al. (2001)
Tensile strength (MPa)	Nacre	70–110		Currey (1977)
	Nacre	160		Currey and Taylor (1974)
	Nacre	170	5	Meyers et al. (2008)
Compressive strength (MPa)	Nacre	540	235	Menig et al. (2000)
	Nacre tablet		2500	Barthelat et al. (2006)
	Nacre tablet		11000	Bruet et al. (2006)

compressive tests on small samples report data that is more representative of the macro-scale performance (Currey and Taylor 1974; Currey 1977).

Many more aspects of the mechanical performance of nacre have been reviewed and discussed (see e.g. Barthelat et al. 2006; Meyers et al. 2008; Chen et al. 2012); however, Figure 9.5 is included here as a good example of the need to consider orientation of test axis when evaluating and reporting mechanical properties in composite materials. Table 9.1 summarises some of the reported strength values for nacre in different test modes and different orientations.

Protection from Crushing and Impact: Mollusc Shells

Nacre is just one layer of the shell of the abalone, mussel or oyster. This lustrous material is typically found on the inner surface, adjacent to the organism. The outer layer is often a different crystalline structure, but formed from the same calcium carbonate mineral and combination of organic components. Marine molluscs require many types of function and protection from their exoskeletons due to the action of waves, tide and movement of the rocky substrate on which they live or the action of predators. The shell must also allow for growth during the lifespan of the organism, and controlled vulnerability during feeding, while providing good resistance against the beaks of predators.

Emerging Nature-Based Materials

Looking at the bivalves, such as mussels (*Mytilus edulis*) and clams (*Mercenaria mercenaria*), this exoskeleton takes the form of a pair of shells that hinge at one location (Figure 9.6a) and can be extended in concentric bands around the perimeter of the shell with each growth event (Figure 9.6b). In the gastropods, such as limpets and whelks, a single exoskeleton provides protection, while contact with the rock or substrate is maintained on the underside. The shell provides a protective role, against predation and environmental events. It also acts as a skeleton, allowing movement by muscle attachment, and prevents sediment from collecting in the cavity (Zuschin and Stanton 2001; Mackenzie et al. 2014). Amongst the molluscs, a great deal of variety in their shell designs can be seen, and further variation can be found when looking at the microstructure of the shell itself.

The shell is composed of calcite and aragonite, which are both forms of the same calcium carbonate mineral. Differences in the crystalline structure of the calcite and aragonite are harnessed within the deposition of the structure, to introduce localised order, influencing the strength and impact resistance of the shell. The aragonite within the nacreous layer provides high stiffness and resistance to crack propagation, while in many shells, other forms are present in outer layers, for example, in mussel shells a more fibre-like crystal structure can be seen in outer layers (Greisshaber et al. 2013), providing greater crack deflection and strength within a preferred orientation. Cryofracture of the mussel shell reveals an outer layer of rhombohedral calcite, with a core layer of long aragonite crystals, and the inner face of the shell is nacreous. Other studies have shown that the prismatic calcite in the outer layer is obliquely aligned and may help avoid splitting when the shell is placed under perpendicularly aligned loads (Feng et al. 2000). Within the abalone shell, the nacre deposition occurs adjacent to a region of columnar aragonite, then a 'green organic layer', and is followed by spherulitic aragonite prior to returning to the tiled configuration. Development of these layers continues throughout the life of the animal, increasing shell thickness, and diameter (Chen et al. 2012).

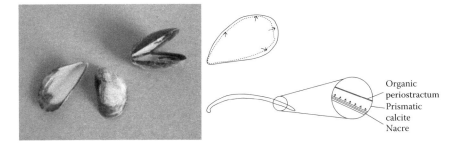

FIGURE 9.6
(a) Shells from blue mussel *Mytilus edulis*. (b) Schematic of shell showing direction of growth over time and cross section indicating the calcitic and nacreous layers. New layers of nacre are deposited to the internal surface within the extrapalial space adjacent to the organism.

There are many forms in which the calcite or aragonite may be deposited in shells. Currey and Taylor (1974) categorised eight shell microstructures for molluscs: columnar nacre (aragonite), sheet nacre (aragonite), foliated calcite, prismatic (calcite or aragonite), crossed lamellar aragonite, cross-foliated calcite, complex crossed lamellar aragonite and homogeneous. The nacre discussed earlier in this section is typically either columnar (with a stack of coins texture) or sheet (with a bricks and mortar texture). Further structures were identified by Kobayashi and Samata (2006) with some differences in nomenclature. Electron microscopy images of five types of nacre microstructure are shown in Figure 9.7. A recent review by Zlotnikov and Schoeppler (2017) has compiled evidence from across many less reported mollusc shells and revealed target type (similar to the columnar type) and spiral type growth fronts in natural shell, and in synthetic growth medium.

Kobayashi and Samata (2006) observed that across the bivalve molluscs, there were only three typical combinations of these morphologies, namely: (1) simple or fibrous prismatic structures that transitioned to nacreous in the internal region of the shell; (2) foliated structures that transitioned to crossed lamellar then to complex crossed lamellar structures and (3) fibrous or composite prismatic structures that transitioned to crossed lamellar or homogenous layers then to complex crossed lamellar or homogenous layers in the interior. They designated these as the nacreous group, the foliated group and the crossed lamellar group. The nature of the protein phase also differs between these three broad groupings of shell type.

Organic matrix proteins form an important part of the nacre structure, with, for example, nacrein, mucoperlin, lustrin A, perlustrin and perlucin occurring in different species (Kobayashi and Samata 2006). The organic matrix has a strong effect on the strength of the shell, and salt-water ageing of mussel shells has been shown to reduce resistance to fragmentation, due to the loss of organic material from within the nacreous layer (Zuschin and Stanton 2001). By contrast, clam shells with lower initial organic content, and a homogeneous brittle crossed lamellar structure did not show loss of strength on ageing in salt water. In addition, an organic layer, named the pereostracum, coats the outer surface of the shell. This is a sclerotinised quinone-tanned protein, which is important in templating the calcium carbonate deposition, as discussed previously. It appears that this layer also enhances the wear-resistance of the shell, as demonstrated by the tribological studies of Wählisch et al. (2014).

The great range of mineralisation strategies in shells hint at there being many mechanisms at play in providing strength, stiffness and toughness to the mollusc shell. Let us first consider the strength, and then the toughness of the nacreous layer. This is one of the most studied and discussed features, and the stress transfer within this relatively simple arrangement of planar hexagonal crystals of aragonite between very thin layers of organic material has changed the approach of research and development in many ceramic composites systems. The recognition of inclusion of organic molecules into

Emerging Nature-Based Materials

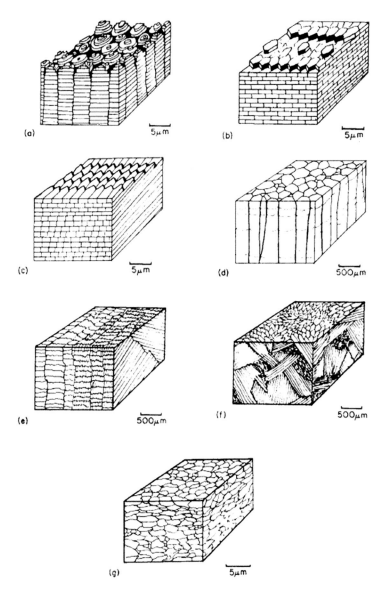

FIGURE 9.7
Scanning electron microscopy images of five mollusc shell microstructures, (a) nacreous (aragonite), (b) crossed lamellar (aragonite), (c) foliated (calcite), (d) prismatic (calcite), and (e) myostracum (aragonite).

Source: **Suzuki and Nagasawa (2013).**

the crystal structure of the aragonite has also inspired an ongoing quest to understand and optimise interaction of mineral and organic phases to create high performance materials.

Strength and Hierarchy in Nacre

Nacre can have a Modulus of Elasticity (MOE) of 30 to 120 GPa (Meyers et al. 2008; Bechtle et al. 2010), and the stiffness, strength and toughness of nacre can be related to its composition using mechanical theory. Here, tensile load is carried by the protein matrix by shear stress, and relates to the particle aspect ratio (length divided by thickness).

Nacre can be considered to be a hierarchical material with two levels of hierarchy. Nano-grains are bonded together to form the platelets (level 1), and the platelets are assembled into the bricks and mortar arrangement (level 2). Bechtle et al. (2010) applied Huajian Gao's hierarchical mechanics model (Gao 2006, 2009) to nacre, alongside other biological hierarchical materials. They demonstrated that with an aspect ratio of 40, the tensile strength (170–230 MPa, Meyers et al. 2008, Table 9.1) and particle pullout seen in failed nacre samples were in acceptable agreement with modelled values (380 MPa). Stress at the hierarchical level can be calculated by Equation [9.1]. The MOE of nacre was well predicted by Equation [9.2], and the study indicated that increased levels of hierarchy in fact tend to reduce the theoretical value of modulus of elasticity

$$\sigma_{n+1} = \left(\sigma_n \psi_n\right)/2 = \left(\tau_p \rho_n \psi_n\right)/2, \tag{9.1}$$

where σ_{n+1} is the average tensile stress at level $n + 1$; σ_n is average tensile stress of hard particle ($\sigma_{max}/2$); ρ_n is hard particle aspect ratio (l_n/d_n); τ_p is shear stress in protein phase; ψ_n is hard particle volume fraction. Strength of the material can be determined as either the shear strength of the soft protein phase or the tensile strength of the hard particle phase, whichever is smaller:

$$1/\left(E_{n+1}\right) = 1/\left(\psi_n E_n\right) + 4\left(1 - \psi_n\right)/\left(\psi_n^2 \rho_n^2 G_n^P\right), \tag{9.2}$$

where E_{n+1} is elastic modulus at level $n + 1$; ψ_n is mineral content; ρ_n is aspect ratio; E_n is particle elastic modulus; and G_n is shear modulus of protein at level n.

A simple man-made composite that had behaviour analogous to nacre is the glued laminated timber and cross laminated timber, which are increasingly used in construction. The relatively rigid timber elements of 20–30 mm thickness are separated by comparably thin layers of organic adhesive (approximately 1 mm) with lower stiffness. The load transfer at each interface occurs by the same shear lag mechanism, with end-to-end joints between timber members of one layer placing localised strain on the adhesive system, which is ultimately transferred due to the rigidity of the timber below or above the jointed layer. Few other composite systems present such a high level of reinforcing element relative to the low percentage of adhesive or matrix as is demonstrated in nacre.

Emerging Nature-Based Materials 231

Toughness in Nacre

Nacre is also frequently considered for its toughness. The work of fracture of nacre was reported to be 3000 times higher than that of pure aragonite by Jackson et al. (1988), and 1700 times higher by more recent researchers (Chen et al. 2012). We have already noted the creation of interfaces by crack deflection within the bricks and mortar structure, and the mechanism of platelet pullout. This increases energy consumption in generating larger crack surface area than if the crack propagated in a straight line, giving a higher work of fracture. It relies on the relative weakness of the organic interface. Let us now consider the thin organic inter-layers between the layers of hexagonal aragonite tiles in more detail.

The organic layer comprises a layer of chitin sandwiched between two layers of silk hydrogel and hydrophilic proteins (Addadi et al. 2006; Meyers et al. 2010). The organic sheets are present before the mineral phase is deposited, and their composition is instrumental in the nacre deposition process, both in nucleating mineralisation, and in providing horizontal boundaries between layers, which gives rise to this tablet structure. The central layer of β-chitin provides a network of long strong fibrils, but also contains holes of approximately 50 μm diameter, allowing mineral to pass through during deposition. Mineral material has been observed to be continuous from tiles on one side to the other, providing mineral bridges between tiles on opposing layers (Meyers et al. 2010). The terraced cone model of biomineralisation seen in *Haliotis* spp. demonstrates this sequential build-up of nacre in each platelet. The mineral that passes through these small holes creates nano-asperities, offering small rigid connections between tablets.

The organic matrix layer is thought to nucleate crystal formation, with variation in the protein composition, which may help template the stack of coins model (Addadi et al. 2006; Meyers et al. 2010). A second theory is that the crystal that nucleated in one layer propagates through the holes in the organic matrix allowing continuation of the same crystalline structure within the newly formed layer. However, the PIC X-ray images for nacre from different species revealed that while in some species (e.g. *Nautilus pompilius*), the angle of neighbouring plates was similar though the stack; in other species, the angles were only continued for a short number of layers (e.g. *Mytilus californianus*), or discontinuous (e.g. *Atrina rigida*) (Addadi et al. 2006).

In addition, in studies on the nacre in mussel shells (*Mytilus edulis*) the filaments of protein between layers of aragonite tablets are significant, and clearly aligned perpendicular to the two adjacent tablet faces under tension (Figure 9.8b, Greisshaber et al. 2013). It was shown by etching that the organic matrix forms a substantial part of the tablet itself, and could be hypothesised that the protein within the hydroxyapatite plays an additional role in toughening. Incorporation of the organic component into the crystalline order of mineral phase itself ensures better load transfer and greater interfacial shear stress than would be possible if the organic layer relied on surface adhesion

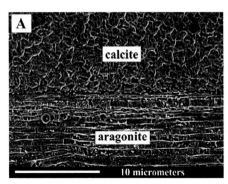 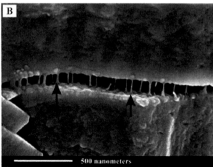

FIGURE 9.8
(A) Outer and inner layer of mussel shell, seen in cross section revealing needle-shaped crystals of calcite arranged in fibrillous structures, and tablets of aragonite in the nacreous layer. (B) Fibrils of protein between two aragonite tablets in the nacre of mussel shell.

Source: **Greisshaber et al. (2013).**

alone. As crack propagation typically occurs by platelet pullout, it is likely that both the mineral bridges between tablets (nano-asperities) and the bridging effect of protein inclusions passing from the tablet into the organic phase both enhance toughness.

There is increasing recognition of the role of the organic components in regulating biomineralisation, controlling nucleation, crystal polymorph (type), texture and morphology. The role of the β-chitin and the silk components of the organic layer are yet to be fully understood; yet, the control exerted by the silk proteins on mineralisation is likely to be significant (Levi-Kalisman et al. 2001; Nudelman 2015). The highly aligned chitin in the core of the organic sheet is water insoluble and provides structural support. The alanine and glycine rich silk proteins are also relatively hydrophobic, but additional proteins with higher water solubility are present, containing aspartic acid or glutamic acid, and these are implicated in controlling the crystal nucleation and morphology. The silk proteins and β-chitin provide a three-dimensional framework in which the mineralisation occurs, i.e. contributing to load transfer between layers. We will revisit this topic in 'Self-assembly and Biotemplating' section in relation to self-assembly in bio-based and bio-inspired materials.

The many toughening mechanisms in nacre were summarised by Mayer (2006) as follows:

- Creation of new surface area by delamination and/or fracture
- Crack diversion
- Pullout of the ceramic phase from the minor organic component
- Asperities on the platelet surfaces may also be involved in the pullout process, as well as may provide resistance against pullout of the platelets

Emerging Nature-Based Materials

- Hole formation at the ends of the displaced ceramic phase elements, which may be related to stress-whitening
- High level of anchoring of the organic (adhesive) phase(s)
- Ligament, or filament, formation in the organic phase(s) that are viscoelastic, as well as highly resilient
- Crack bridging by ligaments of the organic phase(s)
- Strain-hardening of the organic phase(s), accompanied by prior chain extension, breaking of cross-links, etc.

The difference between this biomineral composite and man-made ceramics clearly rests not only with the fine texture of the platelets, but is significantly benefited by the action of the organic components, despite their low abundance. The importance of the viscoelastic character of the organic phase was demonstrated in lab-scale assemblies to mimic the failure mechanisms of nacre in larger components. Further research in this area is likely to yield new approaches in the development of composite materials, whether at the macro-, micro- or nano-scale.

Hierarchy in Bone

In the mineralised protein tissues, the wide range of mechanical properties achieved can be large. All bone contains collagen and hydroxyapatite, yet samples may vary considerably in their strength and hardness. The MOE of trabecular (spongy) bone is typically 0.8–14 GPa, while the harder cortical bone found on the outer portion of most bones has a range of values approximately double this, 6–20 GPa (Currey 2002; Chen et al. 2012). Even comparing samples of cortical bone taken from the femur, a range of MOE and hardness values have been shown between the shaft (femoral diaphysis, 19 GPa MOE, 0.6 GPa hardness, respectively) and femoral neck (14 GPa, 0.45 GPa), the difference between femoral bone and lumbar vertebra was greater, due to different structural motifs (Hoffler et al. 2000). This study, by nanoindentation, was intended to increase understanding of these variations between structural elements of bones from different regions of the body, to increase the ability to match mechanical properties of implants or bone therapies to improve success.

The orientation of samples, but also location within the bone, govern the level of strength or stiffness observed. The trabecular bone in the core is regularly lower in strength than the denser cortical bone at the surfaces of the bone. Location of the strongest material at the surface as a strong cylinder provides great bending resistance. Another benefit of the variation in strength is lightweighting – bird bones have a low-density core and this reduces body mass significantly, assisting flight. In the tibia and fibia, and other long bones within limbs, the dominant strong axis being the longitudinal direction intuitively makes sense. Many more examples could be selected, but this anisotropic behaviour is achieved by combining the constituent parts of the bone as a composite material with several levels of hierarchy (Figure 9.9). Much can be learnt by considering these in turn.

It is the hierarchical structure of the bone that provides the great strength, low density and optimised balance of properties for the specific role within the organism. It allows some regions to be highly compacted and resistant to compression, and others to be cellular with elongated filaments and relatively thin walls to resist only the anticipated mode of applied stress. Seven layers of hierarchy can be identified, starting from (i) the amino acids within polypeptide chains, (ii) their combination into tropocollagen triple helices, (iii) then the aggregation of this tropocollagen with hydroxyapatite particles into collagen fibrils. The mineralised collagen fibrils assemble into (iv) fibril arrays, and (v) these fibrils co-align in lamellae in which the fibrils of the lamella have common orientation, but the adjacent lamella will have a different fibril alignment. Several other types of lay-up of fibrils are possible indifferent bone tissues, for example, radial fibril arrays in dentin, and have different resulting properties as described by Weiner and Wagner (1998). (vi) The lamellae form concentric cylinders around the Haversian canals of osteons and (vii) the bone tissue combines these osteons into compact bone tissue (Weiner and Wagner 1998; Launey et al. 2010; Ritchie 2010).

Anisotropy, i.e. different degrees of strength or stiffness in different orientations, results from bone's highly optimised structure, and the alignment of the constituent components at different hierarchical levels. In addition, as the structure is essentially a composite of fibrils, fibril arrays and layers, differences in strength from different tests are seen, even within the same orientation. For example, in the longitudinal direction, compressive strength (200 MPa) and tensile strength (150 MPa) differ because of the composite hierarchical structure (Sherman et al. 2015). The achievement of a balance between strength, stiffness and toughness is another attribute. Furthermore, bone is a lightweight material, achieved by density variation across its macro structure.

The compact bone is often called cortical bone and is located at the outer portion of the bone. It can be several tenths of a millimetre to several millimetres thick, depending on the bone type. A second type of tissue, the spongy bone, cancellous bone or trabecular tissue is found in the centre of many bones and contains struts of approximately 100–300 μm thick. In a slightly different set of seven hierarchical levels presented by Weiner and Wagner (1998), the relative location of the cancellous bone near the centre, and the compact bone at the surface provided the final level of hierarchy of the complete bone for mechanical optimisation (Currey 1984; Weiner and Wagner 1998).

Few man-made composites incorporate the number of levels of hierarchy seen in bone. However, a recognition of the composite character at each level of hierarchy and the effect of forming fibrils, fibres and spherical or cylindrical layered structures akin to the osteons and Haversian canals may offer

Emerging Nature-Based Materials 235

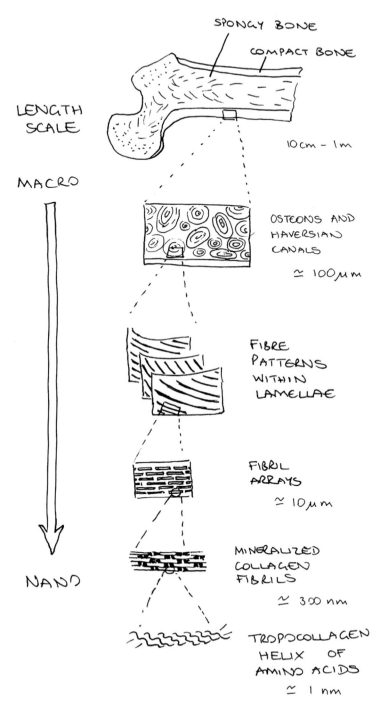

FIGURE 9.9
Six of the levels of hierarchy in bone, adapted from Launey et al. (2010).

design benefits in generating tough materials. Also in the quest to create self-healing structures, structures have been formed with encapsulated reserves of resin or reagent that could be released into cracks or delaminations whilst in continuing service.

Foams and honeycombs are frequently used in synthetic composites and engineered structures. These architected cellular materials can reduce weight while retaining stiffness within a wide range of products. While foams are relatively easy to form, and use within complex sections, designers often return to consider the network of struts, or geometry of the cells, which occur within bones such as from bird wings, in order to better understand the action of tensile forces and structural efficiency (Schaedler and Carter 2016). Cellular materials will be discussed further in 'Form and Structure' section.

An important feature of bone is the presence of the osteons, which in living tissue continually resorb and redeposit bone tissue, allowing remodelling in response to mechanical stress. The osteoclasts remove old material, while the osteoblasts deposit new tissue. While this process falls outside the scope of this chapter, it provides inspiration for self-healing in materials, and provides important additional insight into the ageing related changes in bone strength and toughness. The steady increase in the number of osteons and the alteration of mechanical properties of the collagen are increasingly recognised as contributing to altered toughness in bone with age (Nalla et al. 2006; Ritchie 2010).

Toughening Mechanisms in Bone

Having established the hierarchy of bone, it is now possible to recognise that the toughening processes within bone occur at multiple length scales. Each level of the structural hierarchy contributes to the total toughness (Launey et al. 2010). Fracture behaviour is the result of mutual competition between intrinsic damage mechanisms happening ahead of the crack tip as it propagates, and extrinsic shielding mechanisms which act behind (but also sometimes ahead of) the crack tip (Figure 9.10). Within bone deflection and crack bridging are extrinsic shielding mechanisms, and plasticity at different length scales acts as an intrinsic toughening mechanism.

Intrinsic Toughening Mechanisms

Several mechanisms take up energy during deformation of bone. At the smallest length scale, the collagen itself undergoes molecular stretching, and unwinding of the tropocollagen helix. This involves the breakage of hydrogen bonds and competes with intermolecular sliding. All these molecular processes combine to allow large plastic strain to occur without brittle failure Differences in the unfolding processes have been demonstrated for simulations of model collagen molecules under different strain rates, reflecting the differences in modulus observed experimentally

Emerging Nature-Based Materials

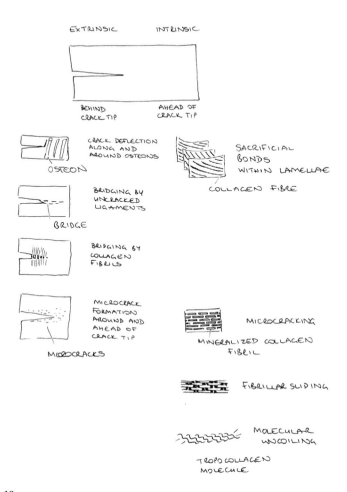

FIGURE 9.10
Intrinsic (right hand side) and extrinsic (left hand side) toughening mechanisms in bone, moving from the macro to the micro to the nano length scale. Adapted from Launey et al. (2010).

(Gaultieri et al. 2009). Modelled values for Young's modulus of hydrated collagen fibrils ranged from 300 MPa to 1.2 GPa at small and large deformation, the deformation of tendon (an unmineralised collagenous tissue) was discussed in Chapter 3, where proteinaceous structural materials were introduced.

The second main intrinsic toughening mechanism in bone is the fibrillar sliding of mineralised collagen fibrils. The presence of the hydroxyapatite crystals increases the stiffness of the collagen relative to unmineralised material, but the presence of hydroxyapatite–collagen interfaces is thought to contribute to continuous glide between the collagen fibrils and the mineral. This enables significant energy dissipation, once stress is sufficient to initiate sliding, effectively increasing the resistance to fracture (Buehler 2007).

The onset of plastic deformation occurs at higher strains for mineralised collagen fibrils than for native collagen, and the work of fracture (area beneath the stress strain graph) is several fold greater for the mineralised fibril in tension (Buehler 2007). To achieve this energy dissipation mechanism Buehler et al. (2008) point out that the adhesion energy between the collagen and the hydroxyapatite must be in a critical regime: sufficiently large to make initiation of slip more difficult than in native collagen, but sufficiently small that it occurs before covalent bonds within the collagen itself are broken. The ionic interactions between the hydroxyapatite and the collagen are of the correct order of magnitude for this to be the case, thus contributing significantly to toughness in bone (Buehler et al. 2008).

It is also proposed that fibrillar sliding occurs within arrays of collagen fibres, where the mineralised collagen fibrils are twisted into collagen fibres and bound by a thin layer of extrafibrillar matrix of non-collagenous proteins. At this level of the hierarchy, the strain can be envisaged as being carried by the fibres as tensile load, and by the extrafibrillar matrix that must resist slippage under shear (Launey et al. 2010). The strength of the matrix is likely to be sacrificial, i.e. lower than the strength of the collagen fibres, to allow deformation at a fraction of the load required to break the collagen fibres themselves. Finally, Launey et al. propose that microcracking at any of the length scales (from sub-micrometer to tens of micrometers) would provide a mechanism of micro-scale deformation (Zioupos and Currey 1994; Launey et al. 2010).

It is clear that the hierarchical structure of the bone contributes interfaces at many levels of the structure, which have potential to consume energy in slippage, small deformation and crack deflection, contributing to the total work of fracture of the material.

Extrinsic Shielding Mechanisms

Crack-tip shielding works as an extrinsic mechanism at the micro- and the macro-scale to provide toughness in cortical bone. The mechanism relates to deflection of the crack path during crack propagation, rather than resisting the crack initiation process. Three potential options for crack propagation at a simple interface are shown in Figure 9.11; importantly, the delamination ahead of the crack tip may greatly reduce the energy available for crack propagation, and arrest crack development. Bone includes many interfaces, at different hierarchical levels in its structure, so crack deflection and crack blunting due to delamination may occur. One major contributor are the osteons, in which lamellae provide many microstructurally weak interfaces along which cracks may travel preferentially; however, the cylindrical nature of these interfaces and the alignment of the osteons longitudinally may significantly deflect the crack propagation route. At the largest scale, this weak interface occurs between the cement lines between the osteons and the bone matrix, or along the hyper-mineralised boundaries between primary osteons in antler.

Emerging Nature-Based Materials 239

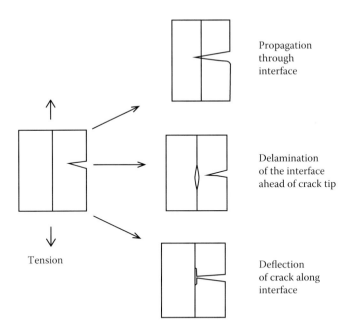

FIGURE 9.11
Schematic presenting three different behaviours when a crack through a material approaches an interface. When interfacial bonding is strong, the crack may propagate directly to the material on the other side. When interfacial bonding is weaker than the stress ahead of the advancing crack tip, delamination may occur. In other cases, the relatively weak interface may deflect the path of the crack along the interface leading to delamination.

Fracture toughness in bone is anisotropic, with crack propagation in the transverse direction of long bones being up to five times higher than crack propagation in the longitudinal direction (Koester et al. 2008). Crack deflection is a significant toughening mechanism arising from the occurrence of microcracking; the other significant mechanism is crack bridging (Launey et al. 2010). Here, the mechanism relates to formation of small cracks ahead of the crack tip when the crack is propagating in the longitudinal direction, so-called mother and daughter cracking. The microcracks ahead of the main crack coalesce but may lead to small, unfractured regions that are now behind the crack tip, providing a bridge across the crack, allowing some residual load to be carried.

These extrinsic mechanisms contribute further capacity for consuming energy, decreasing the likelihood of catastrophic failure of the bone itself. As a result, bone is a highly efficient composite material, with toughness that is far superior to a simple combination of its protein and mineral phases.

Age and Toughness

Despite its high toughness, it is well known that bone tends to become more brittle with age, i.e. toughness is lost. Some of the decrease in strength of bone with age is attributed to a change in mineralisation, measured as bone mass density (BMD), with decreased BMD being linked to brittleness, and osteoporosis (Launey et al. 2010). However, some of the decrease in strength in bone with ageing, especially the reduction of toughness, is related to degradation of the collagen network (Zioupos et al. 1999; Zioupos 2001). The degree of cross-linking within the collagen increases with age, and at high cross-link densities, there is less scope for fibrillar sliding under load. Instead, extension of the collagen fibrils involves molecular fracture and breakage of cross-links, giving rise to brittle behaviour. Other factors such as the change in number and structure of osteons have also been discussed and our understanding of the ageing processes in bone is still advancing.

Antler

In the case of antler, a material that also contains hydroxyapatite in fibrillar collagen, the structure is optimised for the impacts associated with its use in social displaying and defence. Antlers have no marrow and contain cancellous bone that is distributed throughout the entire antler, with good alignment. Again, the hydroxyapatite enhances strength, stiffness and rigidity, while the incorporated collagen provides toughness, and a greater proportion of collagen leads to higher toughness in antler. The mineral content of antler is typically lower (36%) than that of bovine femur bone (43%) (Currey 2002). While strength and MOE tend to increase with increasing mineral content (i.e. are higher in bone), it is known that this also decreases the work of fracture.

The tensile strength of bovine femur is between 100 and 140 MPa; however, strain at failure and work of fracture was 4–5 times higher for antler (Figure 9.12, Currey 1979; Zioupos et al. 1996). In antler, the post yield region

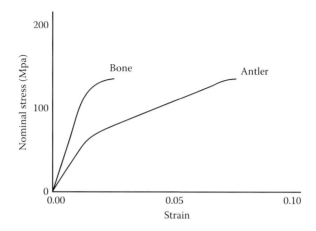

FIGURE 9.12
Schematic showing typical stress–strain behaviour of bovine femur and antler in uniaxial tensile test.

Emerging Nature-Based Materials 241

of a typical stress–strain graph extends considerably longer than for bone, resulting in an ultimate tensile stress, which is similar to that seen in bone, but at considerably greater strain (8%–9% compared to 1.5% in bone). The sustained performance antler at a lower modulus of elasticity above the initial yield point allows for a continuum dynamics model in fatigue tests, which is not found in bone (Zioupos et al. 1996).

The predominance of primary osteons in antler, and their high degree of alignment, achieves very high toughness when the direction of crack propagation is perpendicular to the osteon axis. When antler is loaded transversely, J values of approximately 60 kJ/m^2 are seen over the first 600 µm of crack extension (Launey et al. 2010), which is double the toughness of human cortical bone (approximately 30 kJ/m^2, Koester et al. 2008). When tested in the other alignment, antlers are considerably less tough (approximately 4–5 kJ/m^2, but this is still higher than the equivalent alignment of test in human cortical bone).

Microstructure of Organic Composites

Hierarchy is not unique to the mineralised natural materials. In Chapter 3, the strength of plant bast fibres was discussed, relating to the well aligned long, stiff microfibrils of crystalline cellulose. The cell wall of wood fibres was seen to contain layers in which the alignment is controlled to achieve different roles, the primary wall being randomly aligned, with orthotropic properties, and the secondary wall layers S1, S2 and S3 having specific orientations of the microfibrils. In fact, several levels of hierarchy can be identified in plant fibres or in wood. This is not only between the molecular and the microfibril level, but the lamellae of microfibrils, the cell wall layers, and, at a higher level, the action of the cell as a hollow cylinder and of the tissue comprising individual cells joined by a lignin-rich matrix material forming a foam.

The cellular appearance, and clearly visible differences between different planes of the wood also hint at anisotropy, and large differences are seen in the strength between longitudinal and transversely loaded samples. This relates to the shape and alignment of the cells, and their wall structures. The anisotropy at each level of hierarchy contributes to the complexity of modelling mechanical properties and toughness. Above the cellular structure, the presence of growth rings, and the contribution of minor tissue types such as rays and vessels form the higher tiers of hierarchy contribute to the macro-scale property differences between woods of different species.

Even materials that appear relatively uniform, such as tendon (which is made of collagen), contain hierarchy at the molecular level. The differences in strength and elasticity seen different tissue types – for example, skin or tendon – relates to the manner in which collagen fibrils are combined into macroscopic tissue with different roles within the body.

Hierarchy in Collagen

Fibrillar collagen can be considered to have four broad levels of hierarchy; the first three of that have been introduced in the hierarchy of bone: molecular scale (the combination of polypeptides into a left-handed tropocollagen triple helix, approximately 1.6 nm diameter, Figure 9.13a,b), fibrillar scale (the five tropocollagen unit or microfibril, 2–3 nm diameter, Figure 9.13c,e), micro-scale (the collagen fibre, ranging in diameter from 100 to 500 nm, Figure 9.13d) and the macro-scale (response of the tissue) (Sherman et al. 2015). In addition, Haut (1986) also considered both the microfibril, the subfibril at the fibrillar scale, and the fascicle (aggregates of co-aligned fibrils) as an intermediate before tissue scale.

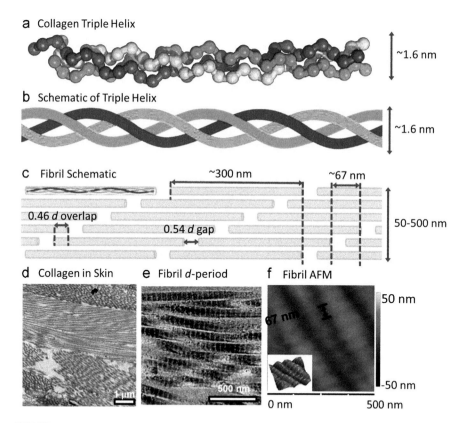

FIGURE 9.13

Hierarchy in the structure of collagenous tissues. (a) Helical left handed procollagen chains form a right handed triple helix of approximately 300 nm length. (b) Schematic representation of a triple helix of collagen. (c) Arrangement of triple helices into fibrils in a staggered manner with a gap (0.54d) and overlap (0.46d) region. (d) Layers of collagen fibrils in a cross section of skin. (e) TEM image of collagen fibrils, showing the characteristic banding pattern. AFM image of hydrated collagen fibrils in a scale from arapaima fish.

Source: (a–e) Sherman et al. (2015), (f) Yang et al. (2014).

Emerging Nature-Based Materials

In cartilage and tendon, the elasticity of collagen is an important characteristic, and the molecular uncoiling, taking up relatively high strain in a recoverable manner is an important mechanism. These molecular motions occur in several ways, providing the typical J-shaped curve when testing collagen in tension. In addition, slippage of microfibrils is an important mechanism within the viscoelastic response of collagen, and this sliding of microfibrils relative to each other within individual collagen fibrils was demonstrated by Yang et al. (2012). Studies indicate that fibrillar slippage in non-mineralised collagen takes up less energy than mineralised collagen fibrils, with plastic deformation beginning at 5% strain rather than 6.7% strain (Buehler 2007). In addition, non-mineralised collagen is less stiff than mineralised collagen fibrils, 4.59 GPa rather than 6.23 GPa for small deformations (Buehler et al. 2008).

Sherman et al. (2015) reviewed the reported experimental and modelled values for strength at each of the four levels of hierarchy and showed that modulus decreases at the higher levels of hierarchy. The modelled values for modulus at the molecular level ranged from 0.35 to 12 GPa; however, at fibril and microfibril level, values ranged from 0.2 to 1.2 GPa (excluding one bead and string modelling study by Buehler (2006, 2008) with 4.4–38 GPa depending on strain rate). At the fibre level, differences between tissue of origin were seen, and values ranged from 30 to 1100 MPa. At the tissue level skin, cornea and valve tissues are considerably lower modulus (0 to 50 MPa) than tendon (1 GPa). At the molecular level, Sasaki and Odajima (1996) used changes in molecular pitch (of the triple helix) during tensile testing under X-ray diffraction to estimate the modulus of tendon collagen as 3.0 GPa. Modelling has been used to determine values ranging from 2.4 to 7 GPa (Gaultieri et al. 2011).

Collagen shows different stiffness at different levels of applied strain, giving a typical J-shaped curve in testing, with a toe, heel and linear region. For example, Gaultieri et al. (2011) reported a modulus of approximately 300 MPa at small strain (below 10% strain) and 1.2 GPa above 10% strain for wet collagen microfibrils. They inferred that below 10% strain the molecule was being straightened within the microfibril, losing its kinked arrangement, whereas above 10% strain, the molecules were undergoing stretching, i.e. the covalent bonds within the now largely straightened molecules were under strain. Differences seen in dry collagen fibrils (which have a higher modulus of 1.8 GPa, and near linear relationship) indicated that the presence of water has a strong effect on the mechanical behaviour of collagen. Elimination of water from between tropocollagen helices is thought to result in a reduction in intermolecular spacing, and a reduction in ability for intermolecular sliding, leading to stiffer more brittle response, for example, in dry skin compared to hydrated skin.

Time-dependent behaviour is an important aspect of collagen's mechanical role within the organism. The ability to transfer load rapidly when running and jumping relates to elastic load transfer and recovery, whereas sustained loading periods may induce viscoelastic behaviour within the tendon, with hysteresis in the recovery – the stain relaxation being partially elastic and partially time-dependent. Viscoelasticity is generally regarded as the time-dependent strain, which is recoverable over time, whereas elastic strain is recovered instantaneously on removal of the load. Many biological macro-molecular materials exhibit a combination of elastic and viscoelastic behaviour in a similar way.

The stress–strain response of rabbit skin also differs with strain rate, achieving higher stress and greater strain when strain rate was higher (1000 s^{-1}) than 10 s^{-1}. When compared with the behaviour of wavy steel wires with circular segments, the elastic component of this deformation behaviour could be accounted for, and the viscous component, based on the Maxwell model, could be related to disruption of hydrogen bonding between collagen fibrils, and their time-dependent sliding (Yang et al. 2015).

Architecture plays an important role in the macro-scale mechanical properties of collagenous tissues. This is especially significant in skin, where assembly varies from organism to organism. For example, the rhinoceros has a tightly woven structure of collagen fibrils within a thick dermis, to provide protection. The collagen fibres in rhinoceros skin are straighter than in most mammals, resulting in a much less prominent toe region of the stress–strain graph, and it lacks the tear resistance of normal mammalian skin. This is thought to allow greater dissipation of energy by superficial gashes across the skin in a fighting animal (Sherman et al. 2015). By comparison, the New Zealand white rabbit has skin that is better adapted to resist tearing if snagged by branches of the undergrowth, and the collagen fibrils are extremely wavy, and arranged in a mesh with many different orientations. On stretching, a large deformation can be taken up with very little stress due to straightening and reorientation. Tearing occurs at 50% or 150% strain, transversely (parallel to the Langer lines) or longitudinally (perpendicular to the Langer lines) (Sherman et al. 2015; Yang et al. 2015).

Skin in birds also contains a layer of dermis that embeds the bases of feathers securely. Beneath the spongy dermis, there are two layers of compact dermis in which densely packed fascicles are aligned in one of two orientations. The thicker of these two layers corresponds with the direction that provides greatest support for the feathers. The fascicles themselves comprise wavy fibrils, introducing the strain absorbing behaviour and leading to observation of an extended toe region within the stress–strain graph (Sherman et al. 2015).

The above examples provide just three ways in which collagen fibrils and fibres can be combined to provide different mechanical properties in skin. In other tissue types, there may be greater alignment, for example, in tendon, or the tissue may show complex lay-ups with different fibrillar orientations, for example, in fish scales. Collagen-based tissues clearly demonstrate the benefits of hierarchical structures on mechanical properties, and the great potential for performance optimisation by simple adjustment of these composite architectures to achieve the properties required in a given tissue.

Plant Cell Walls as Composites

Plant fibres are likely to have provided the model for almost all fibre composites. The harvest and use of long plant fibres from whichever native plants were best suited in ropes, textiles and many artisan products harnesses the inherent longitudinal strength of the aligned cellulose within the cell wall. The dominance of strength in a single orientation, i.e. anisotropy, of plant fibres allowed later products such as linoleum to harness the strong axis of the textile reinforcement long before the term composite became widespread in modern materials science.

Much later in history, parallels have been drawn between fibre composites and the plant secondary wall microstructure itself (Mark 1967; Mark and Gillis 1970; Ansell 2015). Here, we consider the cellulose microfibrils to be the fibre element, and the hemicellulose, pectin and lignin to be the matrix. The orientation of these crystalline microfibrils in the primary and the S1, S2 and S3 secondary wall layers of lignified cells was introduced in Chapter 3. It is important to recognise that discussions of cell wall architecture frequently focus on either lignified fibres with prominent secondary cell walls such as the tracheids of wood, or on non-lignified cells with primary cell walls, such as the parenchyma, which are relevant to food science. There is much that can be learnt from both disciplines (Harris 2006).

In the primary wall, the microfibrils are laid into a network with multiple directions within the plane of the cell wall. This gives near orthotropic mechanical properties, and some very interesting work has considered the effect of cellulose, hemicellulose and pectin on the mechanical properties. Chanliaud et al. (2002) created networks of cellulose fibrils harvested from cultures of bacteria. When the cellulose sheets were tested they showed stiff elastic behaviour, and modelling indicated a Young's modulus of up to 500 MPa. The incorporation of pectin amongst the cellulose fibrils reduced the modulus of these orthotropic sheets to around 120 MPa but they retained largely elastic properties at strains of up to 4.8%. The interesting effect was that incorporation of hemicellulose had a profound effect on modulus, which dropped to the range 2–10 MPa, but the behaviour became non-linear, indicating either plastic or viscoelastic time-dependent behaviour. The viscoelastic effect of hemicellulose within the cell wall has long been acknowledged, but the magnitude of its effect on this experiment was unexpected. Clearly there is much to learn from the composite nature of the primary cell wall, with its approximately random fibre alignment. More recent studies have used combinations of cellulose fibrils with other hemicelluloses (e.g. arabinoxylans, xyloglucans, β-glucans, galactoglucomannans) and revealed a range of behaviour that is revealing the nature of interaction between microfibrils and other cell wall polysaccharides (Mikkelson et al. 2015; Prakobna et al. 2015) and improving modelling capability for these viscoelastic materials (Bonilla et al. 2016). The bio-inspired hemicellulose hydrogel cellulose microfibril composites have potential in packaging, electronic devices and other applications due to the high modulus and strength (Teeri et al. 2007; Henriksson et al. 2008, Prakobna et al. 2015). We will consider the mechanical effects of non-lignified cells with primary walls in 'Form and Structure' section, in particular with the motion effects, which they provide for the plant.

When considering the wall as a composite, the alignment, or microfibril angle, within each layer of cell wall become of great interest, to consider the properties achieved within the quasi-cylindrical structure of the cell. The combined effect of each layer has a profound effect on the behaviour of the cell as a whole; and on the next level of hierarchy, the combined cell mechanical responses governs the overall response of the plant tissue, or the textile fibre, or the piece of wood (Hofstetter and Gamstedt 2009; Salmen and Burgert 2009). Microfibril angle can have a strong effect on the tensile or compressive response of the living cell, such as reaction wood in branches or inclined trunks, resisting the action of gravity using the helical alignment of the fibrils (Fratzl et al. 2008).

Recent advances in microscopy and imaging have allowed the biosynthesis of cellulose, and its deposition within the cell wall with preferred orientation, to be studied (Vincent 1999; Doblin et al. 2002). The process is mediated by enzymes, with enzyme rosettes progressing around the cell membrane, depositing fibrils of cellulose. These are approximately 5 nm in diameter and aggregate into the microfibrils recognised as the structural component of the cell wall (Doblin et al. 2002). Axial, transverse, helical, crossed and random textures of microfibrils can be laid down, controlled by cellulose synthase enzyme complexes located within the cell membrane (Emons and Mulder 2000; Ansell 2015). The selected angle and degree of co-orientation of microfibrils is related to the function of the cell and its architecture. The microfibrils are encased in a sheath of hemicellulose, and a wide range of models has been proposed for the nature of interaction between these components (Carpita and Gibeaut 1993; Cosgrove and Jarvis 2012).

There is emerging evidence indicating that different hemicelluloses are preferentially located adjacent to the cellulose or at a greater distance in the intra-fibrillar spaces. There is potential for epitaxial alignment of xylan alongside the cellulose, slotting together on a two-fold helical screw (Simmons et al. 2016). In Figure 9.14, the xylan is combined with lignin forming a complex within the woody tissue of a conifer. By contrast, other hemicelluloses such as glucomannans have some segments that align with the cellulose microfibrils, but tend to occupy the spaces between microfibrils, forming flexible bridges between these rigid elements (Figure 9.14, Cosgrove and Jarvis 2012). Pectins and structural proteins such as extensin may also be present (not shown, Carpita and Gibeaut 1993). Water is known to be present within the hemicellulosic regions, and may assist the viscoelastic behaviour or adjustment and re-adjustment of hydrogen bonds between hemicelluloses, which allows phenomena such as creep. The activity of enzymes in permitting extension during cell expansion, then inhibiting excessive stretch once the developed cell is under internal turgor pressure is crucial, and has led to ongoing revision of the cell wall model (Cosgrove and Jarvis 2012).

Emerging Nature-Based Materials

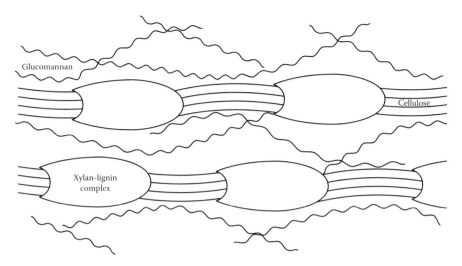

FIGURE 9.14
Simplified diagram of cellulose microfibrils within a hemicellulose matrix, where the xylan component is closely associated with the cellulose, and the glucomannan contains segments that adhere by hydrogen bonding to the cellulose, forming a bridge between cellulose aggregates. Based on a schematic proposed by Cosgrove and Jarvis (2012) for the secondary cell wall of conifer wood. In the cell wall, the components would be compacted together and available space lower than shown.

Many different plant cell walls with differing compositions are known, for example, the distinct difference in hemicellulose composition of monocotyledons compared to primary cell walls of dicotyledons or the secondary cell walls of woody dicotyledons; however, the principle of embedding stiff cellulose microfibrils in an amorphous matrix of various polysaccharide components provides strength, and flexibility to the developing plant cell. During cell wall deposition the hemicellulose allows the microfibrils to align in the same manner as liquid crystals in the nematic phase, this may be guided and aligned by cortical microtubules (Neville 1993; Emons and Mulder 2000). We have already noted that the cellulose microfibril alignment within different cell wall layers may be highly aligned to maximise unidirectional strength, or near random to allow orthotropic strength. As a result, the plant is able to control the strength and stiffness of each cell according to the performance requirements of that tissue.

Strength in Compression and Tension: Wood and Plant Stems

The tracheids make up the xylem of the tree, which has a role in bearing the weight of the trunk, branches and canopy during the life of the tree (Figure 9.15). Considering the simplest case, the tracheid, where the S2 layer of the tracheid forms the majority of the cell wall (up to 80% of the tissue volume in latewood fibres) and the microfibril angle within this tissue is

FIGURE 9.15
Radial section of Norway spruce wood, showing tracheids (vertical) and short sections of radial parenchyma (horizontal). Bordered pits in the tracheids are larger than the piceoid cross-field pits in the parenchyma cell walls.

approximately 8° as introduced previously. The process of secondary cell wall deposition occurs once the tracheid has finished expanding, and deposition defines the cell's role as a tracheid (Kerr and Bailey 1934). Lignification of the cell wall occurs as the final process, during programmed cell death, at which point the cell contents are removed from the cell to leave the tracheid effectively acting as a hollow tube, designed for conduction of sap within the tree stem under negative internal pressure (Tomos and Leigh 1999; Ménard et al. 2015).

The xylem, or woody part of the tree, must support the canopy, meaning that these quasi-cylindrical structures (tracheids) are loaded longitudinally in compression. In engineering terms, we must consider the resistance of these elements to buckling failure. When one cylindrical element (fibre) buckles, it induces stress in its neighbouring fibres. Typically, the buckling is limited to small diagonally oriented features within the wood, recognised as compression creases (Figure 9.16). Similar features are observed when unidirectional fibre composites are loaded in axial compression (André et al. 2014). As the wood cell wall is also dominated by its unidirectional composite structure within the S2 layer, a similar mechanism of dislocation development may also act on the micro-scale within the wall, forming kink bands that are visible under polarised light microscopy (Hughes et al. 2000; Hughes 2012), in addition to the macro-scale compression creases observed by eye (Desch and Dinwoodie 1981). As a result of localised formation of compression creases, the tree trunk is able to absorb energy when load exceeds the strength of the cell wall, for example, in older tissues at the base of the tree, or in stems that experience high wind loading. This can avoid or delay large-scale failure until severe overloading occurs and splitting or other mechanisms begin.

It is only after the tree is harvested, sawn and dried that the high tensile strength properties of this composite structure become of interest. The tensile

Emerging Nature-Based Materials

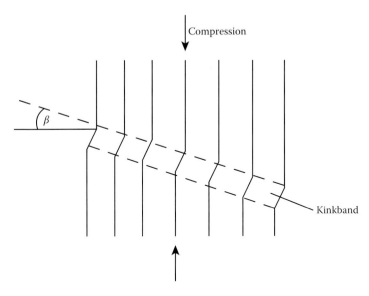

FIGURE 9.16
Schematic of a compression crease as seen within a unidirectional fibre composite, and as observed in the tangential longitudinal orientation of wood.

strength of wood is generally greater than its strength in compression, as the alignment of long microfibrils of high stiffness cellulose within the length direction of the timber provides enormous resistance to deformation. For example, Norway Spruce wood has a tensile strength of 104 MPa, but compressive strength of 36.5 MPa parallel to the grain (Desch and Dinwoodie 1981). The interfaces between cells (the highly lignified middle lamella), and between cell wall layers (where microfibril angle changes and hemicellulose and lignin transfer load), and between microfibrils (predominantly hemicellulose), each contribute to accommodate the stress of shear forces acting between the neighbouring elements. These interfaces, at each level of hierarchy, contribute to the overall strength and stiffness of the wood, as will be considered in 'Hierarchy in Wood' section.

Preferential alignment of fibrous cells to achieve strength can also be seen in plant tissues, such as the bast fibre plants – hemp, jute, kenaf and flax – which have great relevance in making natural fibre composites, as discussed in Chapter 3. Annual plants and grasses and many others also selectively utilise cells with prominent secondary cell walls containing highly aligned microfibrils, in order to contribute the tensile and compressive resistance needed to elevate their stems, flowering structures or leaves. In monocotyledon stems such as wheat, grass, miscanthus and bamboo, sclerenchyma contribute to the structural rigidity of the stem. Here, the stem has a cylindrical form (Figure 9.17a), most easily recognised in bamboo, where long barrel like elements between the nodes act as a structural whole. The hollow cylinder requires the combined tensile and compressive strength to resist

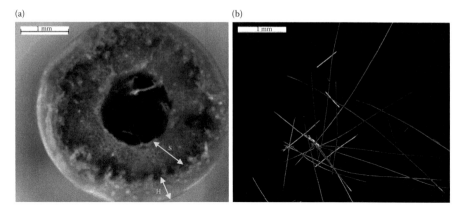

FIGURE 9.17
(a) Cross section of a stem of *Miscanthus sinensis*, where the hypodermis (H) contains many long needle-like sclerenchyma fibres, and the stele (S, or pith) contains vascular bundles within parenchyma ground tissue. (b) Macerated sclerenchyma fibres from *Miscanthus* stem.

Source: **William Turner.**

bending moments applied by wind or animal loading. The location of stiff sclerenchyma (Figure 9.17b) within vascular bundles dispersed throughout the culm provide sufficient stiffness to the cylinder of plant tissue. Another region of fibrous tissue is located around the outside of the stem in the hypodermis. In both cases the cellulose microfibrils in these needle-like cells are aligned near-longitudinally to form a stiff structural component (Turner et al. 2017).

Hierarchy in Wood

Wood is possibly the most studied plant tissue, due to its many structural and decorative applications. The mechanical properties can largely be related to the long tracheids or fibre tracheids, which form the bulk of the tissue.

It has been demonstrated that the hierarchical structure of wood may be considered as five levels of organisation (Hofstetter et al. 2005):

i. The stem cross section contains a succession of concentric annular structures, whether this is earlywood and latewood in temperate timbers or induced by seasons within tropical species.
ii. The earlywood and latewood are both made up of cells, of which the majority will be tracheids (in softwoods) or fibre tracheids (in hardwoods) and have significantly greater length than width.
iii. The cell wall consists of sequentially deposited layers, with differing thickness, and differing composition, as well as different helical angles of winding about the long axis of the cell. In most cases, the layers are the middle lamella (between neighbouring cells); the

Emerging Nature-Based Materials

primary cell wall; the S1, first secondary cell wall; the S2, second secondary cell wall; and potentially S3, third component of the secondary cell wall, followed by a tertiary cell wall, and in some cases a warty layer. The S2 layer may comprise 80%–90% of the total cell wall material for a softwood tracheid.

iv. The cell wall layers contain fibres of approximately 50–200 nm diameter, containing microfibrils of crystalline cellulose, with amorphous regions of cellulose. The angle of winding of the microfibrils within a cell wall layer contributes to the mechanical strength.

v. The crystalline cellulose within the microfibrils adopts a cellulose I conformation, with parallel chains in a $0.84 \times 1.03 \times 0.79$ nm sided prism, with angle 84°. Only the core of the microfibril is crystalline, and additional cellulose on the surface being amorphous in structure, this also allows the inclusion of some additional sugar monomers such as mannose and xylose in these outer regions. The cellulosic microfibrils are set in an amorphous matrix of hemicellulose (also a polysaccharide) and lignin (an aromatic polymer with high degree of cross-linking).

The crystalline cellulose regions of the microfibrils extends over a length of 600 nm, with amorphous cellulose continuing between these crystalline segments. The adjacent microfibrils transfer load across the amorphous regions. As a result, the tensile modulus of the cellulose microfibril has been measured as 134 GPa (Salmen and Burgert 2009). Next, the microfibrils are embedded in hemicellulose, which has a typical tensile modulus of 5–8 GPa. These aggregated layers of polysaccharide are encased in lignin, a stiff well cross-linked polymer, acting as an adhesive and resisting lateral forces.

When we return to the macro-scale and consider the properties of a timber such as sugar maple, we see it has a tensile modulus of 16.5 GPa (Schneider and Phillips 1991). The modulus is much lower than the cellulose component, but it is the result of load transfer between each individual element at each level of hierarchy. This will include shear at the interfaces within the cell wall, and between cell layers, and between neighbouring cells. It will include twisting of the microfibrils towards a state that is closer to full parallel alignment with the axis of loading. This torsion will lead to shear forces between individual microfibrils. It will also include the fact that this cellular material contains voids (the cell lumena), and if the cells are sealed, this air may exert a small resistance to change in volume relating to the loading.

However, while xylem accounts for much of the strength and toughness of wood, it must be noted that timber has a great diversity of anatomical features depending on species. Xylem physiology, i.e. the study of the anatomy of woods, can yield far greater explanation of anisotropic strength and physical properties than is possible to capture here. Classical texts such as Panshin and DeZeeuw (1970) or Desch and Dinwoodie (1981) can provide a primer for the interested reader. One important distinction is between the softwoods (from coniferous trees) which have a simpler anatomy than the hardwoods (from broadleaved trees). In addition to the xylem (tracheids), all timbers contain rays, which transport material radially within the stem. Hardwoods additionally contain vessels, which vary greatly in diameter, both between species (the small vessels in maple are barely visible in the grain, whereas the prominent texture of oak is due to large vessels in the earlywood. This is mentioned here as a caveat that in addition to the general model of hierarchy proposed by Hofstetter et al. (2005); it is possible to identify additional contributions to hierarchical behaviour, depending on species and xylem structure.

Natural Fibre Composites to Mimic Plant Fibres

Various researchers have attempted to use fibre composites to mimic the helical reinforcement, which is abundant in nature. The helical alignment of microfibrils in a plant fibre, or of fibrils within collagen, provides an excellent model for components under tension. Indeed, rope, string and yarn are simple examples of the same efficacy of design we recognise in the collagen triple helix, or the pultruded fibre composite, both described earlier, in Chapter 3. When the helical alignment is included in a cylindrical component, the system may also provide excellent properties in longitudinal compression, when layers of different winding angles are combined.

Recently, filament wound cylinders of hemp fibre yarns and thermoset polyester matrix were produced by Węcławski et al. (2014) and tested in compression. Their results clearly demonstrate the influence of winding angle on compressive properties, with Young's modulus increasing from 1.5 to 5.6 GPa as winding angle decreased from 90° to 10°. The stress at maximum load and the modulus of rupture also increased as the winding angle became shallower (i.e. closer to the longitudinal axis of the cylinder). While the values seen in compressive loading were lower than those typically observed in aligned natural fibre composite (e.g. compared with pultruded rods tested in tension: 122 MPa max stress and 16.84 GPa Young's modulus, Peng et al. 2011), the performance in compression clearly shows the mechanism by which plant fibres perform so well in compressive loading, as is typical in tree trunks, bamboo, maize stems or many other plants.

Flax fibre reinforced cylinders were produced by Yan and co-workers, with fibre arranged in weaves with longitudinal and circumferential alignment. Excellent properties were demonstrated in hollow tubes (Yan and Chouw 2013), and further increased when the tubes were filled with polyurethane foam (Yan et al. 2014). Hybrid flax and carbon (80:20) fibre cylindrical forms have been used in manufacturing bicycles, where the presence of flax fibre provides a vibration-damping role, working alongside the carbon fibre (Museeuw 2017). Here the helical alignment is achieved using weaves of the two fibres. The strength of flax fibre reinforced epoxy cylinders has also been

Emerging Nature-Based Materials 253

used in development of reinforcing cuffs for concrete. The flax fibre reinforcement enhances peak strength, strain, fracture energy and ductility of the concrete cylinders (Wang 2016).

In the tree trunk, once the tissue has been lignified, the compressive force of the canopy above it is the most significant mode of loading. The failure modes in buckling described by Węcławski et al. (2014) may hold clues for differences in buckling of earlywood (thin walled) and latewood (thick-walled) tracheids when wood is loaded in compression. The work of Fratzl et al. (2008) introduced in Chapter 3 also considered winding angle, but to evaluate net compressive and tensile effects arising from changes in internal pressure. In the plant cell example, the obvious cause would be hydrostatic pressure of the sap. This becomes an important consideration if applying the concepts described here to the case of annual plant stems. Both concepts support bio-inspiration, and we will return to the hydrostatic examples in 'Hydrostatic Support' section. Helically wound cylinders have also been successfully investigated for pressurised cellular materials to achieve plant-inspired structures for morphing aircraft wings and robotics (Li and Wang 2017). These will be further discussed with other hydraulic mechanisms in 'Cellular Materials for Shape Modulation and Motion' section, but the use of helical winding is a clear parallel to plant cell wall structure, and alteration of winding angle results in a range of deformations on pressurisation – allowing flexure and movement of the complete system.

Surfaces and Textures

Texture and Superhydrophobicity

Surface texture has been shown to have significant influence over the selection and use of materials. One of the most commonly known examples of biomimetics is the application of the lotus leaf effect to increase hydrophobicity of glass, and to create self-cleaning properties, whereby the texture naturally sheds dirt. This was a revolutionary approach within surface science. Traditional approaches relied on coating the substrate with highly hydrophobic compounds – waxes, silanes or PTFE – and in terms of natural materials carnauba wax is highly prised for car polishes. The contact angle (θ) of a droplet on an ideal surface (Figure 9.18) is related to the surface tension of the liquid and gas, and liquid and solid by the Young equation (Equation 9.3). Hydrophilic surfaces have contact angles of less than 90°, while hydrophobic surfaces, for example, silanes and fluorinated polymers exhibit contact angles of over 90°:

$$\gamma_{SG} = \gamma_{SL} + \gamma_{LG} \cos\theta. \tag{9.3}$$

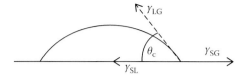

FIGURE 9.18
Schematic of a droplet of water on a smooth surface, indicating the contact angle θ between the droplet and the surface. Surface tension of the liquid (γ_{LG}), interfacial tension between the liquid and solid (γ_{SL}), and surface free energy (γ_{SG}) are also indicated.

However, even with hydrophobic polymers prepared as smooth or highly polished surfaces, it is only possible to achieve contact angles of up to 74° (carnauba wax), 107° (silane) or 116° (siloxane) (Cheng et al. 2006; Lim et al. 2008). Instead, it was discovered that the lotus leaf uses surface texture to increase contact angle to well over 90°, i.e. to a point where water forms easily shed beads on the surface (Figure 9.19). When the contact angle increases to greater than 150° the surface is referred to as superhydrophobic (Darmanin and Guittard 2015). This state generally also has a low sliding angle (less than 5°) i.e. water droplets readily roll off the surface even when it is nearly horizontal.

FIGURE 9.19
Superhydrophobicity causes the water droplets to bead and run off the leaves of *Alchemilla mollis*.

Emerging Nature-Based Materials

Lotus Leaf Effect

The effect of texture on water repellency was first identified in the lotus leaf (*Nelumbo nucifera*), where high resolution scanning electron microscopy showed that the surface has closely spaced nodules protruding 3–10 μm above the surface, 20–40 μm apart (Barthlott and Neinhuis 1997; Neinhuis and Barthlott 1997). There are also smaller particles (70–100 nm in size) of a hydrophobic wax material deposited on the surface (Guo et al. 2011). In fact many plant leaf surfaces that show superhydrophobicity have similar textures, based on rods, nodes, lobes or papillae (Guo and Liu 2007). Various contact angle values reported from the literature are shown in Table 9.2. It is interesting that Cheng et al. (2006) report that the wax of the lotus leaf is actually hydrophilic (74°), but the superhydrophobic properties are generated by the texture.

Contact angle theory described by Cassie and Baxter (1944) describes the effect of this trapped air on the ability to shed water droplets. A second theory by Wenzel (1936) presents a related model in which surface texture is considered without trapped air, and some surfaces demonstrate each behaviour. Figure 9.20a shows the Wenzel model, in which the water droplet interacts with the surface texture, and Figure 9.20b the Cassie–Baxter state, where micro capillaries in the texture retain air, preventing wetting of the surface.

The Wenzel equation (Equation 9.4) may lead to an increase interaction with the roughness parameter r

$$\cos\theta = r\cos\theta^{Y}. \tag{9.4}$$

In the Cassie–Baxter equation (Equation 9.5), the solid fraction (ϕ_s) and air fraction ($1 - \phi_s$) recognise the entrapment of air between the fine texture of the nodules on the surface. When in the Cassie–Baxter state, the increase in solid–vapour interface contributes to low hysteresis (H) and low sliding angle (α):

$$\cos\theta = \phi_s(\cos\theta^{Y} + 1) - 1. \tag{9.5}$$

TABLE 9.2

Contact Angle for Different Leaves Showing Superhydrophobicity

Plant Leaf		Contact Angle	Structure
Lotus leaf	*Nelumbo nucifera* Gaertn.	162°	Protrusions
Rice leaf	*Oryza sativa* L.	157°	Papillate
Taro leaf	*Colocasia* spp.	159°	Elliptic within caves
Canna leaf	*Canna generalis* Bailey	165°	Hierarchical
Ramee leaf	*Boehmeria iongispica* Stead.	164°	Fibrous
Chinese watermelon leaf	*Bennincosa hispida* Cogn.	159°	Fibrous

Source: Guo and Liu (2007).

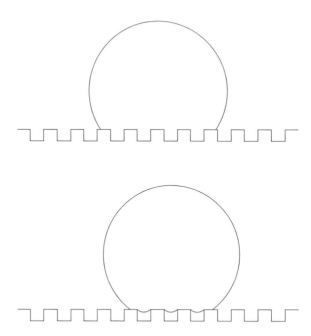

FIGURE 9.20
Schematic showing the (a) Wenzel, (b) Cassie–Baxter states for a droplet of water on a textured surface.

If the texture allows air to be trapped within its fine structure, and the Cassie–Baxter model holds, it is possible to achieve superhydrophobicity even with materials that are naturally hydrophilic, i.e. would have contact angle below 90° in the smooth state. The so-called re-entrant structures of the surface can induce a negative Laplace pressure difference, changing the liquid–vapour interface from concave to convex, impeding liquid penetration, and favouring the Cassie–Baxter state. Similarly, the theory can be applied to generating superoleophilic materials. The technique is in great demand for generating anti-icing coatings, liquid-repellent surfaces and many other applications.

In addition, the rose petal effect has been observed in which hierarchical texture allows high contact angle, but high adhesion of the water droplet is achieved by the secondary ridge-like texture (Bhusan et al. 2008; Bhusan and Her 2010). Various combinations of wetting and filling behaviour within the microstructure and the nanostructure have now been identified and described (Figure 9.21, Bhusan and Nosonovski 2010). Plant leaves may combine superhydrophobic regions with other regions where preferential collection and aggregation of water droplets occurs, to induce drop rolling towards the stem of the plant (Shirtcliffe et al. 2009). This has potential for passive management of water droplets on surfaces, such as solar panels or glazing.

Emerging Nature-Based Materials

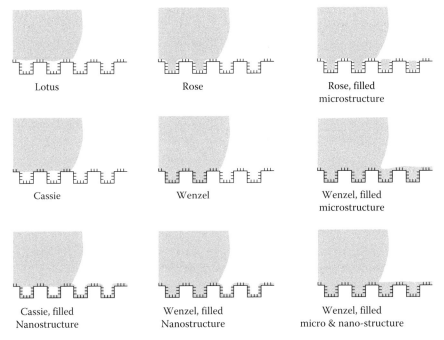

FIGURE 9.21
Schematic showing the droplet–surface interface for nine potential wetting scenarios for surfaces with hierarchical roughness.

Lotusan paint was the first product to capitalise on the biomimetic potential of the lotus leaf. Self-cleaning glass was another significant example, greatly reducing operating costs in tall buildings. Over the past decade, many biomimetic surfaces have been created, demonstrating the role of texture in superhydrophobicity. Onda et al. (1996) showed the difference in water contact angle between a fractal surface of alkyl ketene dimer (174°) and a flat surface of the same material after mechanically flattening it (109°). Since then, many textures similar to those described for plant leaves have been created by wet chemical etching methods, electrochemical methods, lithography, self-assembly, sol–gel methods, plasma etching and vapour deposition, and other approaches (see e.g. the review by Guo et al. 2011). Contact angles of over 150° are achieved by these nano-textured surfaces by inclusion of air pockets within the surface. There are still a great many different examples of textures in nature, both on plants and animals, which capture or manipulate the location of water droplets; many examples are reviewed by Darmanin and Guittard (2015).

One problem that has been encountered when designing superhydrophobic surfaces is that while droplets falling as rain may land in the Cassie state and show superhydrophobicity, droplets formed by condensation may be nucleated within the texture and adopt the Wenzel state. This is less desirable in high performance materials such as anti-fogging surfaces on lenses. The Wenzel state is the thermodynamically stable state, and the Cassie–Baxter state is meta-stable (Boreyko and Chen 2009). Work to overcome the tendency to adopt the Wenzel state in dew formation has looked again at the lotus leaf for inspiration. Dew formation on lotus leaves in nature is generated in the form of rolling Cassie type drops whereas when adhered to cold plates for laboratory condensation experiments Wenzel type drops were formed. The superhydrophobicity of these lotus leaves was restored using vibration, liberating the droplet from intimate interaction with the texture, and restoring the air layer. Transition between the Cassie–Baxter and the Wenzel behaviour has been demonstrated under pressure or vibration (Boreyko and Chen 2009; Bormashenko et al. 2007).

Texture and Adhesion

The principle of controlling surface microtexture to increase or decrease 'stickiness' can be applied to many other applications, whether in increasing or decreasing the adhesion or wettability of a surface. There has been a great deal of investigation into gecko feet, leading recently to the development of adhesive materials, which can be applied and then removed from surfaces such as walls. In the surface of a gecko's foot, nearly half a million microscopic hairs (or setae) provide texture. Each seta is 30–130 μm long and made of keratin, but it is only a tenth of the diameter of a human hair (Autumn et al. 2000). On the surface of each seta, hundreds of even smaller projections terminate in spatula-shaped structures, which are 0.2–0.5 μm wide. While the ability of geckos to climb seemingly smooth walls has long been known, and the anatomy of the foot studied for nearly a century (Hora 1923; Ruibal and Ernst 1965), the measurement of the adhesive force achieved by this structure was not achieved until more recently (Autumn et al. 2000). 100 mm^2 of gecko footpad area can achieve an adhesive force of 10 N, so each seta is thought to contribute 20 μN, with an average stress of 0.1 Nmm^{-2} when loaded parallel to the wall–seta interface, i.e. in shear.

Autumn et al. (2000) considered the gecko's behaviour in attaching and detaching the foot from the wall, and observed that orientation of the setae, and the degree of pressure applied during initial contact influenced the efficacy of adhesion. Use of increased preload led to higher adhesive force being observed in the same test mode; however, the preload required two modes – initial pressure into the surface, followed by small sliding force parallel to the surface. This achieved over ten times the perpendicular adhesive force achieved simply by preload (13.6 μN in place of 0.6 μN). In testing parallel to the surface loads for individual setae of 194 μN were seen, greatly exceeding the estimated values from whole foot calculations mentioned above (Autumn et al. 2000).

The reversible nature of this adhesion is of great interest for synthetic materials. Synthetic gecko adhesives have been manufactured from polymers and from carbon nanotubes (Geim et al. 2003; Sitti and Fearing 2003; Northern and Turner 2005; Yurdumakan et al. 2005). However, ability to re-use these for multiple cycles remains to be developed, and interactions with water present challenges (Lee et al. 2007). Attempts to overcome these limitations include combining the nano-texture of gecko adhesive with wet adhesive systems such as 3,4-dihydroxy-L-phenylalanine (DOPA), a protein-based adhesive occurring in nature as excreted by mussels (see 'Texture and Adhesion' section). The resulting combined system showed great improvement of adhesive strength both in air and in water (Lee et al. 2007).

Texture for Other Applications

Researchers have identified the unique wettability of spider silk, for example, the cribellate spider capture silk has a unique texture, formed at spindle-knot structures. The curvature of these structures and change in wettability results in designed water collecting properties (Zhang and Zheng 2016). The silk has an initial structure of puffs and joints, which is deformed on wetting to create the spindle-knot structures, by a process of wet rebuilding. This material shows a combination of functional texture and hygric response, resulting in a structure that becomes highly efficient at holding water only when water is present. Several synthetic techniques have been reported that can replicate the spindle-knot structure (dip-coating, fluid-coating, coaxial electrospinning, wet assembly and microfluidic technology) and synthetic fibres can be used for water collection or for long-distance water transport (Zhang and Zheng 2016). It has even been demonstrated that temperature responsive effects can cause the spindle-knot dimensions to change, allowing, for example, droplet collection at one temperature and shedding at higher temperature (Hou et al. 2013; Zhang and Zheng 2016). Photo-responsive and water-responsive spindle-knot structures have also been demonstrated (Feng et al. 2013).

The transition of flowers of *Diphylleia grayi* from white to transparent on contact with water (Ong 2014) has been mimicked by researchers seeking superoleophobic surfaces for underwater diving goggles and other materials. Again, the texture of the surface is the property, which control performance, and femtosecond lasers have been used to create lacunae similar to those that naturally occur on the surface of the petals. The whiteness of the petals relates not to pigment but the diffraction at these air–cytolymph interfaces, and on wetting the water within the intercellular spaces (which has a similar refractive index to the petal substance) creates a water–cytolymph interface with increased light transmission, resulting in transparency. For silica glass irradiated by femtosecond laser, the surface takes on a misty appearance, which becomes transparent on wetting. The wetted surface has also demonstrated superior oleophobicity (Yong et al. 2015). Further examples of the influence of texture on properties are likely to follow.

Chemistry and Adhesion

Chemical bio-inspiration can be found in development of adhesives, where, for example, the high strength adhesive excreted by mussels (*Mytilus edulis*) has been studied. This protein-based adhesive shows high adhesion even when wet, as it must retain a strong bond at all states of the tide in the mussel's natural habitat. Other protein-based adhesives found in biological organisms such as barnacle and oyster adhesive utilise different proteins, and may incorporate mineral content, increasing the range of potential properties and applications, which could be addressed with sufficient study (Brubaker and Messersmith 2012). Protein-based adhesives can be traced back throughout human history, with egg, milk, hide, blood and other proteinaceous materials being used since at least the Greco-Roman period, and examples from ancient Egypt being animal glues and casein adhesives. Improvement in understanding of the protein composition is likely to unlock new potential in the field of protein adhesives. In particular, the chemically interesting aspect of mussel foot protein is its strength in the wet state.

The mussels excrete a protein-based adhesive from their byssal plaques, which they use to adhere to their substrate on the coastline. The bond must resist high rates of water flow, and high forces relating to the acceleration of waves breaking within the intertidal range. Studies of Mefp (*Mytilus edulis* foot protein) fractions conducted by Waite and co-workers have revealed that several related proteins are present, and that all are rich in DOPA units (Figure 9.22, Waite and Tanzer 1981; Taylor et al. 1994a,b; Vreeland et al. 1998; Waite and Qin 2001). Mefp-2 is most abundant by weight (40%) while Mefp-1 has attracted much attention for its strong adhesion in vitro. Mefp-1 features a ten-peptide repeat pattern, which includes two DOPA units, two hydroxyproline units and one dihydroxyproline unit, giving a strong basic character and the potential for quinone cross-linking and complexation of metal ions such as Fe^{3+}. Up to eighty repeats of this pattern form a long relatively rigid protein (Laursen 1992). Mefp-3 is believed to act as a primer between the byssal pad and the substrate, and this protein is relatively short but even more rich in DOPA units.

The catecholic amino acid DOPA engages in hydrogen bonding, metal complexation, π-π and π-cation interactions (Vreeland et al. 1998). The adhesive

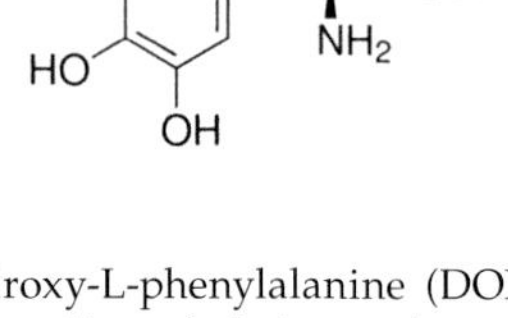

FIGURE 9.22

Chemical structure of 3,4-dihydroxy-L-phenylalanine (DOPA). The two phenolic hydroxyl groups contribute to hydrogen bonding, chelation and coordination with metallic surfaces, while the aromatic ring allows π–π electron sharing with aromatic surfaces and π–cation interaction that enhances adhesion to charged surfaces.

Emerging Nature-Based Materials

proteins of other marine mollusc species have also been studied and found to contain DOPA. Many synthetic DOPA-rich adhesives have been investigated since the first reports of Mefp composition (Yu and Deming 1998; Yu et al. 1999). Novel strategies also include direct functionalisation of surfaces, the use of catechol-functionalised initiators and genetic engineering to create suitable peptides (Forooshani and Lee 2017).

Form and Structure

Biomimetics is not limited to material composition, or to tribology (surface science). Many examples of bio-inspiration or biomimicry in fluid dynamics, design geometry and system engineering can be found. The physical form of a material can do much to increase performance, whether this be the mechanical effect of cross section or shell curvature, or the alteration of aerodynamics, reflection or heat capture with curvature and texture. Form can also increase or decrease resistance to drag, needed for motion. In architecture and design where materials and structures are applied to real life challenges and solutions, we see additional benefits of bioinspiration.

Shape and Form

The shapes and complex forms of many biomineralised materials have also fascinated biologists for many years. Structures such as sea sponges, corals and molluscs gained attention for their new and unusual forms during 19th century exploration voyages, allowing natural historians to catalogue and study organisms from newly discovered regions. The intricate design and impressive mechanical strength of the sea sponge spicule (Figure 9.23), formed using minimal quantities of materials, is a particularly striking example of the skeletons and structures reported by the Challenger expedition of 1873–1876 (Figure 9.23a). Modern biologists and materials scientists have looked again at these structures to gain insights into efficient use of materials in complex hierarchical structures (Weaver et al. 2007).

Shell collectors and taxonomists studying molluscs and other marine life also identified many beautiful symmetries and mathematical relationships, such as within the spirals of nautilus shells (Figure 9.24b). Spirals based on the golden ratio, or the Fibonacci spiral that closely approximates it (Figure 9.24a), have often been used by architects, commenting that the inspiration was from the spirals of a nautilus shell or from phyllotaxic spirals seen in plants (Figures 9.25–9.27). Leaf placement around a stem or the arrangement of scales or seeds in cones and flower heads often follow interesting mathematical relationships, and use space in a highly efficient manner.

FIGURE 9.23
(a) Botanical drawing of *Euplectella aspergillum*, or Venus flower basket, from the Challenger Reports, Volume XXI, collected in the Philippine Islands at 1000 fathoms.
(*Source:* NOAA Central Library Historical Collections).

(b) A golden crab contemplates a spectacular group of Venus flower basket glass sponges.
(*Source:* NOAA Okeanos Explorer Programme, Gulf of Mexico, 2012).

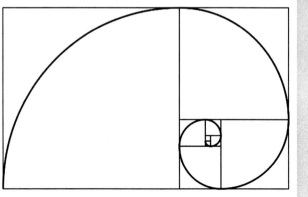 

FIGURE 9.24
(a) Fibonacci spiral. (b) Biological illustration revealing the spiral pattern seen in the cross section of a pearly nautilus.

The benefits of phyllotactic principles based on plant leaf arranegemtn can be used to maximise the balcony area for apartments within a building. One proposed example by Saleh Masoumi of VERK studio was the Phyllotaxy Tower (World Architecture News 2012). Plants have optimised leaf placement to maximise light capture and access to carbon dioxide; so too the balconies achieve daylight and green space for building occupants throughout the tower. A similar concept was proposed for the 'Urban Cactus' building in Rotterdam by UCX Architects. Another architect who has championed

Emerging Nature-Based Materials 263

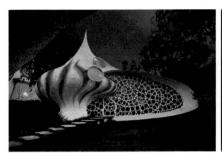

FIGURE 9.25
Nautilus by Javier Senosiain of Arquitectura Organica. Image used with permission.

Source: http://www.arquitecturaorganica.com/nautilus.html.

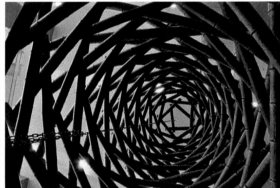

FIGURE 9.26
Bamboo Fibonacci Tower Catenary, constructed by Akio Hizume based on phylotaxic principles for the Tanabata Festival in Japan 2012. Images used with permission, (a) Arata Ikura, (b) Tomoko Ninomiya,

Source: http://starcage.org/tanabata/tanabata.html.

integration of green space within buildings is Ken Yeang. His recognition that buildings must be influenced by the physical and climatic conditions of the site led to his Compass projects (recognising wind conditions) and Sunpath projects (recognising solar orientation) (Yeang 2000; John et al. 2005). The resulting bioclimatic skyscrapers, such as Mesinaga Tower, Malaysia (1992) and Solaris Tower, Singapore (2010), can be optimised for local light and wind conditions. In hot-humid zones, this entails avoidance of excessive solar irradiation and provision for sufficient moisture evaporation by local breezes (Yeang 1992). Much can be achieved by the physical form of the building, and recognition of its interaction with local air flow.

In terms of structural design, there are many benefits in imitating successful structures in nature. The examples above mimic plant stems and mollusc shells directly on the appearance and shape of the structures; however, many examples can be found where the design principles are extracted and utilised in developing new approaches to construction. Some date back several hundred years, for example, the use of suspension in bridge design, later developed into support structures for stadia such as Munich Olympic stadium or the Worker's stadium in Beijing. Others have become possible with modern approaches to model the stress distributions and to form the materials, for example, the concrete structures of Per Luigi Nervi (Palazetto dello sport and Norfolk Scope) or the folding structure demonstrated in Milwaukee Art Museum by Santiago Calatrava. Wang and Wang (2007) proposed many structural forms that are demonstrated in nature and applied in architecture or engineering (Table 9.3).

Many of the tallest skyscrapers of the modern age are based on the efficiency of cylindrical structures, even if modelled as cubic cross sections, as first demonstrated in the Sears Tower (now Willis Tower, Chicago). Continued improvement on this concept has allowed successes such as Taipei 101 to mimic bamboo more literally in its ability to sway to accommodate seismic activity without causing discomfort to the occupants, but also to resist wind loading from frequent typhoons. The double cylindrical structure uses steel and concrete elements, both in specially selected grades, but also a tuned mass damper to oppose the action of wind loading (Poon et al. 2004).

TABLE 9.3

Bioinspired Building Forms and Examples

Biological prototype	Structure model	Example	
Bamboo	Barrel structure		Willis Tower Image: Daniel Schwen, Wikimedia commons

(Continued)

Emerging Nature-Based Materials

TABLE 9.3 (*Continued*)

Bioinspired Building Forms and Examples

Biological prototype	Structure model	Example	
Fruit bunch	Suspension structure		BMW Tower, Munich, 1972 Image: Markus Matern, Wikimedia commons
Spider's web	Suspended-cable structures and suspension bridges		Munich Stadium, 1972 Image: Wikimedia commons
Bird's nest	'Cross the door' shape steel trusses structure		Beijing National Stadium, 2008 for Beijing Olympic Games Image: Tom Nguyen, Wikimedia commons

(*Continued*)

TABLE 9.3 (Continued)
Bioinspired Building Forms and Examples

Biological prototype	Structure model	Example	
Animal bone	Folding structure		Quadracci Pavilion at Milwaukee Art Museum, 1994 Image: Wikimedia commons
Soap bubble	Inflatable membrane structure		Fuji Pavilion of Osaka World Expo, 1970 Image: Wikimedia commons
Eggshell	Thin-shell structure		Oceanographic Park, Valencia, 2003 Image: Rauenstein, Wikimedia commons
Honeycomb	Assembling type spherical grid		Montreal Biosphere (US Pavilion at Expo 67), 1967 Image: Guilherme Duarte Garcia, Wikimedia commons

(Continued)

Emerging Nature-Based Materials

TABLE 9.3 (*Continued*)

Bioinspired Building Forms and Examples

Biological prototype	Structure model	Example	
Sunflower	Reinforced concrete grid structure		Norfolk Scope, 1970 Image: Faithless the wonderboy, Wikimedia commons

Source: Adapted and developed from Wang and Wang (2007), Yuan et al. (2017).

Shells provide more than just the mathematical relationships mentioned above. Thin-shell structures in architecture utilise materials with remarkable efficiency. The combined forces within a thin-shell structure are resolved in such a way that large spans of the roof can be supported only at the perimeter. Curvature and symmetry allow efficient transfer of load to withstand externally applied forces (Yuan et al. 2017). In nature, the curvature of shells provides maximum volume with only few regions of high stress concentration, offering maximum protection from the impact of bird beaks or other predator strategies. In architecture, the curvature of thin-shell structures offers resistance to wind loading and other applied forces (Figure 9.27).

Not only do trees provide a useful structural material, timber, their various growth forms have led to many advances in architecture over the centuries. A good review of dendriform architecture is provided by Rian and Sassone (2014). Notably, the high gothic experimentation with arches and fan vaulting draws on the shape of the tree with many branches (Figure 9.28), and the Dougong brackets used in Chinese traditional structures show tree-like cantilevering of wood blocks of increasing length, until sufficient span is reached, mimicking the tree branches (Figure 9.28). A wide variety of vaulting systems using multiple arches to support the roofs of cathedrals demonstrate tree-like structures in Western architecture, culminating in the elegance of fan vaults, such as Kings College Cambridge (Figure 29a). In the 20th century, Antonio Gaudí drew heavily on natural forms for his any structures, imagining the structure of the Sagrada Familia as having the structure of a forest (Figure 9.29b), with tree-like columns dividing into different branches to support the intertwined hyperboloid vaults of the ceiling (Huerta 2006).

Modern architects have demonstrated the potential of increasing the span of dougong inspired designs while retaining the simplicity of jointing, for

FIGURE 9.27
(a) Simplified representation of the variety of shell structures that can be formed from arches, parabolas and hyperbolas. The sinusoidal slab structure gains stiffness from the curvature of the wave pattern. (b) Ridges in the structure of the cockle shell contribute additional stiffness to the thin structure of the shell.

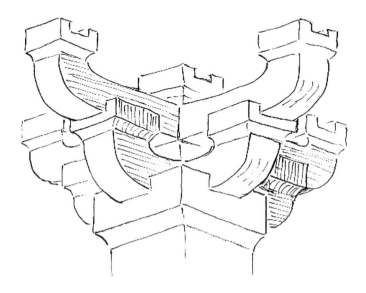

FIGURE 9.28
Dougong joints in traditional Chinese and Far Eastern timber architecture (dou = cap; gong = block).

Emerging Nature-Based Materials 269

FIGURE 9.29
(a) Fan vaulted ceiling at Kings College Cambridge. (b) Vaulted ceiling in the Sagarada Familia, Barcelona.

example, in the bridge constructed to join two buildings of the Yushuhara museum. In constructing, the Dougong Cube a different modern approach was used. The cube was built for Tsinghua University utilised laminated boards of birch to re-approach the same cantilevered system using CNC cut profiles and modular construction for a contemporary structure. Both are examples of timber architecture being extended into new applications. Another development, not based on dougong, but able to trace its origins to the crossed orientation of layers within the plant cell wall, is the crossed laminated timber (CLT) in use within many of the tall timber buildings of the early 21st century. While the design retains relatively traditional concrete core construction, Brock Commons in Vancouver utilises cross-laminated timber floor elements supported by glued laminated timber columns. Lower rise multi-storey timber structures use the CLT within walls as well as floor elements, allowing greater parallel to be drawn to the cell wall on which the panels are based.

Cellular Materials

There are many examples in nature where cellular materials are used to achieve lightweight at the same time as high stiffness and strength. The most obvious examples can be found in birds, where flight requires weight minimisation, and the structural tissues such as bone and the feather shaft or rachis still require high stiffness. Many birds have bones that have apparently hollow centres, more accurately containing struts or membranes to define a cellular interior space. The interior of the rachis of a feather, or interior of a porcupine quill are composed of cellular foams. Many annual plant stems also contain a central cellular pith. In these cellular structures, the internal walls or struts assist in transferring load between the outer surfaces.

The lightweight and mechanical efficiency of cylindrical structures, rather than solid cross sections is well known; however, the incorporation of cells or struts within such forms may further improve efficiency. Cells may be open or partially open, with connections between cells, or closed, with full separation of each cell from its neighbour. The effect of both open celled and closed cell systems is well reviewed by Gibson and Ashby (1988). The mode of action changes in closed-cell systems, often achieving greater stiffness than in open-celled structures of similar form – due to the internal gas pressure in the closed cell system, which resists changes in volume and cell dimensions. An introductory review by Schaedler and Carter (2016) provides many insights into the different systems being made possible today based on biologically inspired or mathematically optimised systems.

In the above examples, foams or honeycombs may be seen to be transferring load between opposing faces of a stronger stiffer material. It is also worth commenting that foams have useful properties in their own right, such as the low density and high thermal resistance achieved by retarding thermal conduction and convection processes.

Cork is a natural insulator, having a very low density (120 to 240 kg/m^3), and good thermal insulation (approximately 0.07 Wm^{-1} K^{-1}). The insulation provided by cork is not limited to thermal effects, but also electrical, vibration damping and the retardation of oxygen migration, which has led to its use in protecting wine from the atmosphere. Robert Hooke was the first scientist to use the term 'cell' in relation to cellular materials, while observing cork cells through a microscope. He commented that they resembled 'a great many little Boxes, separated out of one continued long pore, by certain Diaphragms' (Hooke 1665). The cellular nature of wood, which has already been considered in great detail regarding the cell wall structure, is responsible for its lightweight and high specific strength properties. In this context, the large range of densities seen between different timbers (near 150 kg/m^3 for balsa to over 1260 kg/m^3 for lignum vitae) reminds us that foams may cover a wide range of stiffness, hardness or compliance, depending on the thickness of the cell walls and the geometrical features (Table 9.4).

Relative density (ρ^*/ρ_s) can be calculated for these timbers, assuming a value of 1500 kg/m^3 for the wood cell wall material. This is the density of the foam (ρ^*), divided by the density of the wall material (ρ_s). As you can see, in timbers such as African blackwood (used in making musical instruments such as clarinets and oboes due to its density and texture) and lignum vitae (used for crown green bowling balls due to its weight), the foam is very dense, indicating very thick cell walls, and relatively small pores. In addition, there are a considerable weight of extractives and gums present within the timber, which may explain why the Young's modulus of these two timbers has not continued to increase in line with the trend seen in the data for the less dense woods. In other examples such as oak, although the wood is relatively dense its anatomy leads to variation in observed strength or stiffness, deviating from the main trend. Hardness however increases consistently throughout the selected list.

Emerging Nature-Based Materials 271

TABLE 9.4

Typical Mechanical Properties of Selected Species of Timber with a Range of Different Density Values

	Density (kg/m³)	Relative Density	Young's Modulus (GPa)	Janka Hardness (N)
Balsa	150	0.1	3.71	300
Obeche	380	0.25	6.69	1910
Norway spruce	405	0.27	9.70	1680
Scots pine	550	0.37	10.08	2420
European larch	575	0.38	11.80	3290
Caribbean pine	625	0.42	12.03	4920
Oak	675	0.45	10.60	4980
Jarrah	835	0.56	14.70	8270
Greenheart	1010	0.67	24.64	11260
African blackwood	1270	0.85	17.95	16320
Lignum vitae	1260	0.84	14.09	19510

Source: The Wood Database, Meier (2016).
Young's modulus in bending. Janka hardness is resistance to indentation perpendicular to the grain of the wood.

Honeycombs and Foams

When cells are arranged in a two-dimensional array of prismatic cells, they are referred to as a honeycomb. The most frequent and efficient tessellating cell is the hexagon, as demonstrated by honey bees within the hive (Figure 9.30a); however, others based on triangles, squares, diamonds or combinations can be considered. Polyhedral cells, arranged into a three-dimensional structure, form a foam (Figure 9.30b). Studies on the optimal number of faces within the polyhedron, and the corner angles for optimisation have been reported (Gibson and Ashby 1988).

Honeycomb-based cellular structures have been used for many years in lightweighting, both in domestic furniture, door construction and in high performance panels for aviation and other transport sectors (Figure 9.31). Honeycombs made of resin impregnated paper, or aluminium are relatively easily formed, and can be used to design panels with the prescribed stiffness, using second moment of area and modulus data for the upper and lower skins. When loaded in bending, one face will become stretched, under tension, and the opposite will become compressed, so the honeycomb must provide separation of the two skins and resist lateral shear forces. Similar effects can be achieved with foams of different density, to form sandwich composites between two stiff outer lamellae. In both examples, the contribution of the honeycomb or foam to the total stiffness is relatively minor, so long as shear stress is not exceeded, and overall stiffness is achieved by increasing distance between the two skins or adjusting skin thickness and composition.

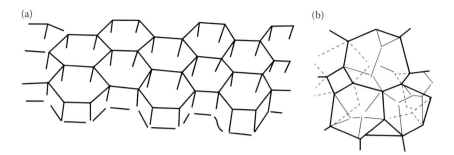

FIGURE 9.30
(a) Honeycomb based on tessellating hexagonal cells. (b) Cluster of polyhedral cells typical of a foam.

FIGURE 9.31
This simple honeycomb cored composite with hardwood skins achieves lightweight and high stiffness.

Cylindrical forms filled with foams are an efficient mechanical structure. We see these demonstrated in porcupine quills or in some plant stems. Porcupine quills have a solid cortex, with an internal foamy core; in some species of porcupine, the quills also contain radially aligned stiffeners, for example, *Hystrix* spp. (Yang and McKittrick 2013). The keratin within a porcupine quill is in the α-form, and tensile modulus of 1.9–2.3 GPa, and strain at break of 63–170 MPa is reported (McKittrick et al. 2012).

Feathers, which also have a foam-like medulla, show similar to higher values of modulus and failure stress due to the presence of a mixture of

Emerging Nature-Based Materials

α-helix and the β-sheet forms of keratin, for example, 1.8–3.8 GPa tensile modulus in the swan feather rachis, and 74 to 188–240 MPa failure stress depending on location within goose feather rachis (Bonser and Purslow 1995; Bostandzhiyan et al. 2006; McKittrick et al. 2012). In addition, the cross-sectional dimensions, profile and composition of the feather spine changes along the length (Figure 9.32), clearly much can be learnt from the design of each segment within this profile of hollow and cellular tissue (Purslow and Vincent 1978). For example, Wang and Meyers (2017) discussed the transition from a cylindrical or ellipsoidal section to a square section, which is seen in feather moving from the proximal (near body) to the distal end. They reported that square sections give higher bending rigidity than circular, as the circular sections tend to distort to an oval shape on flexing. In addition, fibres within the wall of the seagull and crow feather rachis showed preferred orientation that differed between the lateral sides (crossed angles) and the dorsal and ventral (upper and lower) sections (axial alignment). The exception was in the proximal dorsal section, where two distinct layers were seen, the inner being longitudinally aligned, and the outer being circumferential. In the calamus, this outer layer of circumferential fibres fully wrapped the feather shaft, providing resistance to axial splitting (Figure 9.33).

In terms of composition, the keratin within the feather occurs in both the α- and the β-form, with the β-form being laid down later in feather development, and in the hard surfaces such as the calamus or barbs (Alibardi 2007).

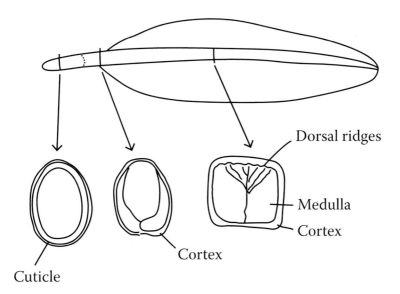

FIGURE 9.32
Simplified structure of a feather, showing the cross section at three locations along the shaft. The oval section from the calamus is hollow, whereas the square section from the barbed region of the shaft has a foam-like medulla in the centre.

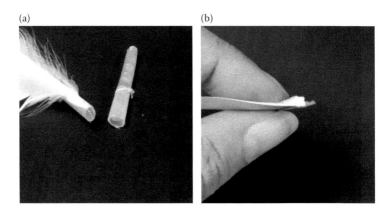

FIGURE 9.33
(a) Cross section of a seagull feather just above the calamus, where the profile is a hollow ellipse. (b) Cross section nearer the distal end, where the cross-section is quasi-rectangular.

The innermost sheath layers do not contain β-keratin; cells are more likely composed of an α-keratin with properties more similar to hair or wool. The typical stress–strain curve of wool keratin shows an initial linear elastic region, followed by a yield region in which plastic deformation occurs, and finally a post-yield region in which the gradient increases again. It is widely recognised that the α-form can be transformed to the β-form when stretched and once the protein chains have been aligned into sheet-like conformation in the β-form, a higher tensile modulus is observed. In flying birds differences in strength and stiffness have tended to show higher stiffness and lower failure stress furthest from the body (McKittrick et al. 2012). This is also consistent with a greater degree of cornification in the older portion of the feather further from the body.

Hedgehog spines have much in common with porcupine quills, and Karam and Gibson (1994) proposed that this foam filled cylindrical structure provided a model for bio-inspired cylindrical structures that are optimised to resist buckling loads. Since this time, further studies on porcupine quills loaded in tension and compression have further demonstrated the superior properties of these simple foam cored structures, especially in buckling (Yang and McKittrick 2013; Yang et al. 2013; Torres et al. 2014).

In many successful natural examples, however, the form achieved is not a simple plate or sheet, but a much more complex profile, such as a bone or a beak. Curvature of the outer profile is accompanied by varied textures of internal cellular dimensions and shape. Some regions of the bone will contain smaller cells with thicker dividing walls to provide necessary rigidity in many dimensions, other regions, possibly those of long bones with quasi-cylindrical shape will have considerably more open architecture, possibly with only a relatively small number of struts, or partial membranes remaining.

Emerging Nature-Based Materials

Seki et al. (2006) investigated the cellular foam within a toucan beak, which is a well-known example of a lightweight structure. Despite its prominent size and colour, the beak is sufficiently low density to not hinder the bird's flight. Although the form is more complex than a quill or spine, this beak utilises foam to increase strength. In the toucan beak, the shell was constructed of hexagonal plates of keratin, which were 30–60 μm diameter and 2–10 μm thickness. Inside the beak, a foam structure, made up of closed cells coats the inside of the shell, with a hollow central space (Seki et al. 2006).

The keratin of the beak is different to that in the feather spines and is less stiff than feather keratin. The shell of the beak may be melanised (by deposition of melanin granules, providing a dark pigment) depending on species and time of year (Bonser and Witter 1993). The melanised beak of a starling (during breeding season) had an indentation hardness that was almost double the winter-spring yellow coloured beak, which is without melanin. In addition, Seki et al. (2006) analysed the protein amino acid composition of the foam within the toucan beak, finding a high glycine content, potentially indicating that this foam was collagen. Within the foam, bony struts were seen, further indicating the foam could be a bone structure. Hardness values from the shell keratin and the fibrous or bony struts of the foam were comparable, but slightly higher for the foam material (0.55 GPa) than the shell keratin (0.50 GPa). Compression tests on the foam showed a prolonged plateau of crushing failure relating to collapse of the cell walls, after this (at a strain of about 0.9) densification of the material begins, and a higher modulus is seen.

The performance of the foam as a structural element was demonstrated by modelling, using Gibson – Ashby constitutive equations. For all foams, the relative density (ρ^*/ρ_s), i.e. the density of the foam, divided by the density of the wall material, is a significant factor in governing strength. For a closed cell foam, applied deformation results in a change of internal pressure in the cell, which influences mechanical response (Equation 9.6). In an open cell foam, where air is able to pass freely from one cell to the next, the relationship between plastic collapse stress of the foam (σ_{pl}) and yield stress of the wall material (σ_{ys}) is simpler, and relates more directly to the relative density of the foam (Equation 9.7) (Gibson and Ashby 1988):

$$\frac{\sigma^*_{pl}}{\sigma_{ys}} = \frac{C_5(\varphi\rho^*)^{3/2}}{\rho_s} + \frac{(1-\varphi)\rho^*}{\rho_s} + \frac{p_0 - p_{at}}{\sigma_{ys}}, \tag{9.6}$$

$$\frac{\sigma^*_{pl}}{\sigma_{ys}} = \frac{C_5(\varphi\rho^*)^{3/2}}{\rho_s}. \tag{9.7}$$

Parameter C_5 is a constant relating to the mode of failure, and has been experimentally obtained as 0.3 for plastic collapse and 0.2 for brittle crushing (Gibson and Ashby 1988). φ is the ratio of the volume of face to the volume of edge. p_0 is the initial fluid pressure inside the cells, and p_{at} is atmospheric pressure. When modelled for data from the toucan beak, the predicted envelope between open and closed cell foams correlated quite well with observed values (Seki et al. 2006).

Interestingly, the mechanism of tensile failure seen in the shell of the toucan beak was similar to that seen in nacre, due to the rigid plates embedded in a more plastic matrix (Seki et al. 2006). The matrix allows some degree of slippage between plates, until the stress exceeds the UTS of the plate keratin, in which case fracture of the plate occurs. Depending on strain rate and applied stress, a mixture of plate pullout and plate fracture occurs.

The use of struts offers different possibilities than the closed cells. The simplest example where all bracings are under tension; yet, incredible stiffness is achieved with minimum weight is the bicycle wheel – all spokes are tightened to maintain the central location of the wheel hub, and when loaded any compressive force is resisted by tension across all spokes, creating a very rigid structure. The use of tension within internal struts in hollow materials increases the performance dramatically. Near the base of a feather, in the calamus, the internal space is not cellular but reinforced by septa, acting as struts and located at suitable intervals to resist buckling of the hollow cross-section. The bending load imposed on the feather spine during flight results in tensile and compressive forces on opposite sides of the cylinder, but the design is efficient in resisting the resulting internal shear forces (Purslow and Vincent 1978).

In engineering, these principles can be harnessed in roofing trusses, where some internal bracings provide tensile resistance, and others transfer compressive load, resulting in a structure that crosses a long span using minimal material. Similarly, in many lightweight designs for use in aviation and high tech applications. Milwich points to the example of the truss-like structure of cactus wood that inspired the design of the Rotex robotic arm, in which fibres were aligned in composite struts and rods to form a tubular lattice with bending and torsional resistance (Milwich et al. 2006). Other research has sought to emulate the lightweight cross section of the horsetail (*Equisetum hyemale*), which has a double ring of structural tissue (Figure 9.34a), with large voids between the two rings separated by pillars of a nearly T-shaped

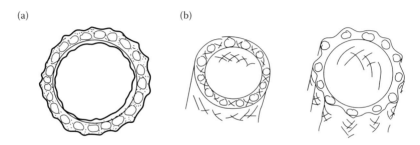

FIGURE 9.34
(a) Cross section of a horsetail stem, typically 6 mm diameter. (b) Two technical plant stems formed by braiding and pultrusion to mimic the lightweight and stiffness of the plant stem.

Emerging Nature-Based Materials

cross section to communicate stresses and ensure structural performance. Various braided tubes with designed in voids were trialled as technical plant stems (Figure 9.34b), using polyurethane as a matrix and pultrusion to form the hollow composite (Milwich et al. 2006).

Schaedler and Carter (2016) report octet lattices in which tension is the dominant mode of resistance, allowing the strength and stiffness achieved to be proportional to the solid volume within the product, known as ideal linear scaling. The difficulties of manufacturing some of these complex geometrical patterns within truss-like materials have recently been overcome using additive manufacturing and 3D printing techniques. These technologies are also likely to help in achieving the complex designs and non-uniform profiles of future cellular designs, such as aircraft wing sections or furniture, allowing greater experimentation with bio-inspired designs.

Cellular Materials for Shape Modulation and Motion

The importance of pressure within closed cell cellular structures was mentioned in 'Cellular Materials' section. Great progress has been made to harness the stiffness induced by pressure in designs based on different cellular geometries. These may also utilise pressure differences to achieve curvature or motion. Li and Wang (2017) reviewed many of these and categorised them based on the geometry and effect achieved.

Honeycomb: In a honeycomb, if the walls are rigid to resist deformation in one or more directions, but able to accommodate shape change in another, pressurisation of the cells can lead to shape change or dimensional change, or expansion in the work direction. Honeycomb celled pouches can be used to increase thickness, or to expand laterally (Li and Wang 2017). Similar orthotropic behaviour can be achieved using diamonds or other tessellating polygons. The system has been developed to working prototype stage for use in shape morphing wings in aircraft (Vos and Barrett 2010, 2011).

Synthesized polygon bilayers: Cellular structures with two or more distinct cell layers have been proposed. The bilayer form acts in a similar manner to the hygromorphic bilayer discussed in the passive actuators section – curving in response to pressure, relating to different extensibility of the two cell shapes when in the pressurised state. A simple example is shown in Figure 9.35, based on hexagonal and pentagonal layers (Pagitz et al. 2012). It is proposed that assembling more than two layers with correctly designed geometry three-dimensional shape morphing may be achieved for adaptive wing foils and car seats (Li and Wang 2017). In addition, by allowing variable stiffness or altered geometry of the cells within the bilayer, it is possible to form materials that deform with non-zero Gaussian curvature (Pagitz and Bold 2013).

Topologically optimised geometry: A similar concept was achieved using specialised topologies by Vasista and Tong (2012, 2013). The moving iso-surface threshold topology (MIST) is combined with mixed finite element techniques to optimise design. For example, cells with segments of concave

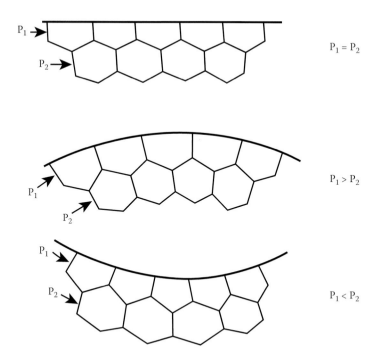

FIGURE 9.35
Effect of pressure in synthesized polygon bilayers. Schematic shows the cross section of polygon tubular structure in which the upper layer cavities are connected (with pressure P_1), and the lower later cavities are connected (pressure P_2), allowing control of the pressure difference between the two layers to cause flexure of the panel.

and convex profiles can be used, such that the thin compliant concave section flattens out on pressurisation, extending to produce bending in the adjacent lever.

Inflatable soft cylinders: The use of materials fabrics and flexible plastics rather than rigid metals and hard plastics can be used to create inflatable structures that function when pressurised and pack away to a small volume when uninflated. Examples for inflatable aircraft wings are based on large aperture cylinders, which are rigidified by adjacent small pressurisable cells on one face. The smaller pneumatic cells were designed such that their shape change with different pressures would induce curvature in the total wind structure (Cadogan et al. 2004; Li and Wang 2017).

Flexible cylinders with helical fibre reinforcement: The use of helically wound fibres to reinforce cylindrical cells was introduced in 'Microstructure of Organic Composites' section. The helical winding provides efficient reinforcement against hoop stresses and longitudinal stresses depending on winding angle. Control of winding angle in cylindrical cells for inflation has been investigated within the fluidic flexible matrix composite (F²MC) proposed by Bakis and Rahn and co-workers (Shan et al. 2006; Philen et al. 2007;

Emerging Nature-Based Materials 279

Zhu et al. 2012). When pressurised, the cylinder may expand or contract laterally, and extend or shrink longitudinally, depending on the winding angle (Figure 9.36). The neutral angle is generally around 55°, an angle lower than this will cause the cylinder to contract when pressurised, and an angle greater than this will cause elongation. The system has been investigated for robotics, morphing aircraft wings, and orthopaedic or prosthetic devices, among other applications (Li and Wang 2017). Combination of these systems with a hydrogel core to generate the internal pressure offers further similarities to the hydraulic activation of similar systems in plant tissues. Dicker et al. (2014) report the solar tracking of the leaves of Cornish mallow (*Lavatera cretica*) by adjusting pressure in the pulvinus as the inspiration for hydrogel actuators.

Foldable origami cells: A range of origami cells was demonstrated by Martinez et al. (2012). These were based on origami tubes with a bellows pattern (or Yoshi-Miura pattern), the linear actuator was able to lift loads 120 times its weight, while similar designs were used to bend in controlled manner by adjustment of the pattern or the extensibility of their pleats. Pressurisable Miura-Ori cells were proposed by Cheung et al. (2014), this achieved stiffness in the z-direction and flexibility in the x and y directions, offering predictable anisotropic behaviour when pressurized. A similar fluidic origami concept was proposed by Li and Wang (2015), resulting in a three-dimensional cell structure that can allow fluid flow and strategically controlled pressurisation, similar to the regulated turgor pressure in plant tissues. Sophisticated crease patterns are being designed and are likely to achieve complex shape morphing (Li and Wang 2017).

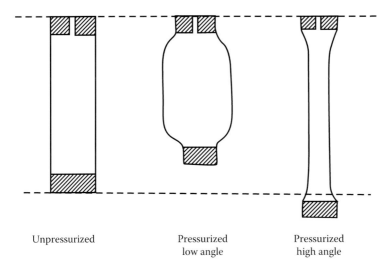

FIGURE 9.36
Schematic indicating the behaviour of helically reinforced flexible cylinders (a) unpressurised, (b) with a low winding angle, when pressurized, (c) with a high winding angle when pressurised.

Regardless of the geometry used by the cellular systems described above, pressure is required to generate the motion or shape change. In order to achieve these pressure differentials various techniques can be used, such as pneumatics (Martinez et al. 2012). However, two strategies for motion show bio-inspiration. The first is active pressure generation, utilising proton pumps and sucrose cotransporters, as found in living tissues to control cell pressure. The second is passive pressure generation, based on osmotic water diffusion across the cell membrane (Li and Wang 2017). Many of these designs and strategies find application in soft robotics where motion or grip requires adaptability.

Fluid Dynamics

Organisms show a wealth of adaptation strategies to regulate their internal temperature and conditions (for example, water balance) by controlling their interaction with their environment. As the majority of life exists on land (in air) or in fresh or marine water (in liquid), these strategies rely to a greater or lesser extent on fluid dynamics. Thus many examples of bio-inspiration or biomimicry can be found where design geometry and fluid dynamics combine to optimise performance. The field of biomechanics has led to great advances in aviation, sporting performance and robotics, much of which falls outside the scope of this chapter, but the reader is referred to an excellent review of the biomechanics of solids and fluids (Alexander 2016).

Temperature and Humidity Control

One example of biomimicry in fluid dynamics is the use of air flow within termite mounds as the inspiration for passive cooling mechanisms in high-rise buildings for warm climates (Figure 9.37). This is best known in the Eastgate Centre, designed by Mick Pearce for construction in Harare, Zimbabwe. The building opened in 1996. The same design concept has also been demonstrated in Portcullis House, London, and Council House Two (CH2) in Melbourne (French and Ahmed 2010). The building design allows for central upward movement of warm air, exploiting natural convection flows, while drawing cool air in and through the occupied spaces. Once heated by passing through the building this air exits through large central chimneys, and the induced air flow maintains the cycling of air throughout the building for occupant comfort.

Since the construction of the Eastgate Centre, researchers Turner and Soar (2008) continued to study termite mounds and have reviewed the two theories of ventilation in the context of the termite structures: thermosiphon flow (Figure 9.37a) and induced flow (or the stack effect, Figure 9.37b). While the principle effect – air circulation – does occur in the building, and helps regulate short term spikes in temperature, they comment that a large component of the success of the Eastgate Centre is the heat sink effect provided by the building walls, and the use of fans to mechanically circulate air at night.

Emerging Nature-Based Materials

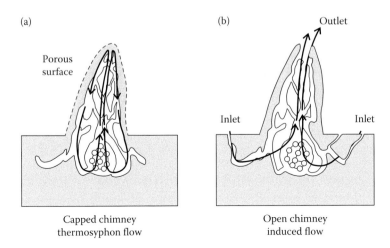

FIGURE 9.37
Schematic showing the two traditional models for air flow inside a termite mound. (a) Thermosyphon flow. (b) Induced flow (or the stack effect).

This third principle, using the heat capacity of the building's thermal mass to regulate temperature, is also seen in the termite mound – in the thermosiphon model, where air passes close to the surface of the mound to exchange moisture and respiratory gases with the exterior, and returns cooled to the base of the mound (Badarnah and Fernández 2015). Their investigations led to a new proposal; that termite mounds act like lungs (Turner and Soar 2008). The many pillars and ridges on the outermost surface of the termite mound, and the location of air tunnels at a very shallow depth below the surface maximises the available area for thermal transfer in a very similar manner to the lung. As a result Turner and Soar proposed that buildings might also be designed as lungs, where walls are designed not as barriers but as adaptive interfaces, 'where fluxes of matter and energy … are not blocked but are managed by the wall itself'. This concept will be further considered in the context of bio-inspired actuators in 'Hydrostatic Support' and 'Responsive Forms – Actuators' sections.

Consideration of maximising surface area and introduction of functional surfaces within buildings is an emerging area with great potential. On possibility is for passive scavenging of pollutants, such as the titanium dioxide nanoparticle functionalised screen installed at Manuel Gea Gonzalez hospital in Mexico City. The titanium dioxide substrate acts as a catalyst for reaction of pollutants in the presence of ultra-violet (UV) light, converting nitrous oxides from polluted urban air into more environmentally benign soluble nitrate (Doudrick et al. 2012). Titanium dioxide paints offering similar benefits are marketed for buildings and masonry. A titanium dioxide concrete blend has also been demonstrated in the biodynamic panels of the Palazzo Italia at Milan's World Fair in 2015 (Nemesi and Partners).

Functionalised surfaces for the interior environment have also been proposed. One area is the passive regulation of humidity, which is becoming recognised in timber interior fittings, or the beneficial effect of timber panelling in high humidity environments such as saunas, to intensify the heat experience (Kraniotis and Nore 2015). In addition, moisture buffering of wood and bio-based materials in buildings, as well as breathability of buildings is increasingly recognised, and there are likely to be continued experimental designs seeking to utilise bio-based materials in this area for their inherent porosity and moisture buffering capabilities (Cerolini et al. 2009; Kraniotis et al. 2015; Kraniotis and Nore 2017). The development of panels incorporating scavengers to regulate volatile organic compounds (VOCs) has also been explored (Mansour et al. 2016; Stefanowski et al. 2016). Wool proteins, for example, show great potential for selective adsorption of formaldehyde, and other VOCs of differing polarity have been investigated, to reflect the full range of interior atmosphere pollutants arising from modern urban life. The re-thinking of building envelopes to adapt a process view of nature can be extended to integrate multiple functions and sensors, offering many opportunities of bio-inspired structure and properties in service (Soar 2015a,b).

Conduction, Convection and Radiation

All fur and feathers serve as protection from heat gain, by their insulative effect trapping a layer of low thermal conductivity air in the spaces between fibres or feathers. This also protects against heat loss in colder climates (Badarnah and Fernández 2015). The coat structure affects this protection, for example, due to moulting black bear fur in summer is 52% less effective as an insulator than the bear's winter coat (Schmidt-Nielsen 1997). The structure of penguin feathers allows a near complete layer of air to be trapped within the coat, minimising conduction, but the feather design, with a short stiff quill section (30–40 mm) and a prominent after-feather (20–30 mm), allows feathers to be locked down to exclude air during diving avoiding unnecessary buoyancy (Dawson et al. 1999; John et al. 2005). Radiative heat loss is also minimised by the feather design and the entanglement of the after-feather barbs and cilia (Dawson et al. 1999).

The use of fibrous insulation products is widespread, and insulation products based on sheep's wool or plant fibres such as flax have been developed using the same principle (Hill et al. 2009; Loxton et al. 2013; Ormondroyd et al. 2017). Treatment of wall systems and insulation as a complete unit of many layers, similar to the structure of animal fur, allows the effect of reduced convection and conduction to be combined with reflection or refraction to return heat, and to maintain some ventilation to prevent build-up of condensed moisture. Problems may occur in simple insulation systems when used in cold wet maritime climates if they neglect to consider dew point. It is possible that investigation into fur structure and biomimicry may lead to advances in the performance of fibre-based insulation systems, for example,

Emerging Nature-Based Materials

by incorporating radiation scattering or emission mechanisms while continuing to suppress conduction and convection processes (Badarnah 2015).

Reflection of light or heat radiation is also demonstrated in nature, in locations where solar gain could be excessive, leading to stress within the organism. The white colour of polar bear fur, which hides the dark skin of the bear beneath allows selective reflectance, with the black body effect of the dark pigmented skin to gain heat while reflecting many wavelengths of light. The greater magnitude of radiative losses than conductive losses in the model of polar bear fur was discussed by Simonis et al. (2014). They proposed that scattering of radiation by the fur is a mechanism that facilitates heat retention. The different thicknesses of fur at different levels within the coat providing the necessary multiple interfaces for retro-diffusion of heat.

In architecture, Craig et al. (2008) designed a roof structure for a building in which the same spectral selection principle was used with a honeycomb cored panel to allow transmission of longwave infrared heat energy, while reflecting shortwave infrared back to the atmosphere. The loss of longwave radiation from within the building was a cooling mechanism for temperature regulation. Just over half of the temperature gained from solar energy is due to shortwave infrared, and many natural materials, including bird's eggs, desert snail shells and leaves, reflect this range of electromagnetic radiation to maintain acceptable temperatures (Vogel 2005). While the system proposed by Craig and co-workers relied partially on the effect of white and black pigment, the pore dimensions of the honeycomb panel provided the selective porosity necessary to allow the longwave radiation to pass.

Other examples of bio-inspired design for thermal regulation in buildings include counter-current heat exchange, as demonstrated by penguin feet and the circulatory systems of whale and dolphin flippers (Scholander et al. 1950; Scholander and Schevill 1955; Badarnah and Fernández 2015). Counter-current systems form the basis of heat exchangers, and opportunities to bring bioinspired design into this technology to improve building heat and ventilation control may continue.

Small-Scale Fluid Dynamics and Diffusion

Plants have adapted to regulate the level of cooling provided by transpiration through stomata openings (Figure 9.38a). The location of stomata on leaves, with guard cells that open and close controlled by turgor pressure (Figure 9.38b), allows transpiration of water to be regulated and excessive moisture loss on hot dry days to minimised. They are also required to allow carbon dioxide into the leaf to support photosynthesis.

The location of the majority of stomata on the underside of the leaf is an obvious design choice to prevent excessive moisture loss, but additional features such as the shape of the guard cells, and presence or absence of auxiliary cells, and the design of the stomatal pores offers further control. In some cases, epithelial cells or waxy deposits are present, which may further

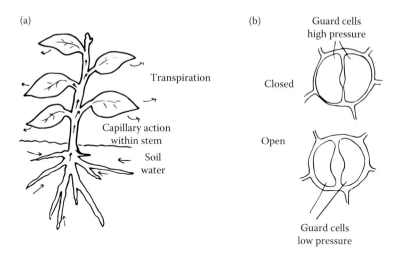

FIGURE 9.38

(a) Simplified schematic of transpiration processes in a plant, by which soil water is drawn up through the roots and stem under negative pressure due to evaporation in stomata on the underside of the leaves. (b) Close up of simple stomata, where pairs of guard cells open when internal pressure drops. Some plants use a more sophisticated system with companion cells adjacent to the guard cells, allowing greater control.

regulate air flow or minimise negative vapour pressure effects (Parlange and Waggoner 1970; Field et al. 1998; Franks and Farquhar 2007). In some plant leaves, the location of stomata together within stomatal crypts allows the plant to optimise relative humidity retention at this crucial interface between the interior and the exterior of the plant. Modelling and comparison between stomatal design of different plant species have shown that fluid dynamics, and the effect of shape on the mobility of air is a crucial design criterion for regulating water loss while permitting CO_2 entry. The many designs of stomatal pores seen in different vegetation may provide optimal CO_2 exchange, while offering protection from surplus water, or control of limited water resources (Roth-Nebelsick et al. 2013).

Various designers have commented that selective or tunable porosity would be desirable in buildings. John et al. (2005) referred to the traditional Bedouin tents observed by McDonough in Jordan. The responsive nature of the tent allows efficient regulation of heat despite high external temperatures. Here, the loose weave of goat hair in the canvas is said to allow passage of air, and the development of convection currents within the tent, providing cooling. An effect that is not seen in the closed texture canvas of synthetic fabric tents. It was commented that on wetting the goat hair swells, closing the texture, and providing a tight barrier against rain (McDonough 2004; John et al. 2005). The idea of passive regulation of ventilation has been picked up and developed by various researchers. Later, we will discuss actuators based on bio-based materials and hydrogels; in addition, researchers have

Emerging Nature-Based Materials

investigated textiles with responsive properties, such as switching between hydrophilic and hydrophobic behaviour. It is likely that buildings of the future will be able to utilise one or more of these strategies depending on the requirements of the local environment.

The tuneable wettability of fibres with spider silk inspired textures has been introduced earlier (Hou et al. 2013; Zhang and Zheng 2016), and would make an ideal candidate material for passive moisture scavenging or regulation of humidity. Other researchers have looked to cactus spines for bioinspiration, as the contact angle on these fine conical structures influences moisture motion (Guo and Tang 2015; Tan et al. 2016). A good review of the theories underpinning development of surfaces for water capture, or other applications for controlled wettability is given by Song and Zheng (2014).

Moisture Harvesting

Other researchers have proposed strategies such as moisture harvesting surfaces to regulate temperature and humidity in buildings. Various examples of moisture harvesting structures are known in nature, including the various fog basking desert beetles found in Namibia (*Physosterma cribripes, Stenocara gracilipes, Onymacris unguicularis, O. laeviceps* and *O. bicolor*). The texture of the beetle's back (or the top fused wings known as 'elytra') has a fog collecting structure, allowing the water droplets to be drunk by the insect (Parker and Lawrence 2001; Malik et al. 2014). In buildings, water harvesting surfaces have been considered to reduce the humidity, and to provide passive cooling. In the Las Palmas Water Theatre (Grimshaw Architects) on the Canary Islands, a large net structure slows air from the sea, and evaporates sea water in the presence of plentiful sunshine so that the air becomes warm and humid, causing condensation when the air reaches cold pipes in the next layer of the structure. The condensed water that has been distilled and salt free is gathered into recycling facilities, making the building self-sufficient, and allowing water to be supplied to nearby buildings (Yuan et al. 2017). While this process is a temperature driven condensation process, other researchers have considered the potential of surface texture based on insect backs, frog and lizard skin, spider cribellate silk or plants such as cacti to provide similar cooling effects (Malik et al. 2014).

Hydrostatic Support

Cells: Containing and Using Hydrostatic Pressure

The basic construction of the plant cell wall was introduced earlier ('Plant Cell Walls as Composites' section). Let us now consider how these cells act together within the plant tissue to utilise turgor pressure. Hydrostatic pressure is capable of supporting vast slender structures against lateral forces such as wind. In many annual plants, or wilting plants, which are to a great extent unlignified, the flower heads and leaves are held aloft simply due to hydrostatic pressure (Figure 9.39a). In design and engineering, the principle finds its way into the use of pneumatic pressure in everything from bouncy castles and rigid inflatable boats (RIBs, Figure 9.39b) to large structures such as sports halls and greenhouses (Figure 9.39c,d). We will also later consider modern hydrostatic structures for high tech applications.

While at first glance, this is a simplistic example, the design element within the plant tissue is surprisingly complex. Hydrostatic pressure is actually an essential element in the plant's structural design, and control of turgor pressure allows rotation and adjustment of angles in leaves, flowers or growing shoots. Wilting under drought conditions or infection demonstrates the results of insufficient turgor pressure.

Wainwright (1970) observed that inserting a needle into parenchyma tissue of a nasturtium, a small droplet of water will be observed to emerge through the hole – confirming that this tissue is under positive pressure. These walls of these cells are providing lateral resistance to constrain the pressure of fluid within – a typical hydrostatic system. A second observation

FIGURE 9.39
The head of a daffodil is supported purely by the hydrostatic pressure of cells in the stem. A RIB relies on pneumatic pressure to retain its shape in demanding conditions. The large span and great height of the greenhouses at the Eden Project, Cornwall, are formed from inflated hexagonal cells.

Sources: **(a, b) Morwenna Spear, (c, d) Lynne Pugh.**

Emerging Nature-Based Materials

by Wainwright is more surprising in this context, the same needle if inserted into xylem cells, and a droplet appears when the needle is withdrawn, this droplet is rapidly drawn back into the tissue, as these cells are under negative pressure. These cells are performing the more commonly recognised role of transporting fluid up through the stem of the plant. Fluid flow in plant tissue performs multiple functions, and the plant tissues have been optimised for their respective roles.

In terms of segregating the roles within the plant, this division of labour, or segregation of flow upwards and downwards in the stem, makes good sense. However, from the point of view of supporting the stem in a vertical position a further observation is needed. The tissue under negative pressure (collenchyma) is located on the periphery of the stem, providing a ring of cells that are under negative pressure, constantly under potential to collapse. These exert a net tension effect around the margin of the stem. The parenchyma cells, on the other hand, which are under positive pressure, are located in the centre of the stem, providing the positive pressure in all directions (Wainwright et al. 1976). The parenchyma can be said to be acting as hydrostats – providing an inflating, or outward support role to the stem. Note that the system as a whole also acts hydrostatically, with internal pressure being resisted by a stiff peripheral tissue under tension. The combination of this resistance to compressive forces in the core, and the state of tension on the outer portion provides the net hydrostatic force at a second hierarchical level within the total stem. Similar states of external tension and internal compression can also be observed in a carrot root: if you cut it in half longitudinally, then slice this in half along the mid line. Segments of outer tissue will bow outwards and shrink under the now unopposed tension, while the storage tissue at the core will expand now that internal pressure is unresisted.

In engineering, this principle is applied in hydraulic systems to operate lifts and motors, and in the 19th century, hydraulic power networks were developed for many cities, for example, London, Liverpool, Birmingham, Glasgow and Manchester. One still operates in Bristol Harbour today. There are also examples of architecture inspired by pneumatic or hydraulic forces to form the structure or maintain the building shape, for example, Frei Otto's Pneu structures (Otto 1995). More recently, Khire et al. (2006) have used designs based on honeycomb cell structures to form rigid inflatable structures for housing. Sports halls and greenhouses based on inflatable diamond or hexagonal cells are also well known. The National Aquatics Centre in Beijing used transparent ethylene tetrafluoroethylene membranes to form cushion units within a grid structure inspired by the space-filling nature of foam to create the wall and roof panels. Here, the air becomes the structural element, generating the necessary tensile forces in the membranes to allow the whole panel of inflated cells to act as a continuous structural element (Yuan et al. 2017). Within the cells of the grid, external and internal cells allowed for thermal insulation and ventilation.

Turgor Pressure and Motion

Many biological materials are not static, whether this is muscle-controlled motion in animals, or moisture actuated movement such as the opening and closing of flowers, or angling of leaves in plants. In many cases, these rely on live tissue to respond to metabolic signalling. Plant movement can be nastic (response to non-directional stimulus) or tropic (such as the phototropic growth response towards directional light, Figure 9.40a). Nastic movements include the thigmonastic response to touch of mimosa, where the plant defines the direction of movement, or the whirling, searching movement made by growing vine seedlings until the find a support to twine around (Figure 9.40b). The thigmotropic reaction of the seedling to the presence of a support is defined by the contact with the support. Turgor pressure is an important factor in the majority of plant motions, whether this is localised to a few cells, such as in the control of stomata, or readjusting the whole stem in response to light or gravity (Forterre 2013). Plant cells are commonly at high pressures of 0.4 to 0.8 MPa, when fully hydrated, and can reach 4 MPa in specific tissues such as the guard cells of stomata (Tomos and Leigh 1999; Franks et al. 2001; Taiz and Zeiger 2002). Cells can be considered to act as hydrostats, as was introduced in 'Cells: Containing and Using Hydrostatic Pressure' section.

Many plants show temporary or reversible motion, and two of the best known examples may be the touch response of *Mimosa pudica* leaves, or the rapid closing of the Venus flytrap (*Dionaea muscipula*) when a fly enters the carnivorous flower. Turgor pressure is an important mechanism in these nastic responses. Special cells are tuned to respond rapidly to stimulus, reducing pressure and causing motion. In the Mimosa leaf support structure, a localised tissue in structures named pulvini, which are formed of parenchyma cells, are able to respond to electrochemical signals, decreasing

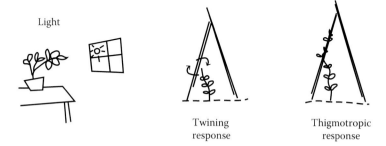

FIGURE 9.40
(a) Phototropism is most obvious when plants in low light grow towards a distant light source, such as a house plant at a distance from a window. (b) The twining response of many beans and climbing plants is demonstrated when seedlings make circular motions with their shoots until they make contact with a cane or nearby plant stem. This is a nastic response as the direction of circling is governed by the plant. (c) After contact is made with the cane, a thigmotropic (touch induced) response leads to the helical winding of the growing shoot around the support.

internal pressure of pulvinus cells on the lower side, while cells above the central axis increase in volume. This allows the leaf and the leaflets to droop under gravity (Volkov et al. 2010). Reversing the process requires longer as turgor pressure of the lower cells must be restored, while the upper cells decrease in volume.

In the Venus flytrap, a similar process occurs; however, the design of the flower is such that the curvature and in-built stress within the two main trap lobes causes an acceleration as they start to move in response to change of turgor pressure, causing them to snap shup sufficiently fast to catch the visiting insect. The system relies on bi-stability, i.e. a design that can be stable in either of two states, with a snap-through mechanism acting to give fast movement between the two states (Forterre et al. 2005; Li and Wang 2017). The open state is maintained by hydrostatic pressure difference between layers of a bi-layer of cells, allowing a large amount of elastic energy to be stored for release on trap closure (Markin et al. 2008).

Other plant tissue also utilise turgor pressure to move. Growth of shoots in an upward or lateral direction is often determined by auxins or other growth regulating hormones, where lateral growth is required the auxins initially stimulate differential pressures on opposing sides of the stem. Unlike the nastic movements mentioned above, which are short term and initiated by temporary stimulus, these tropic movements result in long-term changes in growth direction, and are reinforced by structural changes in the cell walls to fix the newly adopted shape to some degree once cells have fully differentiated (Iino 1990). A frequently mentioned and more pronounced example is the heliotropic movement of the sunflower (*Helianthus annuus*). This is sometimes misrepresented as being the motion of the sunflower heads following the path of the sun, but is only truly demonstrated by the developing shoots, and later in plant growth by the upper portion of the stem and flower bud, with the motion ceasing by the time the flower opens (Kutschera and Briggs 2016). The flower heads adopt a fixed position facing East, and after flowering the seed head finally droops in this direction. Various researchers have studied the motion of the apical bud, or the leaves and their orientation, reporting that the shoot tracks 12° behind the location of the sun, relating to approximately 48-minute lag (Shell and Lang 1976), and that the nocturnal reorientation (approximately 26°/hour) occurs more rapidly than the solar tracking during daylight (approximately 15°/hour, Koller 2011) (Figure 9.41).

The solar tracking of the sunflower shoots and leaves demonstrated in sunflower plants has provided inspiration to architects and designers seeking to provide temporary adjustment to external factors such as angle of incident light. Photonastic movements in plants also occur, such as leaf angle change between day and night based on non-directional light, or evading damage due to excess solar irradiance by altering angle to reduce photon capture (Koller 2011). Both phenomena have been taken as inspiration for adaptive façades for buildings, either to introduce variable shading or ventilation. Many strategies have been investigated, drawing on different plant-inspired

FIGURE 9.41
Phototropism in sunflowers stops when the flower head develops. Source: Deborah Newton-Perks.

techniques (Fiorito et al. 2015; López et al. 2017). One example is the use of mechanised louvres set within the glazing void of the Ernst and Young building in Sydney harbour. The architect sought the texture and patina of timber rather than tinted glass, yet required a design that would provide shade from high levels of solar irradiance. The use of Accoya (an acetylated wood with high UV stability and dimensional stability in fluctuations of relative humidity) allowed a system of synchronised opening and closing louvres to be formed, giving the appearance of a timber façade when louvres are closed (Figure 9.42).

Other examples include Flectofin, a concept based on the torsional opening of the bird of paradise flower (*Strelizia reginae*) when a bird lands on a perch structure at the entrance to the flower. The weight of the bird depresses

FIGURE 9.42
The natural wood appearance of the Ernst & Young building, Sydney, Australia when the louvres are closed. (b) Stability of the Accoya timber louvres ensures minimum distortion in the changing temperature and humidity of the enclosed space of the building facade. Images: Accsys Technologies plc.

Emerging Nature-Based Materials

the perch causing the two opposing curved segments of the flower to open, allowing pollination. The opening of the laminae of the perch is caused by a torsional buckling mechanism. The concept was transferred to the design of structures that can be installed as passive mechanical louvres in windows or façades by the team at the Institute of Building Structures and Structural Design (ITKE) in the University of Stuttgart (Lienhard et al. 2011; Masselter et al. 2012). Here, the laminae were manufactured from glass-reinforced plastic, which has a highly predictable deformation response to stress. The open and closed positions from a demonstration model are shown in Figure 9.43, with an artist's impression of the dynamically changing façade possible in a building.

Many other adaptive façades demonstrate bio-inspiration in their design for adjustable light or ventilation, using a wide range of materials (López et al. 2017). A construction using bimetal laminae with curvature initiated by thermal gain to open the vents of the structure under hot conditions was demonstrated by Doris Kim Sung in the Bloom installation, at the Los Angeles Materials and Application Gallery in spring 2012 (Furuto 2012). In the same year, the One Ocean Pavilion at the EXPO in Yeosu, South Korea, showcased reinforced polymer lamellae that could be controlled individually to regulate light intensity inside the building (Badarnah 2017). For creative ripple effects, the shades can be opened and closed in sequence and illuminated (Figure 9.44a,b, ArchDaily 2012). The façade system was developed by Soma with Knippers Helbig Advanced Engineering and contributed to the building's design intent, to reflect the Living Ocean and Coast theme of the Expo. It is also possible to utilise bio-based materials to provide the actuator component of the façade – as will be described in the next section.

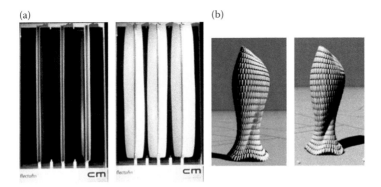

FIGURE 9.43
Flectofin louvres demonstration in the closed and open position, allowing and preventing light through the façade. Artist's impression of Flectofin façade at different angles of incident sunlight. All stills are from video presentation by ITKE, University of Stuttgart, available on https://vimeo.com/48374174 [Accessed 8/5/17].

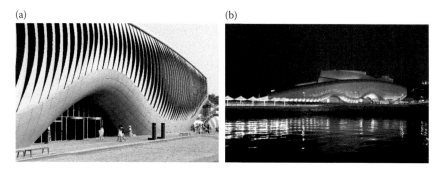

FIGURE 9.44
The One Ocean Thematic Pavilion façade, by SOMA Architecture for the 2012 EXPO in Yeosu, South Korea. (a) Deformable lamellas control the shading and day-lighting. (b) At night the closed lamellae form a sleep profile. Images used with permission, Copyright soma.

Responsive Forms – Actuators

In the plant motion examples given above, live plants respond to stimuli and activate changes in turgor pressure to initiate movement. Other biological materials show movement that is triggered by change of moisture content, such as the opening of dried seed pods, or the opening of mature pine cones in hot weather. These passively responsive tissues have also inspired advances in materials science and design, in the field of actuators.

In the seed pod of a bean, or the pinecone (Figure 9.45a), the tissue is no longer alive (Dawson et al. 1997), but the loss of moisture results in contraction of one portion of the tissue, exerting a force on the scale of the pinecone, or breaking the seal of the seed pod (Burgert and Fratzl 2009; Reyssat and Mahadevan 2009; Guiducci et al. 2016; Shtein et al. 2016). The motion is

FIGURE 9.45
(a) The scales of a pine cone open due to differential swelling in the two dominant layers of cells within the scale. (b) Bilayers of wood veneer form the leaflets within a living wall created by Henri Judin of Aalto University.

Emerging Nature-Based Materials

caused by hydroscopicity of the tissue, and build-up of tensile and compressive forces in opposite faces of the structure due to differential swelling of the tissue. Often this anisotropic swelling is caused by alignment of mircofibrils or cells restricting motion to the plane perpendicular to the fibril alignment (Figure 9.46b). In a pine-cone scale (Figure 9.46a), the tissue on the upper side is fibrous sclerenchyma cells, 150–200 μm long, 8–12 μm diameter, aligned approximately parallel to the scale length and with a low angle of microfibril winding in the cell wall. Meanwhile, the tissue on the lower side comprises shorter sclerid cells (80–120 μm long, and 20–30 μm diameter). Within the sclerid cells, the angle of winding is very great, 74° compared to 30° in the sclerenchyma, and although both cell types take up equal moisture on change of humidity, the hygroscopic expansion of the sclerids was approximately three times greater than the sclerenchyma from the upper surface (Dawson et al. 1997).

Various examples of hygromorphic materials exist. Holstov et al. (2015) report the traditional use of plain-sawn planks in boathouses in Nordmore, Norway, as a simple climate responsive façade system. Here the natural tendency of timber to cup when dry (if cut with the tangential direction aligned with the face of the plank) is harnessed to provide greater ventilation in the boathouse during the summer months. The practice is said to date back to the 19th century (Larsen and Marstein 2000).

In the timber, the swelling is caused by uptake of moisture within the amorphous layers of hemicellulose, which occur between the cellulose microfibrils. The swelling in the hemicellulose pushes the microfibrils apart, but it is resisted by the longitudinal axis of the microfibril; so swelling is considerably greater in the transverse directions of the timber than the longitudinal axis (Figure 9.47). Additional anatomical features of the wood, such as the presence of rays (bands of parenchyma, generally several cells deep

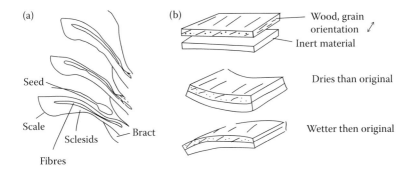

FIGURE 9.46
(a) Section of a pine cone, indicating fibres and sclerids within the scale. Shrinkage of the sclerids on drying is greater than change of dimension in the fibres, so the scale flexes to release the seed. (b) Laminated bi-layer of wood veneer and inert material, demonstrating change on drying and wetting.

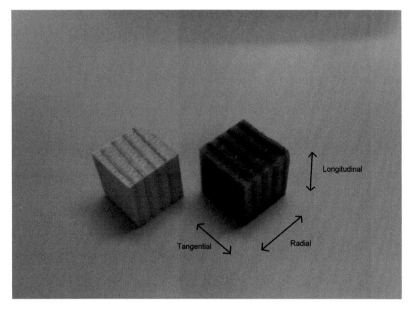

FIGURE 9.47
Dry and saturated blocks of pine wood with identical initial dimensions swelling in the tangential direction was more than double the swelling in the radial direction.

and one or more cells wide, which are aligned radially), offers some resistance to swelling in the radial direction, meaning that tangential swelling is the greatest of the three orientations in wood. As a rule of thumb, tangential swelling is 10x that of the longitudinal and double that of the radial direction, although there is variation between species, all of which is well described by Suchsland (2004).

Several researchers have demonstrated modern hygromorphic materials based upon bi-layer composites, or 'bimorphs', in which the hygroscopicity of wood is harnessed to distort the component by a predictable amount on uptake, or loss of moisture (Cordero and Smith 2013a,b; Reichart et al. 2014; Holstov et al. 2015). Typically, there is a passive layer, such as glass fibre reinforced plastic and various polymers, while timbers including larch, sycamore and walnut have been reported as active layers (Holstov et al. 2015). Such structures can be used for passive ventilation. A living wall design was proposed for shower rooms, with leaflets that open to provide ventilation, or can be watered to close the structure for privacy during showering (Figure 9.45b, Judin 2016). Reyssat and Mahadevan (2009) used paper rather than wood as the active layer to equal effect within a bilayer hygromorphic material. Holstov et al. (2015) liken the concept to that of bi-metallic strips (commonly used on thermostatic sensors) and therefore harness the calculations relating to Timoshenko's theory of bi-metal thermostats (Timoshenko 1925). The formulae utilise the coefficients of hygroexpansion and moisture content change to predict the radius of curvature of the material. Maximum responsiveness can be seen when a tangentially aligned timber veneer is used in the active layer, and an inert material is used for the passive layer.

Holstov et al. (2015) report that there may be benefits in utilising other species with diffuse porous structures such as birch and beech. Here, the relatively uniform distribution of cells may ensure relatively efficient flux of moisture, and avoid regions of high and low responsiveness relating to earlywood–latewood differences, or non-uniform grain such as burls, decorative figure or knots. Aesthetics, such as the colour of the wooden component may also be significant in selection.

The use of a passive system, not requiring electronic or motorised control is of great interest in buildings, for example, in ventilation or moisture control (Menges and Reichart 2012; Holstov et al. 2015). Demonstration of the technology in museum exhibits, and in the Hygroskin meteorosensitive pavilion created by Achim Menges and co-workers from the University of Stuttgart (Figure 9.48) have shown the suitability of hygromorphic systems for built elements. Building physicists have pointed out that the control of ventilation and shading simultaneously using these systems is complex, due to competing effects. Design frequently optimises systems for one function at a time; yet, buildings are exposed to multiple environmental factors and are required to regulate heat, air water and light simultaneously (Badarnah 2017). Development of multiple functionalities into built elements such as adaptive facades presents a challenge in reconciling these parameters. Performance prediction, design optimisation and simulation development is therefore complex (Xing et al. 2017, submitted).

Wood is not the only material that may be used in hygromorphic materials. The use of hydrogels (Erb et al. 2013; Ionov 2013), polyelectrolyte layers (Lee et al. 2013) or conducting polymers (Taccola et al. 2015) is advancing rapidly. Responsive hydrogels are elastic networks of hydrophilic polymer, which are capable of swelling to the desired extent on change of hydration. Hemicellulose and alginates are both naturally occurring hydrogels, with a viscoelastic role in the plant or seaweed tissues where they occur. In animals, proteoglycans combined with protein fibres form hydrogel composites (Calvert 2009). The properties of these hydrogels have been mimicked in synthetic composites; one example using microfibril reinforced hydrogels has demonstrated the potential of these materials to change shape on immersion in water (Gladman et al. 2016). Expansion of the hydrogel between fibres is countered by the linear resistance to movement imparted by the fibres to achieve anisotropic swelling. When such composites combine different preferred alignment of fibres in the separate layers the swelling translates into curvature, allowing the material to fold or unfold in the same manner as described above for wood-based systems. The resulting hygromorphic properties may go beyond simple flexure in one plane, offering twisting or complex modes of flexure, as demonstrated by the '4D printed' flower example (Figure 9.49b, Gladman et al. 2016).

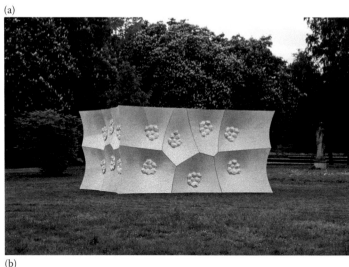

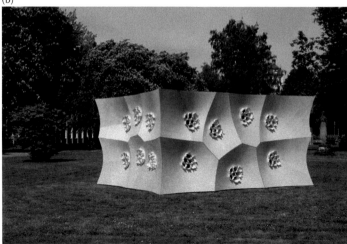

FIGURE 9.48
(a) Hygroskin meterorosensitive pavilion installation in Stadtgarten, Stuttgart, with vents closed. (b) Vents open during a warmer drier climate.

Source: **ICD, University of Stuttgart.**

Other synthetic hydrogels have been used, and a range of motions and folding actions has been developed, many of which have some bio-inspiration in fibril orientation. Ionov (2013) reported the mechanism for controlling twist and coil formation and the potential for combining active and passive regions in complex cross sections. Twist can be introduced by alignment of fibrils within the two layers at a range of opposing angles, for example, 45° and −45°, or 30° and −60° so that shrinkage in the intra-fibrillar material on one

Emerging Nature-Based Materials 297

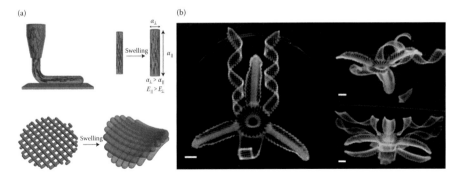

FIGURE 9.49
(a) Principles of 3D printing microfibril reinforced hydrogel actuators that swell when immersed in water to induce movement as the fourth dimension. (b) A 4D printed flower using curling, twisting and curving sections.

Source: Gladman et al. (2016).

face is resisted by the fibril rigidity on the opposite face. This crossed alignment mimics the fibrillary reinforcement within the orchid tree seedpod, as described by Armon et al. (2011). Wu et al. (2013) also demonstrated twist in a single layer hydrogel sheet, with alternate stripes of two hydrogel compositions (Figure 9.50). The angle of stripes relative to the length of the sheet controlled the handedness and pitch of curling observed on soaking in different sodium chloride concentrations. The various strategies and demonstrated results indicate the potential for materials with discrete segments to be designed to flex or to twist. An example of this multimodal modulation was given by Erb et al. (2013) for gelatin sheets with aligned alumina platelets. The programming of platelet alignment in individual regions of the gelatin strip led to twist and both left hand and right hand bending being demonstrated.

One potential application for actuators with self-folding geometries is to encapsulate agents such as yeast inside thermally responsive actuators, so that release occurs only at desired temperature (Stoychev et al. 2011; Ionov 2013). Design of hydrogels to respond to changes in light intensity, pH, temperature, biochemical processes and electric or magnetic fields will open new avenues (Ionov 2013). These hydrogels may find applications ranging from smart lenses (responding to temperature to adjust focal length, Dong

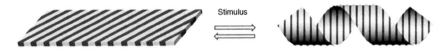

FIGURE 9.50
Hydrogel sheet with alternating stripes of gels with different response, aligned at an angle of 30°, 45° or 60° to the long axis of the strip. On application of stimulus (soaking in salt solution) the sheet twists in relation to the angle of the stripes.

Source: Wu et al. (2013).

et al. 2006) through to adapting surface texture with high or low hydrophobicity in response to relative humidity (Sidorenko et al. 2008) as discussed by Ionov (2013). The Lewis group at Harvard University predict that biomimetic shape morphing materials may be used in smart textiles, soft electronics, medical devices and tissue engineering. Lee et al. (2013) used hygromorphic actuators to demonstrate locomotion along a ratchet track, this offers new potential in passive regulation of systems, in addition to the binary open vs closed control demonstrated in the previous examples.

Biological Systems: Harnessing Natural Processes

One leading scholar in biomimetics and bioinspiration Professor Vincent proposed that there are many levels of biomimicry, ranging from the most direct such as Velcro, to the more generalised and abstract transfer of ideas and concepts. Figure 9.51 shows a pair of maps, indicating this concept and the potential routes for transfer of ideas from biology into engineering (Vincent 2001). The further down the map you move from the original concept, the more powerful the concept can be in providing a new-engineered solution. Thus, it is worth briefly mentioning two fields in which natural materials are being considered in a broader manner to derive larger and more far-reaching benefits in future technologies and materials.

Enzymes for Manufacturing

Enzymes have been identified from natural products and used in cleaning, pulping and material separation processes for many years. The introduction of enzymes to washing powder and stain removers, for example,

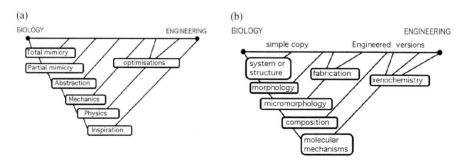

FIGURE 9.51
(a) Biomimetic map indicating increasing abstraction, and the relative ease of extrapolating the idea for application in a new area of engineering. (b) Biomimetic map indicating increasing degrees of fundamental properties and their route to biomimetic engineered solutions.

Source: **Vincent (2001).**

revolutionised cleaning products, allowing lower temperatures to be used, and a decrease in reliance on caustic soda or other harsh ingredients. Other enzymes of selective activity have been used in pre-treatment, pulping and bleaching for paper manufacture; pre-treatment and delignification for bioethanol production form biomass; or fermentation to produce platform chemicals and biopolymers (Saai Anugraha et al. 2016). In many cases, once the enzyme has been identified and isolated, it can be manufactured synthetically, while in others, the prevalence of the enzyme in a culture of a microorganism is sufficient to harvest it or to use the pure culture. Anaerobic digestion, for example, often harnesses communities of microorganisms rather than the isolated enzyme, to achieve the same objective – degradation of biomass into useful fragments that can be harvested.

A wide range of bio-derived platform chemicals is now available, such as volatile fatty acids (acetic, propionic, butyric, caprolic) diacids (succinic, malonic), solvents (methanol, ethanol, acetone) and polyols. Not only these are formed from natural materials such as agricultural residues but also transformed by natural organisms or enzymes harvested from microbial communities. Many of these chemicals are formed by enzymatic fermentation followed by distillation or by enzyme synthesis. For example, digestion of corncobs or other biomass by cellulase from *Trichoderma ressei* is often used to achieve glucose for ethanol production (Sun and Cheng 2002; Chen et al. 2007). In some cases, simultaneous fermentation to utilise the glucose and generate ethanol allows a continuous process to be developed without excess glucose inhibiting the depolymerisation. Similarly, when anaerobic digestion using microbial cultures is used to generate volatile fatty acids for biopolymer synthesis, control of the harvesting of the fatty acids can avoid saturating the liquor with product or inhibition of the reaction. In methanol production, control or inhibition of volatile fatty acid generation is necessary to maintain yield of the desired product (Franke-Whittle et al. 2014).

In nature, however, many natural materials are formed, rather than degraded, using enzymes, so modern research has addressed the use of enzymes to selectively create rather than degrade bonds between components or between monomers. Researchers developing packaging materials based on hemicellulose have recognised the potential of laccase to form new bonds between hemicellulose oligomers, during separation from pulp waste water, to form higher molecular weight polymers of cross-linked galactoglucomannan. The molecular weight can be increased by up to six times, 26 kDa instead of 4 kDa, creating a biopolymer with greater potential (Oinonen et al. 2013, 2016).

The polyhydroxy alkanoate (PHA) biopolymers were introduced in Chapter 3. Polyhydroxy butyrate (PHB) and polyhydroxy valerate (PHV) and their co-polymers (PHBV) can be formed by bacterial cultures, and the form of biopolymer controlled by altering the blend of volatile fatty acids supplied. There is great opportunity for the development of this technology and improving product grade and yield (Kedia et al. 2013; Passanha et al. 2013). Hydrolases may also be used to form the biopolyesters (Uyama and Kobayashi 2002). Ring opening polymerisation of ε-caprolactone can be enzyme catalysed by laccase (Loeker et al. 2004) and peroxidase and laccase can be used in oxidative polymerisation of phenol derivatives to form polymers (Uyama and Kobayashi 2002).

There is interest in training communities of microbes to produce PHA biopolymers from mixed feedstocks in anaerobic digestion processes, although greater control of the feedstock to achieve tightly defined polymer (with target molecular weight and polydispersity, and mix of butyrate–valerate units) may be of greater commercial importance. Enzymatic synthesis often presents a relatively benign manufacturing route, and utilises only moderate temperatures, avoids the need for solvents or high pressure, high temperature synthesis steps. Reclamation and purification of the polymer from the broth after sufficient residence period in the reactor is the main other step, but can be relatively easily achieved by physical separation and dialysis processes.

Self-Assembly and Biotemplating

A significant aspect of many of the natural materials discussed here has been their internal order, and the fact that this order (whether in the form of plates, fibrils, columns, layers, or in complex structures utilising these components to build apertures and scaffolds) is entirely self-assembled. No high tech machinery or craftsman oversee the process, the individual assembly of these minute components is efficiently and replicably performed in situ by the cell or tissue. It is amazing to observe the complexity of structures that can be assembled by this process. The growing sea sponge arranges mineral fibres into spicules with intricate geometry (e.g. the Venus flower basket, Figure 9.23). Aside from the overwhelming intricacy of the cage of the Venus flower basket, consider the elegance and mechanical optimisation of this truss-like structure from incredibly fine silica filaments. At the ultrastructural level, all of this is achieved by controlled nucleation of the mineral on a scaffold of protein, and development of concentric layers of silica to create a tough and strong composite (Woesz et al. 2006).

In modern times, researchers have turned again to the complex forms found within biomineralised shells and skeletons with new eyes. Advances in understanding of the biosynthesis of the structures, and their assembly in relation to proteinaceous filaments, have allowed researchers to mimic the biomineralisation processes in the laboratory. The benefits of harnessing the synthetic processes of nature to achieve these high performance microstructures require considerable time commitment and research (Sarikaya and Aksay 1992), but this research into biosynthesis and biomineralisation is beginning to bear fruit. As a result, a new discipline of biotemplating has emerged, in which the potential of re-creating the structural shapes or hierarchical internal order of natural materials is being explored (Zollfrank et al. 2012). Biotemplating using organic polymers, such as proteins and polysaccharides, allows control over deposition of minerals in a manner that mimics the processes seen in nature.

The self-assembly concept has perhaps been best studied in the context of nacre. The high fracture toughness and strength have long been recognised as being on a par with the most advanced structural ceramics (Sarikaya and Aksay 1992). In 'Toughness in Nacre' section we hinted at the role of proteins in determining the crystalline order of the calcium carbonate mineral phase; and commented that the presence of polymeric material within the aragonite tables providing additional toughening mechanisms in the nacre itself. Many researchers have developed techniques to generate nacre by deposition in vitro, allowing various stages in the process of biomineralisation to be better understood. It follows that these steps can be revisited in order to develop new materials, by self-assembly, with the potential to manipulate and engineer its microstructure.

The biomineralisation process in the mollusc shell includes distinct steps: a layer of biopolymer named the periostracum is formed, which nucleates the growth of calcite. The calcite nucleation is preferentially controlled, by epitaxial effect of the polymer of the organic phase (Lin et al. 2008; Zlotnikov and Schoeppler 2017). Once the calcite layer is sufficiently thick (between 0.5 and 3 mm), nucleation of aragonite is initiated. The aragonite is first in the brick and mortar configuration, then takes a columnar formation, and adopts several other conformations across the thickness of the shell (Lin et al. 2008; Zaremba et al. 1996). Each of these transitions occurs replicably in abalone shells regardless of their region of origin, to achieve a complex composite material that is optimised for its protective role to the mollusc. The mineral content of 95% by volume is achieved by the self-assembly process, which would be difficult to replicate in synthetic composites.

However, biologists seeking to verify the mechanisms that occur in vivo have been successful in replicating various steps of the nacre production. It is a logical extension also to ask, can these self-assembly processes be harnessed to create such multi-layer structures from other materials, for example, to form a sandwich structure, or to create synthetic nacre. For example, growth of 'flat pearls' on glass cover slips inserted into extrapallial fluid in the laboratory have revealed that growth velocity is influenced by water temperature and feeding (Lopez et al. 2011). Analysis of microvilli in the layer of epithelium above the extrapallial space (Nakahara 1991) corresponds well with the arrangement of regular holes in the aligned chitin fibrils that form the layer between aragonite tiles. This chitin layer is excreted from the epithelium, but laid down periodically onto the growing aragonite surface (Lopez et al. 2011), it is proposed that this is to retard growth in the c-direction.

Clearly, replication of this process under laboratory conditions would require considerable engineering; however, many scientists aim to identify and harness processes such as these to generate the nanostructures or microstructures desired to allow a new generation of materials. In the field of nanotechnology, various approaches have used protein scaffolds, or cages, or even viruses to provide the proteinaceous substrate on which deposition of metals, metal oxides or other inorganic materials can be controlled (Whyburn et al. 2008). The early work using ferritin cages gave rise to nanoparticles of controlled shape and size (Douglas et al. 1995). A whole range of biopanning techniques have been developed to help identify the suitable protein structures for specific interaction with the inorganic component, these include phage display systems, cell surface display and ribosome display (Dower and Mattheakis 2002; Whyburn et al. 2008). Once identified, it is possible to manipulate the coat structure of the virus to act as a bioscaffold. Such techniques can produce nano-wires, sheets or heterogeneous materials and multi-dimensional structures. Two-dimensional structures are possibly the most commonly used biotemplates at present, but three-dimensional templating is likely to create new potential in generating nano-lattices or hierarchical materials.

The use of scaffolds to re-generate bone in medicine has also drawn on the recognition of the role of proteins such as collagen in controlling or guiding the deposition of calcium phosphate, or of hydroxyapatite (Cui et al. 2007). While calcium phosphate ceramics have been used for bone repair for several decades, and several systems show improved similarity with the substrate bone giving good results, much work remains to be done, in particular to introduce the hierarchy of natural bone, and to better match the range of properties seen in different bone tissues. Due to the ease of forming the sparingly soluble calcium phosphate from solutions containing the two ions, this has been a popular approach for regenerating hard tissues (Xu et al. 2006; Cai and Tang 2008). At the smallest scale the mineral crystals in bone that are located between the collagen fibrils are 2–10 nm in thickness, with variable lengths (30–50 nm) and widths (15–30 nm), the average being 50 – 25 nm (Wang et al. 2006). In naturally occurring compact bone the crystal orientation is defined in relation to the collagen fibril matrix into which the crystals are deposited, and results in an efficient shear lag mechanism of load transfer within the fine textured composite material produced (Gupta et al. 2006). In an ideal regenerated or repaired bone tissue some of this strength and toughness would be harnessed by mimicking the shape and orientation of the particles and fibrils.

One example of progress towards a biotemplating approach is the use of nano-hydroxyapatite collagen blends to achieve in situ self-assembly (Zhang et al. 2003a). Many other methods of generating mineralised collagen by self-assembly are reviewed by Cui et al. (2007). Type I collagen provides a good substrate for mineralisation, and it has been the basis of most research in the area, although many other forms are known to exist in nature. It has been proposed that the carbonyl oxygen of collagen molecules may chelate calcium ions, thus serving as a nucleation site for the mineral (Zhang et al. 2003b). Further study indicated that the developing hydroxyapatite crystals alter the conformation of the collagen fibrils, and it has indicated a cooperative interaction between the protein and the mineral phases (Cui et al. 2008).

Emerging Nature-Based Materials 303

Biotemplating using polysaccharides to control deposition of inorganic materials may generate nanotubes, or two-dimensional structures (Zollfrank et al. 2012). Another example is the development of high surface area materials for drug delivery by functionalisation of the biosilicate of diatoms (Chao et al. 2014). While this field may move outside bioinspiration, the likely rapid expansion of such strategies is likely to interface with the revelations of self-assembly processes observed in natural organisms, and their influence on development of hierarchical structures. It will certainly offer new opportunities for utilising natural polymers, and synthetically regenerating biominerals in new materials.

Biotemplating may also be possible in the field of organic chemistry and polymer science. Kondo et al. (2002) have been reported to mimic the natural deposition of cellulose microfibrils by bio-directed epitaxial deposition processes. Here they used *Acetobacter xylinum* to excrete cellulose, which was deposited onto a surface of nematic ordered cellulose, with high chain alignment, which had been previously prepared in the laboratory. The *Acetobacter* excreted cellulose ribbons, which were aligned with the chain orientation of the substrate, and demonstrated this by video imaging of the motion of the bacteria relative to the alignment direction of the substrate. They reported that the interaction with the ordered cellulose substrate prevented twisting that normally occurs in the ribbons excreted by *Acetobacter*. The directed will allow highly aligned fibrillar structures to be generated, and it offers the potential to investigate nanostructures and new nanocomposites manufacturing techniques.

Conclusions and Outlook

This chapter has outlined many ways in which the materials scientist can turn to natural materials for inspiration in form, composition, mechanical properties, motion and regulation of environmental conditions. An increasing number of examples combine natural materials with bioinspiration in their use in architecture or the built environment. The natural materials themselves provide a rich palette from which fibres, polymers, and constituent monomers can be taken, processed, developed, refined, polymerised and rebuilt. With the advent of new techniques such as 3D printing or self-assembly new opportunities for using natural fibres or bio-based polymers will emerge, and incorporate greater range of the bioinspired systems mentioned here. The field of biocomposites is broad, and likely to continue to expand as new approaches are investigated.

As there is considerable scope in the field of biomimetics, only those aspects relating directly to material structure or material performance have been considered her. Additional topics and strategies, such as biomimetics in design (e.g. using strategies identified in termite mounds to assist ventilation and heat recovery in architecture) or self-healing materials (e.g. in high fatigue composites) have been excluded. Even in the areas covered, the text serves only as an introduction or appetiser, leaving the reader with suggested references for further reading if this specific facet of the science of natural materials is of interest.

Much can be achieved by combining bio-based materials in composites, and technology to do so has been developing rapidly. The emergence of bio-composites, whether they are formed with long natural fibres, powdered biomass, or nanocellulose, was discussed in Chapter 3. Here we have considered some of the new developments that harness the same elements and utilise bio-inspiration in the re-assembly into functional materials. One example is the 3D printing of fibre reinforced hydrogels to produce actuators that flex to regulate the environment. Other emerging products seek to use biopolymers such as protein or chitosan as substrates for biomineralisation, to achieve superior performance and biocompatibility for bone replacement.

Many more examples of bio-inspiration mentioned in this chapter are formed from synthetic materials, but still offer scope for concept to be recombined with natural bio-based ingredients to deliver the product concept from sustainable sources. The continued exploration of design, form and action in nature, and the advances in bio-based polymers, enzyme synthesis and biotemplating are likely to deliver an expanding range of concepts in future. These fields of study are expanding rapidly, and it has only been possible to skim the surface of the techniques and understanding on which they are developing new materials.

What can be said of the future? It is highly likely that the progress seen over the past twenty years, which could in some areas of study have been an inconceivable pipedream, will continue at a similar pace. There is certainly no shortage of examples in biology to generate inspiration. The concept that gives us mechanical motion in hygromorphic actuators has been brought to fruition by the great progress in engineering and technology (not covered here) with laser etching, 3D printing and high precision computerised systems capable of forming or removing materials to a pre-defined template. When this technology is coupled with the understanding of the roles of individual molecules in a structural, elastic, shape changing or adhering role, high levels of complexity can be designed into materials at an ever increasing level of hierarchy.

All this progress must be coupled with a realistic glance at the environmental sustainability of the process of manufacturing the material. Part of the charm of the existing bio-based materials – wood, bamboo, natural fibre composites – is their naturalness, and their low environmental impact in manufacture. If bioinspiration moves too far into automated systems and highly synthesized precision chemistry without considering natural methods for synthesis, or green chemistry options for material inputs, the field will eventually contribute to decreased sustainability. Life cycle assessment of new materials may be difficult, due to lack of environmental or toxicological data; however, efforts must be made to quantify the benefits of new materials against their environmental burden. However, as research also continues into self-assembly, the use of enzymes, or of combinations of materials as templates to regulate structure development, there is every opportunity to identify further sustainable routes to manufacture.

It is certain that the future will involve a greater degree of responsibility in combining these novel materials into systems that function holistically to deliver multiple functionalities. For example, the buildings of the future are likely to use wall panels into which not only are the electrical and piped services pre-fixed, but also the surface contributes to environmental stabilisation, scavenging VOCs, or regulating humidity. The surface may include pores that open and close to feed the passive ventilation and heat the recovery system, while microtexture within the windows may morph to increase or decrease transmitted and reflected light and allow or inhibit the transfer of heat. With such dreams comes the challenge of modelling and predicting behaviour, in particular avoiding negative effects if system variables move outside the predicted range, leading to over-heating or cooling. Options for human interaction with the systems, or for consumers to understand the technology that is embedded into the fabric of the building, is essential.

Acknowledgements

This chapter incorporates many concepts from architecture, and the author is grateful for many discussions with members of the Plants and Architecture Cluster of the Sêr Cymru funded National Research Network for Low Carbon Energy and Environment (NRN-LCEE) which have prompted this study. In particular Dr Maurice Bosch, Dr Yangang Xing, Dr Peter Wooton-Beard and Dr Durai Prabhakaran whose breadth of expertise has enriched this chapter.

References

Addadi L., Joester D., Nudelman F. and Weiner S. (2006) Mollusk shell formation: A source of new concepts for understanding biomineralization processes. *Chemistry* **12**(4):980–987.

Alexander D.E. (2016) The biomechanics of solids and fluids: The physics of life. *European Journal of Physics* **37**:053001, 43 pp.

Alibardi L. (2007) Keratinization of sheath and calamus cells in developing and regenerating feathers. *Annals of Anatomy* **189**:583–595.

André A., Kliger R. and Asp L.E. (2014) Compression failure mechanism in small scale timber specimens. *Construction and Building Materials* **50**:130–139.

Ansell M.P. (2015) Wood microstructure – a cellular composite. In: *Wood Composites*, Ansell M.P. (Ed.). Woodhead Publishing, Elsevier, Cambridge, pp. 3–26.

ArchDaily (2012) One Ocean, Thematic Pavilion EXPO 2012 / soma. www.archdaily. com/236979/one-ocean-thematic-pavilion-expo-2012-soma. Accessed 14/7/17.

Armon S., Efrati E., Kupferman R. and Sharon E. (2011) Geometry and mechanics in the opening of chiral seed pods. *Science* **333**(6050):1726–1730.

Autumn K., Liang Y.A., Hseih S.T., Zesch W., Chan W.P., Kenny T.W., Fearing R. and Full R.J. (2000) Adhesive force of a single gecko foot hair. *Nature* **405**:681–685.

Badarnah L. (2015) A biophysical framework of heat regulation strategies for the design of biomimetic building envelopes. *Procedia Engineering* **118**:1225–1235.

Badarnah L. (2017) Form follows environment: Biomimetic approaches to building envelope design for environmental adaptation. *Buildings* **7**(40):16 pp.

Badarnah L. and Fernández J. (2015) Morphological configurations inspired by nature for thermal insulation materials. *International Association for Shell and Spatial Structures (IASS) Symposium 2015: Future Visions*, Amsterdam.

Barthelat F., Li C.M., Comi C. and Espinosa H.D. (2006) Mechanical properties of nacre constituents and their impact on mechanical performance. *Journal of Materials Research* **21**(8):1977–1986.

Barthlott W. and Neinhuis C. (1997) Purity of the sacred lotus, or escape from contamination in biological surfaces. *Planta* **202**:1–8.

Bechtle S., Ang S.F. and Schneider G.A. (2010) On the mechanical properties of hierarchically structured biological materials. *Biomaterials* **31**:6378–6385.

Bhusan B. and Her E.K. (2010) Fabrication of superhydrophobic surfaces with high and low adhesion inspired from rose petal. *Langmuir* **26**(11):8207–8217.

Bhusan B. and Nosonovski M. (2010) The rose petal effect and the modes of superhydrophobicity. *Philosophical Transactions of the Royal Society A* **368**:4713–4728.

Bhusan B., Zhang Y., Xi J., Zhu Y., Wang N., Xia F. and Jiang L. (2008) Petal effect: A superhydrophobic state with high adhesive force. *Langmuir* **24**(8):4114–4119.

Bonilla M.R., Lopez-Sanchez P., Gidley M.J. and Stokes J.R. (2016) Micromechanical model of biphasic biomaterials with internal adhesion: Application to nanocellulose hydrogel composites. *Acta Biomaterialia* **29**:149–160.

Bonser R.H.C. and Purslow P.P. (1995) The Young's modulus of feather keratin. *Journal of Experimental Biology* **198**(4):1029–1033.

Bonser R.H.C. and Witter M.S. (1993) Indentation hardness of the bill keratin of the European starling. *Condor* **95**:736–738.

Boreyko J.B. and Chen C.H. (2009) Restoring superhydrophobicity of lotus leaves with vibration-induced dewetting. *Physical Review Letters* **103**(17):174502.

Bormashenko E., Pogreb R., Whyman G. and Erlich M. (2007) Resonance Cassie–Wenzel wetting transition for horizontally vibrated drops deposited on a rough surface. *Langmuir* **23**:12217–12221.

Bostandzhiyan S.A., Shteinberg A.S. and Bokov A.V. (2006) Physicomechanical properties of the bird feather shaft. *Doklady Chemistry* **408**(2):92–95 (original text: *Doklady Akademii nauk SSSR* 408(6):772).

Brubaker C.E. and Messersmith P.B. (2012) The present and future of biologically inspired adhesive interfaces and materials. *Langmuir* **28**(4):2200–2205.

Bruet B.J.F., Qi H.J., Boyce M.C., Panas R., Tai K., Frick L. and Ortiz C. (2006) Nanoscale morphology and indentation of individual nacre tablets from the gastropod mollusc Trochus niloticus. *Journal of Materials Research* **20**(9):2400–2419.

Buehler M.J. (2006) Atomistic and continuum modelling of mechanical properties of collagen: Elasticity, fracture and self-assembly. *Journal of Materials Research* **21**(8):1947–1961.

Emerging Nature-Based Materials

Buehler M.J. (2007) Molecular nanomechanics of nascent bone: Fibrillar toughening by mineralization. *Nanotechnology* **18**:295102.

Buehler M.J. (2008) Nanomechanics of collagen fibrils under varying cross-link densities: Atomistic and continuum studies. *Journal of the Mechanical Behavior of Biomedical Materials* **1**:59–67.

Buehler M.J., Keten S. and Ackbarow T. (2008) Theoretical and computational hierarchical nanomechanics of protein materials: Deformation and fracture. *Progress in Materials Science* **53**(8):1101–1241.

Burgert I. and Fratzl P. (2009) Actuation systems in plants as prototypes for bioinspired devices. *Philosophical Transactions A. Mathematical Physical and Engineering Sciences* **28**(367):1541–1557.

Cadogan D., Smith T., Uhelsky F. and MacKusick M. (2004) Morphing inflatable wing development for compact package unmanned aerial vehicles. In: *Proceedings of 45th AIAA/ASME/ASCE/AHS/ASC Structures, Structural Dynamics and Materials Conference*, April 2004, Palm Springs, CA, USA, no. 1807.

Cai Y. and Tang R. (2008) Calcium phosphate nanoparticles in biomineralization and biominerals. *Journal of Materials Chemistry* **18**:3775–3787.

Calvert P. (2009) Hydrogels for soft machines. *Advanced Materials* **21**:743–756.

Carpita N.C. and Gibeaut D.M. (1993) Structural models of primary cell walls in flowering plants: Consistency of molecular structure with the physical properties of the walls during growth. *Plant Journal* **3**(1):1–30.

Cassie A.B.D. and Baxter S. (1944) Wettability of porous surfaces. *Transactions of the Faraday Society* **40**:546–551.

Cerolini S., D'Orazio M., Di Perna C. and Stazi A. (2009) Moisture buffering capacity of highly absorbing materials. *Energy and Buildings* **41**:164–168.

Chanliaud E., Burrows K.M., Jeronimidis G. and Gidley M.J. (2002) Mechanical properties of primary plant cell wall analogues. *Planta* **215**:989–996.

Chao J., Biggs M.J.P. and Pandit A. (2014) Diatoms: A biotemplating approach to fabricating drug delivery reservoirs. *Expert Opinion on Drug Delivery* **11**(11):1–9.

Chen M., Xia L. and Xue P. (2007) Enzymatic hydrolysis of corncob and ethanol production from cellulose hydrolysate. *International Biodeterioration and Biodegradation* **59**(2):85–89.

Chen P.Y., McKittrick J. and Meyers M.A. (2012) Biological materials: Functional adaptations and bioinspired designs. *Progress in Materials Science* **57**:1492–1704.

Cheng Y.T., Rodak D.E., Wong C.A. and Hayden C.A. (2006) Effects of micro- and nano-structures on the self-cleaning behaviour of lotus leaves. *Nanotechnology* **17**:1359–1362.

Cheung K.C., Tachi T., Calisch S. and Miura K. (2014) Origami interleaved tube cellular materials. *Smart Materials and Structures* **23**:095012, 10 pp.

Cordero S.M. and Smith A. (2013a) Responsive expansion. In: *ACADIA 2013: Adaptive Architecture, Proceedings of the 33rd Annual Conference of the Association for Computer Aided Design in Architecture*, Beesley P., Khan O. and Stacey M. (Eds.), 21–27 October 2013. Cambridge, Ontario, pp. 25–32.

Cordero S.M. and Smith A. (2013b) Responsive expansion. ACADIA 2013 description (online). http://acadia.org/papers/NVEPNX. Accessed 4/12/16.

Cosgrove D.J. and Jarvis M.C. (2012) Comparative structure and biomechanics of plant primary and secondary cell walls. *Frontiers in Plant Science* **3**:204, 6 pp. doi:10.3389/fpls.2012.00204.

Cox H.L. (1952) The elasticity and strength of paper and other fibrous materials. *British Journal of Applied Physics* **3**:72–79.

Craig S., Harrison D., Cripps A. and Knott D. (2008) BioTRIZ suggests radiative cooling of buildings can be done passively by changing the structure of roof insulation to let longwave infrared pass. *Journal of Bionic Engineering* **5**:55–66.

Cui F.Z., Li Y. and Gi J. (2007) Self-assembly of mineralized collagen composites. *Materials Science and Engineering R: Reports* **57**(1–6):1–27.

Cui F.Z., Wang Y., Cai Q. and Zhang W. (2008) Conformation change of collagen during the initial stage of biomineralization of calcium phosphate. *Journal of Materials Chemistry* **18**:3835–3840.

Currey J.D. (1977) Mechanical properties of mother of pearl in tension. *Proceedings of the Royal Society of London B* **196**(1125):443–463.

Currey J.D. (1979) Mechanical properties of bone tissues with greatly differing functions. *Journal of Biomechanics* **12**(4):313–319.

Currey J.D. (1984) The mechanical properties of materials and the structure of bone, Chapter 1, pp. 3–37. In: *The Mechanical Adaptations of Bones*. Princeton University Press, NJ, p. 306.

Currey J.D. (2002) *Bones, Structure and Mechanics*. Princeton University Press, NJ, p. 456.

Currey J.D. and Taylor J.D. (1974) The mechanical behaviour of some molluscan hard tissue. *Journal of Zoology* **173**(3):395–406.

Currey J.D., Zioupos P., Davis A. and Casinos A. (2001) Mechanical properties of nacre and highly mineralized bone. *Proceedings of the Royal Society of London B* **268**:107–111.

Darmanin T. and Guittard F. (2015) Superhydrophobic and superoleophobic properties in nature. *Materials Today* **18**(5):273–285.

Dawson C., Vincent J.F.V. and Rocca A.M. (1997) How pine cones open. *Nature* **390**:668.

Dawson C., Vincent J.F.V., Jeronimidis G., Rice G. and Forshaw P. (1999) Heat transfer through penguin feathers. *Journal of Theoretical Biology* **199**:291–295.

Desch H.E. and Dinwoodie J.M. (1981) *Timber: Its Structure, Properties and Utilization*, 6th Edition. Macmillan, Basingstoke, p. 424.

Dicker M.P.M., Weaver P.M., Rossiter J.M. and Bond I.P. (2014) Hydrogel core flexible matrix composite (H-FMC) actuators: Theory and preliminary modelling. *Smart Materials and Structures* **23**:095021, 13 pp.

Doblin M.S., Kurek I., Jacob-Wilk D. and Delmer D.P. (2002) Cellulose biosynthesis in plants: From genes to rosettes. *Plant and Cell Physiology* **43**(12):1407–1420.

Dong L., Agarwal A.K., Beebe D.J. and Jiang H. (2006) Adaptive liquid microlenses activated by stimuli-responsive hydrogels. *Nature* **442**:551–554.

Doudrick K., Monzon O., Mangonon A. and Hristovski K. (2012) Nitrate reduction in water using commercial titanium dioxide photocatalysts (P25, P90, and Hombikat UV100). *Journal of Environmental Engineering* **138**(8). doi:10.1061/(ASCE)EE.1943-7870.0000529.

Douglas T., Dickson D.P.E., Betteridge S., Charnock J., Garner C.D. and Mann S. (1995) Synthesis and structure of an iron(III) sulfide-ferritin bioinorganic nanocomposite. *Science* **269**(5220):54–57.

Dower W.J. and Mattheakis L.C. (2002) In vitro selection as a powerful tool for the applied evolution of proteins and peptides. *Current Opinions in Chemical Biology* **6**(3):390–398.

Emons A.M.C. and Mulder B.M. (2000) How the deposition of cellulose microfibrils builds cell wall architecture. *Trends in Plant Science* **5**(1):35–40.

Erb R.M., Sander J.S., Grisch R. and Studdart A.R. (2013) Self-shaping composites with programmable bioinspired microstructures. *Nature Communications* 4:1712, 8 pp. doi:10.1038/ncomms2666.

Eschelby J.D. (1957) The determination of the elastic field of an ellipsoidal inclusion and related problems. *Proceedings of the Royal Society A* **241**:376–396.

Eschelby J.D. (1959) The elastic field outside an ellipsoidal inclusion. *Proceedings of the Royal Society A* **252**:561–569.

Feng Q.L., Li H.B., Pu G., Zhang D.M., Cui F.Z., Li H.D. and Kim T.N. (2000) Crystallographic alignment of calcite prisms in the oblique prismatic layer of Mytilus edulis shell. *Journal of Materials Science* **35**:3337–3340.

Feng S., Hou Y., Xue Y., Gao L., Jiang L. and Zheng Y. (2013) Photo-controlled water gathering on bio-inspired fibers. *Soft Matter* **9**(39):9294–9297.

Field T.S., Zwieniecki M.A., Donoghue M.J. and Holbrook N.M. (1998) Stomatal plugs of Drimys winteri (Winteraceae) protect leaves from mist but not drought. *Proceedings of the National Academy of Sciences of the United States of America* 95:14256–14259.

Fiorito F., Sauchelli M., Arroyo D., Pesenti M., Imperadori M., Masera G. and Ranzi G. (2015) Shape morphing solar shadings: A review. *Renewable and Sustainable Energy Reviews* **55**:863–884.

Forooshani P.K. and Lee B.P. (2017) Recent approaches in designing bioadhesive materials inspired by mussel adhesive protein. *Journal of Polymer Science A: Polymer Chemistry* **55**(1):9–33.

Forterre Y. (2013) Slow, fast and furious: Understanding the physics of plant movements. *Journal of Experimental Botany* **64**(15):4745–4760.

Forterre Y., Skotheim J.M., Dumais J. and Mahadevan L. (2005) How the Venus flytrap snaps. *Nature* **433**:41–425.

Franke-Whittle I.H., Walter A., Ebner C. and Insam H. (2014) Investigation into the effect of high concentrations of volatile fatty acids in anaerobic digestion on methanogenic communities. *Waste Management* **34**(11):2080–2089.

Franks P.J., Buckley T.N., Shope J.C. and Mott K.A. (2001) Guard cell volume and pressure measured concurrently by confocal microscopy and the cell pressure probe. *Plant Physiology* **125**:1577–1584.

Franks P.J. and Farquhar G.D. (2007) The mechanical diversity of stomata and its significance in gas-exchange control. *Plant Physiology* **143**(1):78–87.

Fratzl P., Elbaum R. and Burgert I. (2008) Cellulose fibrils direct plant organ movements. *Faraday Discussions* **139**:275–282.

French J.R.J. and Ahmed B.M. (2010) The challenge of biomimetic design for carbon-neutral buildings using termite engineering. *Insect Science* **17**(2):154–162.

Furuto A. (2012) Bloom /DO|SU Studio Architecture. ArchDaily. 11 March 2012, www.archdaily.com/215280/bloom-dosu-studio-architecture. Accessed 8/5/17.

Gao H. (2006) Application of fracture mechanics concepts to hierarchical biomechanics of bone and one-like materials. *International Journal of Fracture* **138**:101.

Gao H. (2009) Mechanical principles of a self-similar hierarchical structure. *Materials Research Society Symposium Proceedings* **1188**:3–14. doi:10.1557/PROC-1188-LL01-01.

Gaultieri A., Vesentini S., Redealli A. and Buehler M.J. (2009) Hierarchical nanomechanics of collagen microfibrils. *Journal of the Mechanical Behavior of Biomedical Materials* **2**:130–137.

Gaultieri A., Vesentini S., Redealli A. and Buehler M.J. (2011) Hierarchical nanomechanics of collagen microfibrils. *Nano Letters* **11**(2):757–766.

Geim A.K., Dubonos S.V., Grigorieva I.V., Novoselov K.S., Zhukov A.A. and Shapoval S.Y. (2003) Microfabricated adhesive mimicking gecko foot hair. *Nature Matter* **2**:461–463.

Gibson L.J. and Ashby M.F. (1988) *Cellular Solids: Structure and Properties*. Pergamon Press, Oxford, p. 367.

Gladman A.S., Matsumoto E.A., Nuzzo R.G., Mahadevan L. and Lewis J.A. (2016) Biomimetic 4D printing. *Nature Materials* **15**:413–418.

Greisshaber E., Schmahl W.W., Singh Ubji H., Huber J., Nindiyasari F., Maier B. and Ziegler A. (2013) Homoepitaxial meso- and microscale crystal co-orientation and organic matrix network structure in Mytilus edulis nacre and calcite. *Acta Biomaterialia* **9**(12):9492–9502.

Guiducci L., Razghandi K., Bertinetti L., Turcard S., Rüggeberg M., Weaver J.C., Fratzl P., Burgert I. and Dunlop J.W. (2016) Honeycomb actuators inspired by the unfolding of ice plant seed capsules. *PLoS One* **11**(11):e0163506.

Guo Z.G. and Liu W.M. (2007) Biomimic from the superhydrophobic plant leaves in nature: Binary structure and unitary structure. *Plant Science* **172**:1103–1112.

Guo L. and Tang G.H. (2015) Experimental study on directional motion of a single droplet on cactus spines. *International Journal of Heat and Mass Transfer* **84**:198–202.

Guo Z., Liu W. and Su B.L. (2011) Superhydrophobic surfaces: From natural to biomimetic to functional. *Journal of Colloid and Interface Science* **353**:335–355.

Gupta H.S., Seto J., Wagermaier W., Zaslansky P., Boesecke P. and Fratzl P. (2006) Cooperative deformation of mineral and collagen in bone at the nanoscale. *Proceedings of the National Academy of Sciences USA* **103**(47):17741–17746.

Harris P.J. (2006) Primary and secondary plant cell walls: A comparative overview. *New Zealand Journal of Forestry Science* **36**(1):36–53.

Haut R.C. (1986) The influence of specimen length on the tensile failure properties of tendon collagen. *Journal of Biomechanics* **19**(11):951–955.

Henriksson M., Berglund L.A., Isaksson P., Lindström T. and Nishino T. (2008) Cellulose nanopaper structure of high toughness. *Biomacromolecules* **9**:1579–1585.

Hill C.A.S., Norton A.J. and Newman G. (2009) Natural fibre insulation materials – the importance of hygroscopicity in providing indoor climate control. In: *Proceedings of the 11th Conference on Non-Conventional Materials and Technologies (NOCMAT 2009)*, 6–9 September 2009, Bath, UK, p. 6.

Hoffler C.E., Moore K.E., Kozloff K., Zysset P.K., Brown M.B. and Goldstein S.A. (2000) Heterogeneity of bone lamellar-level elastic moduli. *Bone* **26**(6):603–609.

Hofstetter K. and Gamstedt E.K. (2009) Hierarchical modelling of microstructural effects on mechanical properties of wood. A review. *Holzforschung* **63**(2):130–138.

Hofstetter K., Hellmich C. and Eberhardsteiner J. (2005) Development and experimental validation for a continuum micromechanics model for the elasticity of wood. *European Journal of Mechanics A/Solids* **24**:1030–1053.

Holstov A., Bridgens B. and Farmer G. (2015) Hygromorphic materials for sustainable architecture. *Construction and Building Materials* **98**:570–582.

Emerging Nature-Based Materials

Hooke R. (1665) *Micrographia: Or Some Physiological Descriptions of Minute Bodies Made By Magnifying Glasses.* Martin and Allestry, London.

Hora S.L. (1923) The adhesive apparatus on the toes of certain geckos and tree frogs. *Journal and Proceedings of the Asiatic Society of Bengal* **9**:137–145.

Hou Y., Gao L., Feng S., Chen Y., Xue Y., Jiang L. and Zheng Y. (2013) Temperature-triggered directional motion of tiny water droplets on bioinspired fibers in humidity. *Chemical Communications* **49**(46):5253–5255.

Huerta S. (2006) Structural design in the work of Gaudí. *Architectural Science Review* **49**(4):324–339.

Hughes M., Hill C.A.S., Sèbe G., Hague J., Spear M. and Mott L. (2000) An investigation into the effects of micro-compressive defects on interphase behaviour in hemp-epoxy composites using half-fringe photoelasticity. *Composite Interfaces* **7**(1):13.

Hughes M. (2012) Defects in natural fibres: Their origin, characteristics and implications for natural fibre-reinforced composites. *Journal of Materials Science* **47**(2):599–609.

Hull D. and Clyne T.W. (1996) *An Introduction to Composite Materials*, 2nd Edition. Cambridge University Press, NY, p. 326.

Iino M. (1990) Phototropism: Mechanisms and ecological implications. *Plant Cell and Environment* **13**(7):633–650.

Ionov L. (2013) Biomimetic hydrogel-based actuating systems. *Advances in Functional Materials* **23**(36):4555–4570.

Jackson A.P., Vincent J.F.V. and Turner R.M. (1988) The mechanical design of nacre. *Proceedings of the Royal Society of London B* **234**(1277):415–440.

Jeronimidis G. and Atkins A.G. (1995) Mechanics of biological materials and structures: Nature's lessons for the engineer. *Proceedings of the Institute of Mechanical Engineering Part C: Journal of Mechanical Engineering Science* **209**(4):221–235.

John G., Clements-Croome D. and Jeronimidis G. (2005) Sustainable building solutions: A review of lessons from the natural world. *Building and Environment* **40**:319–328.

Judin H. (2016) Living wall. In: *Functional Wood*, Cronhjort Y., Hughes M., Paakanen M., Sahi K., Tukiainen P., Tulamo T. and Vahtikari K. (Eds.). Aalto University, Crossover Report 3, pp. 78–79.

Karam G.N. and Gibson L.J. (1994) Biomimicking of animal quills and plant stems: Natural cylindrical shells with foam cores. *Materials Science and Engineering C* **2**(1–2):113–132.

Kedia G., Passanha P., Dinsdale R.M., Guwy A.J., Lee M. and Esteves S.R. (2013) Addressing the challenge of optimum polyhydroxyalkanoate harvesting: Monitoring real time process kinetics and biopolymer accumulation using dielectric spectroscopy. *Bioresource Technology* **134**:143–150.

Kerr T. and Bailey I.W. (1934) The cambium and its derivative tissues. No X. Structure, optical properties and chemical composition of the so-called middle lamella. *Journal of the Arnold Arboretum* **15**:327–349.

Khire R.A., Van Dessel S., Messac A. and Mullur A.A. (2006) Study of a honeycomb-type rigidified inflatable structure for housing. *Journal of Structural Engineering* **132**(10):1664–1672.

Kobayashi I. and Samata T. (2006) Bivalve shell structure and organic matrix. *Materials Science and Engineering C* **26**(4):692–698.

Koester K.J., Ager J.W. and Ritchie R.O. (2008) The true toughness of human cortical bone measured with realistically short cracks. *Nature Materials* **7**:72–77.

Koller D. (2011) Light driven plant movements. *Plant Cell and Environment* **13**(7):615–632.

Kondo T., Nojiri M., Hishikawa Y., Togawa E., Romanovicz D. and Brown Jr. R.M. (2002) Biodirected epitaxial nanodeposition of polymers on oriented macromolecular templates. *Proceedings of the National Academy of Sciences USA* **99**(22):14008–14013.

Kraniotis D. and Nore K. (2015) On simulating latent heat phenomena in a sauna. In: *Proceedings of the 11th Meeting of the Northern European Network for Wood Sciences and Engineering (WSE)*, 14–15 September 2015, Poznan, Poland, pp. 186–193.

Kraniotis D. and Nore K. (2017) Latent heat phenomena in buildings and potential integration into energy balance. *Procedia Environmental Sciences* **38**:364–371.

Kraniotis D., Nyrud A.Q., Englund F. and Nore K. (2015) Moisture buffering, energy potential and VOC emissions of wood exposed to indoor environments. In: *Proceedings of the 8th International Cold Climate HVAC Conference*, Dalian, China, p. 8.

Kutschera U and Briggs W.R. (2016) Phototropic solar tracking in sunflower plants: An integrative perspective. *Annals of Botany* **117**:1–8.

Larsen K.E. and Marstein N. (2000) *Conservation of Historic Timber Structures: An Ecological Approach*. Butterworth Heinemann, Oxford, p. 140.

Launey M.E., Buehler M.J. and Ritchie R.O. (2010) On the mechanistic origins of toughness in bone. *Annual Review of Materials Research* **40**(1):25–53.

Laursen R. (1992) Reflections on the study of mussel adhesive proteins. In: *Structure, Cellular Synthesis and Assembly of Biopolymers, RESULTS 19, Results and Problems in Cell Differentiation*, Case S.T. (Ed.). Springer, NY, pp. 55–74.

Lee H., Lee B.P. and Messersmith P.B. (2007) A reversible wet/dry adhesive inspired by mussels and geckos. *Nature Letters* **448**:338–342.

Lee S.W., Prosser J.H., Purohit P.K. and Lee D. (2013) Bioinspired hygromorphic actuator exhibiting controlled locomotion. *ACS Macro Letters* **2**(11):960–965.

Levi-Kalisman Y., Falini G., Addadi L. and Weiner S. (2001) Structure of the nacreous organic matrix of a bivalve mollusc shell examined in the hydrated state using cryo-TEM. *Journal of Structural Biology* **135**:8–17.

Li S. and Wang K.W. (2015) Fluidic origami: A plant-inspired adaptive structure with shape morphing and stiffness tuning. *Smart Materials and Structures* **24**:105031, 13 pp.

Li S. and Wang K.W. (2017) Plant-inspired adaptive structures and materials for morphing and actuation. *Bioinsiration and Biomimetics* **12**:011001, 17 pp.

Lienhard J., Schleicher S., Poppinga S., Masselter T., Milwich M., Speck T. and Knippers J. (2011) Flectofin: A hingeless flapping mechanism inspired by nature. *Biomimetics and Bioinspiration* **6**(4):045001, 7 pp.

Lim B.C., Thomas N.L. and Sutherland I. (2008) Surface energy measurements of coated titanium dioxide pigment. *Progress in Organic Coatings* **62**(2):123–128.

Lin A.Y.M. and Meyers M.A. (2009) Interfacial shear strength in abalone nacre. *Journal of Mechanical Behavior of Biomedical Materials* **2**:607–612.

Lin A.Y.M., Chen P.Y. and Meyers M.A. (2008) The growth of nacre in abalone shell. *Acta Biomaterialia* **4**(1):131–138.

Loeker F.C., Duxbury C.J., Kumar R., Gao W., Gross R.A. and Howlde S.M. (2004) Enzyme-catalyzed ring-opening polymerization of ε-caprolactone in supercritical carbon dioxide. *Macromolecules* **37**(7):2450–2453.

Lopez M.I., Chen P.Y., McKittrick J. and Meyers M.A. (2011) Growth of nacre in abalone: Seasonal and feeding effects. *Materials Science and Engineering C* **31**(2):238–245.

Emerging Nature-Based Materials

López M., Rubio R., Martín S. and Croxford B. (2017) How plants inspire façades. From plants to architecture: Biomimetic principles for the development of adaptive architectural envelopes. *Renewable and Sustainable Energy Reviews* **67**:692–703.

Loxton C., Mansour E. and Elias R. (2013) Research into natural bio-based insulation for mainstream construction. In: *Portugal SB13: Contribution of sustainable building to meet EU 20-20-20 targets*, 30 October-1 November 2013, Guimarães, Portugal, pp. 631–638.

Mackenzie C.L., Ormondroyd G.A., Curling S.F., Ball R.J., Whiteley N.M. and Malham S. (2014) Ocean warming, more than acidification, reduces shell strength in a commercial shellfish species during food limitation. *PLoS One* **9**(1):e86764, 9 pp.

Malik F.T., Clement R.M., Gethin D.T., Krawszik W. and Parker A.R. (2014) Nature's moisture harvesters: A comparative review. *Bioinspiration and Biomimetics* **9**:031002, 15 pp.

Mansour E., Curling S., Stéphan A. and Ormondroyd G. (2016) Absorption of volatile organic compounds by different wool types. *Green Materials* **4**(1):1–7.

Mark R.E. (1967) Tests of mechanical properties. In: *The Cell Wall Mechanics of Tracheids*. Yale University Press, New Haven, CT, pp. 28–34.

Mark R.T. and Gillis P.P. (1970) New models in cell wall mechanics. *Wood and Fibre* **2**(2):79–95.

Markin V.S., Volkov A.G. and Jovanov E. (2008) Active movements in plants: Mechanism of trap closure in Dionaea muscipula Ellis. *Plant Signalling and Behavior* **3**(10):778–783.

Martinez R.V., Fish C.R., Chen X. and Whitesides G.M. (2012) Elastomeric origami: Programmable paper-elastomer composites as pneumatic actuators. *Advanced Functional Materials* **22**:1376–1384.

Masselter T., Poppinga S., Lienhard J., Schleicher S., Knippers J. and Speck T. (2012) The flower of Strelizia reginae as a concept generator for the development of a technical deformation system for architectural purposes. In: *Proceedings of 7th Plant Biomechanics International Conference*, August 2012, Clermont-Ferrand, France, pp. 389–392.

Mayer G. (2006) New classes of tough materials – lessons from natural rigid biological systems. *Materials Science and Engineering C* **26**:1261–1268.

McDonough W. (2004) Why can't a building be designed like a tree? New Scientist, 20 March 2004, pp. 46–49.

McKittrick J., Chen P.Y., Bodde S.G., Yang W., Novitskaya E.E. and Meyers M.A. (2012) The structure, functions and mechanical properties of keratin. *Journal of Materials* **64**(4):449–468.

Meier E. (2016) The Wood Database. www.wood-database.com/. Accessed 29/6/2017.

Ménard D., Escamez S., Tuominen H. and Pesquet E. (2015) Life beyond death: the formation of xylem sap conduits. In: *Plant Programmed Cell Death*, Gunarwardena A.N. and McCabe P.F. (Eds.). Springer, New York, pp. 55–76.

Menges A. and Reichart S. (2012) Material capacity: Embedded responsiveness. *Architecture and Design* **82**(2):52–59.

Menig R., Meyers M.H., Meyers M.A. and Vecchio K.S. (2000) Quasi-static and dynamic mechanical response of Haliotis rufescens (abalone) shells. *Acta Materialia* **48**:2383–2398.

Meyers M.A., Chen P.Y., Lin A.Y.M. and Seki Y. (2008) Biological materials: Structure and mechanical properties. *Progress in Materials Science* **53**(1):1–206.

Meyers M.A., Lim C.T., Li A., Hairul Nizam B.R., Tan E.P.S., Seki Y. and McKittrick J. (2010) The role of organic intertile layer in abalone nacre. *Materials Science and Engineering C* **29**(8):2398–2410.

Mikkelson D., Flanagan B.M., Wilson S.M., Bacic A. and Gidley M.J. (2015) Interactions of arabinoxylan and (1,3)(1,4)-β-glucan with cellulose networks. *Biomacromolecules* **16**:1232–1239.

Milwich M., Speck T., Speck O., Stegmaier T. and Planck H. (2006) Biomimetics and technical textiles: Solving engineering problems with nature's wisdom. *American Journal of Botany* **93**(10):1455–1465.

Museeuw (2017) Flaxpreg. Museeuw website. www.museeuw-usa.com/technology/flaxpreg/. Accessed 9/5/17.

Nakahara H. (1991) Nacre formation in bivalve and gastropod molluscs. In: *Mechanisms and Phylogeny of Mineralization in Biological Systems*, Suga S. and Nakahara H. (Eds.). Springer, New York, pp. 343–350.

Nalla R.K., Kruzic J.J., Kinney J.H., Balooch M., Ager III J.W., Ritchie R.O. (2006) Role of microstructure in the aging-related deterioration of the toughness of human cortical bone. *Materials Science and Engineering C* **26**:1251–1260.

Neinhuis C. and Barthlott W. (1997) Characterization and distribution of water-repellent, self-cleaning plant surfaces. *Annals of Botany* **79**:667–677.

Neville A.C. (1993) *Biology of Fibrous Composites: Development Beyond the Cell Membrane.* Cambridge University Press, NY, p. 226.

Northern M.T. and Turner K.L. (2005) A batch fabricated domestic dry adhesive. *Nanotechnology* **16**:1159–1166.

Nudelman F. (2015) Nacre biomineralisation: A review on the mechanisms of crystal nucleation. *Seminars in Cell and Developmental Biology* **46**:2–10.

Oinonen P., Areskogh D. and Henriksson G. (2013) Enzyme-catalyzed cross-linking of spruce galactoglucomannan improves its applicability in barrier films. *Carbohydrate Polymers* **95**:690–696.

Oinonen P., Krawczyk H., Ek M., Henriksson G. and Moriana R. (2016) Bioinspired composites from cross-linked galactoglucomannan and microfibrillated cellulose: Thermal, mechanical and oxygen barrier properties. *Carbohydrate Polymers* **136**:146–153.

Olson I.C., Kozdon R., Valley J.W. and Gilbert P.U.P.A. (2012) Mollusk shell nacre ultrastructure correlates with environmental temperature and pressure. *Journal of the American Chemical Society* **134**:7351–7358.

Onda T., Shibuichi S., Satoh N. and Tsujii K. (1996) Super-water-repellent fractal surfaces. *Langmuir* **12**(9):2125–2127.

Ong K. (2014) This flower turns amazingly transparent when touched by raindrops. www.lostateminor.com/2014/11/18/flower-turns-amazingly-transparent-touched-raindrops/. Accessed 6/11/16.

Ormondroyd G.A., Stefanowski B.K., Mansour E., Spear M.J. and Curling S.F. (2017) An industry prioritised survey of thermal, mechanical, hydro and decay properties of natural fibre insulation materials. In: *Proceedings of the International Panel Products Symposium* 2017, 4–5 October 2017, Llandudno, Wales, UK, pp. 179–184.

Otto F. (1995) IL 35. Pneu and Bone. Krämer, Stuttgart.

Pagitz M. and Bold J. (2013) Shape changing shell-like structures. *Bioinspiration and Biomimetics* **8**:016010, 11 pp.

Pagitz M., Lamacchia E. and Hol J.M.A.M. (2012) Pressure-actuated cellular structures. *Bioinspiration and Biomimetics* **7**(1):016007, 19 pp.

Emerging Nature-Based Materials

Panshin A.J. and DeZeeuw C. (1970) *Textbook of Wood Technology: Structure, Identification, Uses and Properties of the Commercial Woods of the United States and Canada.* McGraw Hill, New York.

Parker A.R. and Lawrence C.R. (2001) Water capture by a desert beetle. *Nature* **414**:33–34.

Parlange J.Y. and Waggoner P.E. (1970) Stomatal dimensions and resistance to diffusion. *Plant Physiology* **46**(2):337–342.

Passanha P., Esteves S.R., Kedia G., Dinsdale R.M. and Guwy A.J. (2013) Increasing polyhydroxyalkanoate (PHA) yields from Cupriavidus necator by using filtered digestate liquors. *Bioresource Technology* **147**:345–352.

Peng X., Fan M., Harley J. and Al-Zubaidy M. (2011) Properties of natural fiber composites made by pultrusion process. *Journal of Composite Materials* **46**(2):237–246.

Philen M.P., Shan Y., Prakash P., Wang K.W., Rahn C.D., Zydney A.L. and Bakis C.E. (2007) Fibrillar network adaptive structure with ion-transport actuation. *Journal of Intelligent Material Systems and Structures* **18**:323–334.

Poon D.C.K., Shieh S.S., Joseph L.M., Chang C.C. (2004) Structural design of Taipei 101, the world's tallest building. In: *CTBUH* 2004, 10–13 October 2004, Seoul, Korea. pp. 271–278.

Prakobna K., Kisonen V., Xu C. and Berglund L.A. (2015) Strong reinforcing effects from galactoglucomannan hemicellulose on mechanical behavior of wet cellulose nanofiber gels. *Journal of Materials Science* **50**:7413–7423.

Purslow P.P. and Vincent J.F.V. (1978) Mechanical properties of primary feathers from the pigeon. *Journal of Experimental Biology* **72**:251–260.

Reichart S., Menges A. and Correa D. (2014) Meteorosensitive architecture: Biomimetic building skins based on materially embedded and hygroscopically enabled responsiveness. *Computer-Aided Design* **60**:50–69.

Reyssat E. and Mahadevan L. (2009) Hygromorphs: From pine cones to biomimetic bilayers. *Journal of the Royal Society Interface* **6**(39):951–957.

Ritchie R.O. (2010) How does human bone resist fracture? *Annals of the New York Academy of Sciences* **1192**:72–80.

Rian I.Md. and Sassone M. (2014). Tree-inspired dendriforms and fractal-like branching structures in architecture: A brief historical overview. *Frontiers of Architectural Research* **3**:298–323.

Roth-Nebelsick A., Fernandez V., Peguero-Pina J.J., Sancho-Knapik D. and Gil-Pelegrin E. (2013) Stomatal encryption by epicuticular waxes as a plastic trait modifying gas exchange in a Mediterranean evergreen species (Quercus coccifera L.). *Plant Cell and Environment* **36**(3):579–589.

Ruibal R. and Ernst V. (1965) The structure of the digital setae of lizards. *Journal of Morphology* **117**:271–294.

Saai Anugraha T.S., Swaminathan T., Swaminathan D., Meyyappan N. and Parthiban R. (2016) Enzymes in platform chemical biorefinery. In: *Platform Chemical Biorefinery: Future Green Industry*, Kaur Brar S., Jyoti Sarmar S. and Pakshirajan K. (Eds.). Elsevier, pp. 451–469.

Salmen L. and Burgert I. (2009) Cell wall features with regard to mechanical performance. A review. *Holzforschung* **63**(2):121–129.

Sarikaya M. and Aksay I.A. (1992) Nacre of abalone shell: a multifunctional nanolaminated ceramic-polymer composite material. In: *Structure, Cellular Synthesis and Assembly of Biopolymers*. Springer, Berlin, pp. 1–26.

Sarikaya M., Gunnison K.E., Yasrebi M., Milius D.L. and Aksay I.A. (1990) Seashells as a natural model to study laminated composites. Session 2A Biotechnology and composite materials. In: Proceedings of the American Society of Composites Fifth Technical Conference, Lancaster, PA, pp. 176–183.

Sasaki N. and Odajima S. (1996) Stress-strain curve and Young's modulus of a collagen molecule as determined by the x-ray diffraction technique. *Journal of Biomechanics* **29**(5):655–658.

Schaedler T.A. and Carter W.B. (2016) Architected cellular materials. *Annual Reviews of Materials Research* **46**:187–210.

Schmidt-Nielsen K. (1997) *Animal Physiology: Adaptation and Environment*, 5th Edition. Cambridge University Press, Cambridge, p. 607.

Schneider M.H. and Phillips J.G. (1991) Elasticity of wood and wood polymer composites in tension compression and bending. *Wood Science and Technology* **25**:361–364.

Scholander P.F. and Schevill W.E. (1955) Counter-current vascular heat exchange in the fins of whales. *Journal of Applied Physiology* **8**(3):279–282.

Scholander P.F., Hock R., Walters V. and Irving L. (1950) Adaptation to cold in arctic and tropical mammals and birds in relation to body temperature, insulation and basal metabolic rate. *Biological Bulletin* **99**(2):259–271.

Seki Y., Kad B., Benson D. and Meyers M.A. (2006) The toucan beak: Structure and mechanical response. *Materials Science and Engineering C* **26**:1412–1420.

Shan Y., Philen M.P., Bakis C.E., Wang K.W. and Rahn C.D. (2006) Nonlinear-elastic finite axisymmetric deformation of flexible matrix composite membranes under internal pressure and axial force. *Composites Science and Technology* **66**:3053–3063.

Shell G.S.G. and Lang A.R.G. (1976) Movements of sunflower leaves over a 24-h period. *Agricultural Meteorology* **16**:161–170.

Sherman V.R., Yang W. and Meyers M.A. (2015) The materials science of collagen. *Journal of the Mechanical Behavior of Biomedical Materials* **52**:22–50.

Shirtcliffe N.J., McHale G. and Newton M.I. (2009) Learning from superhydrophobic plants: The use of hydrophilic areas on superhydrophobic surfaces for droplet control. *Langmuir* **25**(24):14121–14128.

Shtein I., Elbaum R. and Bar-On B. (2016) The hygroscopic opening of sesame fruits is induced by a functionally graded pericarp architecture. *Frontiers in Plant Science* **7**:1501, 8 pp.

Sidorenko A., Krupenkin T. and Aizenberg J. (2008) Controlled switching of the wetting behaviour of biomimetic surfaces with hydrogel-supported nanostructures. *Journal of Materials Chemistry* **18**:3841–3846.

Simmons T.J., Mortimer J.C., Bernardinelli O.D., Pöppler A.C., Brown S.P., deAzevedo E.R., Dupree R. and Dupree P. (2016) Folding of xylan onto cellulose fibrils in plant cell walls revealed by sold-state NMR. *Nature Communications* **7**:13902. doi:10.1038/ncomms13902.

Simonis P., Rattal M., Oualim E.M., Mouhse A. and Vigneron J.P. (2014) Radiative contribution to thermal conductance in animal furs and other wooly insulators. *Optics Express* **22**(2):1940–1951.

Sitti M. and Fearing R. (2003) Synthetic gecko foot-hair micro/nano-structures as dry adhesives. *Journal of Adhesion Science and Technology* **17**:1055–1073.

Soar R. (2015a) Part 1: A process view of nature: Multifunctional integration and the role of the construction agent. *Intelligent Buildings International* **8**:78–89.

Soar R. (2015b) Part 2: Pushing the envelope. A process perspective for architecture, engineering and construction. *Intelligent Buildings International* **8**:90–105.

Song C. and Zheng Y. (2014) Wetting-controlled strategies: From theories to bioinspiration. *Journal of Colloid and Interface Science* **427**:2–14.

Stefanowski B.K., Curling S.F. and Ormondroyd G.A. (2016) Evaluating mould colonisation and growth on MDF panels modified to sequester volatile organic carbons. *International Wood Products Journal* **7**(4):118–194.

Stoychev G., Puretskiy N. and Ionov L. (2011) Self-folding all polymer thermoresponsive microcapsules. *Soft Matter* **7**:3277–3279.

Suchsland O. (2004) *The Swelling and Shrinking of Wood*. Forest Products Society, Madison, WI, p. 189.

Sun Y. and Cheng J. (2002) Hydrolysis of lignocellulosic materials for ethanol production; a review. *Bioresource Technology* **83**(1):1–11.

Suzuki M. and Nagasawa H. (2013) Mollusk shell structures and their formation mechanism. *Canadian Journal of Zoology* **91**(6):349–366.

Taccola S., Greco F., Sinibaldi E., Mondini A., Mazzolai B., Mattoli V. (2015) Toward a new generation of electrically controllable hygromorphic soft actuators. *Advanced Materials* **27**(10):1668–1675.

Taiz L. and Zeiger E. (2002) *Plant Physiology*, 3rd Edition. Sinauer Associates, Sunderland, MA, p. 690.

Tan X., Shi T., Tang Z., Sun B., Du L., Peng Z. and Liao G. (2016) Investigation of fog collection on cactus-inspired structures. *Journal of Bionic Engineering* **13**(3):364–372.

Taylor S.W., Luther G.W. and Waite J.H. (1994a) Polarographic and spectrophotometric investigation of iron (III) complexation to 3,4-dihydroxyphenylalanine-containing peptides and proteins from Mytilus edulis. *Inorganic Chemistry* **33**:5819–5824.

Taylor S.W., Waite J.H., Ross M.M., Shabanowitz J. and Hunt D.F. (1994b) Trans 2,3-cis-3,4-dihydroxyproline, a new naturally occurring amino acid, is the sixth residue in the tandemly repeated consensus decapeptides of an adhesive protein from Mytilus edulis. *Journal of the American Chemistry Society* **116**:10803–10804.

Teeri T.T., Brumer H., Daniel G. and Gatenholm P. (2007) Biomimetic engineering of cellulose-based materials. *Trends in Biotechnology* **25**(7):299–306.

Timoshenko S. (1925) Analysis of bi-metal thermostats. *Journal of the Optical Society of America* **11**:233–255.

Tomos A.D. and Leigh R.A. (1999) The pressure probe: A versatile tool in plant cell physiology. *Annual Review of Plant Physiology and Plant Molecular Biology* **50**:447–472.

Torres F.G., Troncoso O.P., Diaz J. and Arce D. (2014) Failure analysis of porcupine quills under axial compression reveals their mechanical response during buckling. *Journal of the Mechanical Behaviour of Biomedical Materials* **39**:111–118.

Turner J.S. and Soar R.M. (2008) Beyond Biomimicry: what termites can tell us about realising the living building. In: *Proceedings of the 1st International Conference on Industrialized, Intelligent Construction (I3CON)*, 14–16 May 2008, Loughborough University, Loughborough, UK, p. 18.

Turner W., Robson P., Bosch M. and Spear M. (2017) A study of miscanthus straw for composites: microscopy and fibre analysis. In: *Proceedings of the International Panel Products Symposium* 2017, 4–5 October 2017, Llandudno, Wales, UK, p. 3.

Uyama H. and Kobayashi S. (2002) Enzyme-catalyzed polymerization to functional polymers. *Journal of Molecular Catalysis* **19**–20:117–127.

Vasista S. and Tong L. (2012) Design and testing of pressurized cellular planar morphing structures. *AIAA Journal* **50**(6):1328–1338.

Vasista S. and Tong L. (2013) Topology-optimized design and testing of a pressure-driven morphing-aerofoil trailing-edge structure. *AIAA Journal* **51**(8):1898–1907.

Velcro S.A. (1955) Improvements in or relating to a method and a device for producing a velvet type fabric. Patent no. 721338, Switzerland.

Vincent J.F.V. (1999) From cellulose to cell. *Journal of Experimental Biology* **202**:3263–3268.

Vincent J.F.V. (2001) Stealing ideas from nature. In: *Deployable Structures. International Centre for Mechanical Sciences (Courses and Lectures)*, vol. 412, Pellegrino S. (Ed.). Springer, Vienna.

Vincent J.F.V., Bogatyreva O.A., Bogatyrev N.R., Bowyer A. and Pahl A.K. (2006) Biomimetics: Its practice and theory. *Journal of the Royal Society Interface* **3**(9). doi:10.1098/rsif.2006.0127.

Vogel S. (2005) Living in a physical world IV. Moving heat around. *Journal of Biosciences* **30**:449–460.

Volkov A.G., Foster J.C., Baker K.D. and Markin V.S. (2010) Mechanical and electrical anisotropy in Mimosa pudica pulvini. *Plant Signalling and Behavior* **5**(10):1211–1221.

Vos R. and Barrett R. (2010) Pressure adaptive honeycomb: mechanics, modelling and experimental investigation. In: *Proceedings of the 51st AIAA/ASME/ASCE/AHS/ASC Structures, Structural Dynamics and Materials Conference*, 12–15 April 2010, Orlando, FL, USA, p. 14.

Vos R. and Barrett R. (2011) Mechanics of pressure-adaptive honeycomb and its application to wing morphing. *Smart Materials and Structures* **20**:094010, 11 pp.

Vreeland V., Waite J.H. and Epstein L. (1998) Polyphenols and oxidases in substratum adhesion by marine algae and mussels. *Journal of Phycology* **34**(1):1–8.

Wählisch F.C., Peter N.J., Torrents Abad O., Oliveira M.V.G., Schneider A.S., Schmahl W., Greisshaber E. and Bennewitz R. (2014) Surviving the surf: The tribomechanical properties of the periostracum of Mytilus sp. *Acta Biomaterialia* **10**(9):3978–3985.

Wainwright S.A. (1970) Design in hydraulic organisms. *Naturwissenschaften* **57**:321–326.

Wainwright S.A., Biggs W.D., Currey J.D. and Gosline J.M. (1976) *Mechanical Design in Organisms*. Edward Arnold (Publishers) Ltd, London, p. 423.

Waite J.H. and Tanzer M.L. (1981) Polyphenolic substance of Mytilus edulis: Novel adhesive containing L-DOPA and hydroxyproline. *Science* **212**(4498):1038–1040.

Waite J.H. and Qin X. (2001) Polyphosphoprotein from the adhesive pads of Mytilus edulis. *Biochemistry* **40**(8):2887–2893.

Wang L. (2016) Plain concrete cylinders and beams externally strengthened with natural flax fabric reinforced epoxy composites. *Materials and Structures* **49**(6):2083–2095.

Wang B. and Meyers M.A. (2017) Light like a feather: A fibrous natural composite with a shape changing from round to square. *Advanced Science* **4**(3):1600360, 10 pp. doi:10.1002/advs.201600360.

Wang X. and Wang L. (2007) Study on the construction mode of bionic architectural morphology. *Urbanism and Architecture* **2007**(8):11–12. In Chinese. Cited by Yuan et al. 2017.

Wang R.Z., Suo Z., Evans A.G., Yao N. and Aksay I.A. (2001) Deformation mechanisms in nacre. *Journal of Materials Research* **16**(9):2485–93.

Emerging Nature-Based Materials

Wang L., Nancollas G.H., Henneman Z.J., Klein E. and Weiner S. (2006) Nanosized particles in bone and dissolution insensitivity of bone material. *Biointerphases* **1**(3):106–111.

Weaver J.C., Aizenberg J., Fantner G.E., Kisailus D., Woesz A., Allen P., Fields K., Porter M.J., Zok R.W., Hansma F., Fratzl P. and Morse D.E. (2007) Hierarchical assembly of the siliceous skeletal lattice of the hectinellid sponge Euplectella aspergillum. *Journal of Structural Biology* **158**:93–106.

Węcławski B.T., Fan M. and Hui D. (2014) Compressive behaviour of natural fibre composite. *Composites Part B* **67**:183–191.

Wegst U.G.K. and Ashby M.F. (2004) The mechanical efficiency of natural materials. *Philosophical Magazine* **84**(21):2167–2181.

Weiner S. and Wagner H.D. (1998) The material bone: Structure-mechanical function relations. *Annual Review Materials Research* **28**:271–298.

Wenzel R.N. (1936) Resistance of solid surfaces to wetting by water. *Industrial and Engineering Chemistry* **28**(8):988–994.

Whyburn G., Li Y. and Huang Y. (2008) Protein and protein assembly based material structures. *Journal of Materials Chemistry* **18**:3755–3762.

Woesz A., Weaver J.C., Kazanci M., Dauphin Y., Aizenberg J., Morse D.E. and Fratzl P. (2006) Micromechanical properties of biological silica in skeletons of deep-sea sponges. *Journal of Materials Research* **21**(8):2068–2078.

World Architecture News. (2012) Open to the elements. http://www.worldarchitecturenews.com/project/2012/21413/verk-studio/phyllotactic-towers.html. Accessed 25/4/2018.

Wu Z.L., Moshe M., Greener J., Therien-Aubin H., Nie Z., Sharon, E. and Kumacheva E. (2013).Three-dimensional shape transformations of hydrogel sheets induced by small-scale modulation of internal stresses. *Nature Communications* **4**:1586, 7 pp.

Xing Y., Bosch M., Ormondroyd G., Donnison I., Spear M. and Jones P. (2017) Exploring design principles of biological and living building envelopes: What can we learn from plant cell walls? *Intelligent Building International* **10**(2):78–102.

Xu H.H.K., Weir M.D., Burguera E.F. and Fraser A.M. (2006) Injectable and macroporous calcium phosphate cement scaffold. *Biomaterials* **27**(24):4279–4287.

Yan L. and Chouw N. (2013) Crashworthiness characteristics of flax fibre reinforced epoxy tubes for energy absorption application. *Materials and Design* **51**:629–640.

Yan L., Chouw N. and Jayaramana K. (2014) Lateral crushing of empty and polyurethane-foam filled natural flax fabric reinforced epoxy composite tubes. *Composites Part B: Engineering* **63**:15–26.

Yang W. and McKittrick J. (2013) Separating the influence of the cortex and foam on the mechanical properties of porcupine quills. *Acta Biomaterialia* **9**(11):9065–9074.

Yang L., van de Werf K.O., Dijkstra P.J., Feijen J. and Bennink M.L. (2012) Micromechanical analysis of native and cross-linked collagen type I fibrils supports the existence of microfibrils. *Journal of the Mechanical Behaviour of Biomedical Materials* **6**:148–158.

Yang W., Chao C. and McKittrick J. (2013) Axial compression of a hollow cylinder filled with foam: A study of porcupine quills. *Acta Biomaterialia* **9**(2):5297–5304.

Yang W., Sherman V.R., Gludovatz B., Mackey M., Zimmermann E.A., Change E.A., Schaible E., Qin Z., Buehler M.J., Ritchie R.O. and Meyers M.A. (2014) Protective role of Arapaima gigas fish scales: Structure and mechanical behavior. *Acta Biomaterialia* **10**(8):3599–3614.

Yang W., Sherman V.R., Gludovatz B., Schaible E., Stewart P., Ritchie R.O. and Meyers M.A. (2015) On the tear resistance of skin. *Nature Communications* **6**:1–10.

Yeang, K. (1992) Designing the Tropical Skyscraper. In: *Mimar 42: Architecture in Development*, Khan H.U. (Ed.). Concept Media Ltd, London.

Yeang K. (2000) *The Green Skyscraper, the Basics for Designing Sustainable Intensive Buildings*. Prestel Verlag, Munich, London, New York, p. 184.

Yong J., Chen F., Yang Q., Du G., Shan C., Bian H., Farooq U. and Hou X. (2015) Bioinspired transparent underwater superoleophobic and anti-oil surfaces. *Journal of Materials Chemistry A* **3**(18):9373–9384.

Yu M. and Deming T.J. (1998) Synthetic polypeptide mimics of marine adhesives. *Macromolecules* **31**(8):4739–4745.

Yu M., Hwang J. and Deming T.J. (1999) Role of L-3,4-dihydroxyphenylalanine in mussel adhesive proteins. *Journal of the American Chemical Society* **121**(24):5825–5826.

Yuan Y., Yu X., Yang X., Xiao Y., Xiang B. and Wang Y. (2017) Bionic building energy efficiency and bionic green architecture: A review. *Renewable and Sustainable Energy Reviews* **74**:771–787.

Yurdumakan B., Raravikar N.R., Ajadan P.M. and Dhinojwala A. (2005) Synthetic gecko foot-hairs from multiwalled carbon nanotubes. *Chemical Communications* **30**:3799–3801.

Zaremba C.M., Belcher A.M., Fritz M., Li Y., Mann S., Hansma P.K., Morse D.E., Speck J.S. and Stucky G.D. (1996) Critical transitions in the biofabrication of abalone shells and flat pearls. *Chemistry of Materials* **8**(3):679–690.

Zhang M. and Zheng Y. (2016) Bioinspired structure materials to control water-collecting properties. *Materials Today: Proceedings* **3**:696–702.

Zhang W., Liao S.S. and Cui F.Z. (2003a) Hierarchical self-assembly of nano-fibrils in mineralized collagen. *Chemistry of Materials* **15**(16):3221–3226.

Zhang W., Huang Z.L., Liao S.S. and Cui F.Z. (2003b) Nucleation sites of calcium phosphate crystals during collagen mineralization. *Journal of the American Ceramic Society* **86**(6):1052–1054.

Zhu B., Rahn C.D. and Bakis C.E. (2012) Actuation of fluidic flexible matrix composites in structural media. *Journal of Intelligent Material Systems and Structures* **23**:269–278.

Zioupos P. (2001) Ageing human bone: Factors affecting its biomechanical properties and the role of collagen. *Journal of Biomaterials Applications* **15**(3):187–229.

Zioupos P. and Currey J.D. (1994) The extent of microcracking and the morphology of microcracks in damaged bone. *Journal of Materials Science* **29**:978–986.

Zioupos P., Wang X.T. and Currey J.D. (1996) Experimental and theoretical quantification of the development of damage in fatigue tests of bone and antler. *Journal of Biomechanics* **29**(8):989–1002.

Zioupos P., Currey J.D. and Hamer A.J. (1999) The role of collagen in the declining mechanical properties of aging human cortical bone. *Journal of Biomedical Materials Research* **45**(2):108–116.

Zlotnikov I. and Schoeppler V. (2017) Thermodynamic aspects of molluscan shell ultrastructural morphogenesis. *Advanced Functional Materials* **27**:1700506, 14 pp.

Zollfrank C., Cromme P., Rauch M., Scheel H., Kostova M.H., Gutbrod K., Gruber S. and Van Opdebosch D. (2012) Biotemplating of inorganic functional materials from polysaccharides. *Bioinspired, Biomimetic and Nanobiomaterials* **1**(1):13–25.

Zuschin M. and Stanton R.J. (2001) Experimental measurement of shell strength and its taphonomic interpretation. *Palaios* **16**:161–170.

10

Bio-Inspired Design – Enhancing Natural Materials

Chris Holland
The University of Sheffield

CONTENTS

References 331

This book is intended for a wide-ranging audience, a mixture of scientists, engineers, designers and artists. As such, we have envisaged the readers represent a diverse set of backgrounds but hopefully a shared desire to work together across disciplines when using natural materials. In this chapter we echo this theme and discuss nature's approach towards developing its materials and how it frequently achieves performance greater than the sum of its parts. We shall cover the concepts of bio-inspiration, natural variation, structural hierarchies and most importantly evolutionary context, arming the reader with a biologist's perspective about material choice. Finally, we will finish on a series of case studies that illustrate how fundamental research has enabled us to tweak the processing of natural materials in order to enhance their properties.

But before we venture into the details of the strategies employed by nature, we should first define bio-inspiration and why it is useful to be bio-inpsired.

Bio-inspiration is simply taking inspiration from nature in order to solve human problems. It is largely synonymous with other terms such as 'bionics', 'biomimesis' and 'biologically inspired design (BID)' with the original concept of 'biomimetics', being coined by physicist Otto Schmit in the mid part of the twentieth century as a result of his work developing an electronic nerve [1] and best introduced to the wider public through Janine Benyus' book on Biomimicry (a book I largely attribute with providing me at the age of 15, the inspiration for my entire research career) [2].

Throughout her works, Benyus outlines nine guiding principles of biomimicry which serve as motivation for bio-inspired design.

321

Nature runs on sunlight
Nature uses only the energy it needs
Nature fits form to function
Nature recycles everything
Nature rewards cooperation
Nature banks on diversity
Nature demands local expertise
Nature curbs excesses from within
Nature taps the power of limits
Janine Benyus' nine guiding principles of biomimicry [2]

The original term of biomimetics and its etymology implies the imitation of a biological model, system or elements. Yet, ever since DaVinci studied bird wings to design a flying machine over 500 years ago [3], successful attempts that have faithfully and fully replicated a system in entirety are few and far between. This can be attributed to several factors, first, a complete replication may not be technically possible because 1) the natural system is not entirely understood or 2) the technology is not available to copy it, 3) the challenge nature solves is not exact the same as the one we need to solve. Thus, with all due credit to the phrase biomimetics, my personal preference is the term bio-inspired, purely because it better manages the expectations of the designer and user.

Whilst defining the human challenge may be straightforward, when looking to nature for the answer, a common pitfall is adaptationalism. Adaptationalism is mistakenly imparting function to every individual trait of an organism. Gould and Lewontin [4] argue that not all traits have a specific purpose; they may be evolutionary relics from previous organisms that employed them (such as the appendix in the human or the vestigial leg bones on some snakes). If the traits do indeed have a purpose, then one must be careful not to infer their specific biological role incorrectly (such as the function of my nose is to support my glasses), and in fact, most traits are multifunctional.

Furthermore traits are always the result of trade-offs [5], compromises in design that arise due to said multifunctionality or limiting resources (i.e. time or energy) and hence to say a feature is 'perfect' or 'optimised' is incorrect. It is perhaps best to think of these traits as just the most competitive solution for that organism at a specific time, with a bit of variation thrown in (*Nature banks on diversity*).

Yet despite these pitfalls, bio-inspiration has helped solve numerous challenges over the years, with a superb list of over 2000 examples to be found on asknature.org, a website developed as part of the Biomimicry Institute (founded by Janine Benyus and Bryony Schwan). Famous examples range from the development of Velcro by Swiss Engineer George De Mestral after taking his dog for a walk and noticing hooked cocklebur seed pods attaching to his trousers [6], the streamlining of the Japanese Shinkansen 500 series bullet train by Eiji Nakatsu after observing how kingfisher birds enter water with practically no ripples [7], to the development of a dry adhesive tape inspired by the tiny hairs present on a gecko's foot [8] (Figure 10.1).

Here it is worth noting that the most prolific examples of bio-inspiration tend to be the result of a 'bottom up' approach: an engineer being inspired by a natural structure. However, these are often argued to be the easiest aspect to copy [9–10] and it is only recently that a 'top down' approach of designers specifying a problem first, then collaborating with biologists to search for natural solutions before employing engineers to help deliver the product. Hence, whilst many engineering applications can be improved by deriving inspiration from biological structures that are visible to the naked eye, there are examples where bio-inspiration may contribute towards human challenges through the mimicry of entire systems, such as the transport efficiency of leafcutter ants for lorry drivers [11] and the increased productivity of plants grown in polyculture for agriculture [12] (*Nature rewards cooperation*).

Yet there is one area that is still somewhat underutilised for inspiration, that of nature's materials. But why do so? A logical progression from the approach of using bio-inspiration to derive or copy the natural process of materials production is to simply get the organisms to make specific materials for us.

FIGURE 10.1
Bioinspiration can be found everywhere, from sticky cocklebur seed pods inspiring Velcro (a) to adhesives based on a gecko's foot (b) to the shape of a kingfishers beak forming the design of the front of the Japanese bullet train (c,d). Photo credits: Flickr CC BY 2.0 changed by cropping and image rotation: (a) B.J. Deming, (b) Bizmac, (c) coniferconifer and (d) Takeshi Kuboki (https://creativecommons.org/licenses/by/2.0/).

Whilst this may sound cutting edge, with the mind wandering towards the areas of genetic manipulation and biotechnology to coerce organisms to generate materials of interest to us, this has in fact been practiced for thousands of years through selective breeding [13] but as well as other examples including adding fluoride to water to reduce cavities in our teeth [14]. Another example may be found in silkworms which produce our commercial textile silk, where it has recently been shown that by changing the temperature and humidity of the environment in which they spin their cocoons, it can improve the quality of the raw material [15–16], which could make it more amenable for a range of different applications beyond textiles, from biomedical implants to holograms (for an excellent overview see [17]).

However, all these systems above rely on the ability and extent to which the natural system can be manipulated (including genetic engineering). As we shall see next, by developing bio-inspired systems, it is possible to extend and even enhance the natural traits desired by humans to solve their problems. But first, any successful bio-inspired application requires understanding, and understanding a biological material requires the confluence and efforts from physicists, chemists, biologists and engineering to determine how structure imparts function, from the atom to the organism.

When assessing the materials that nature creates, one of the first considerations should be the raw materials at hand and the chemistry that can be applied to synthesise new products. Given that nature undertakes nearly all of its materials synthesis in a narrow temperature range of 100°C (~–10°C to 90°C), using water as the solvent and mostly out of carbon, hydrogen, oxygen and nitrogen, when compared to industrial materials synthesis, one may think at first that there is little to learn. But, upon considering the massive functional diversity nature has been able to obtain with such evolutionary constraints, innovation is key (*Nature taps the power of limits*).

This innovation has been driven by a concept of energy management in nature (*Nature uses only the energy it needs*). Simply put, energy spent on anything other than increasing one's fitness/competitiveness to pass on one's genes is energy wasted. This applies to the whole organism but is the sum of the chemical processes within it, and as such, we can assume that most processes that are energetically wasteful are selected against (*Nature curbs excesses from within*). This is in stark contrast to industrial chemical synthesis, which is practically unconstrained with respect to energy input. However, in this current era of environmental awareness, increasing fuel prices and dwindling resources, now is the perfect time to be looking to develop new bio-inspired routes for our industrial materials production whilst also perhaps being able to satisfy consumer demand for sustainable, lightweight, high performance materials.

When compared to industrially produced materials, nature excels at being able to combine material properties that were once thought to be mutually exclusive, specifically strength (the force/load a material can withstand before failure) and toughness (how much energy a material can dissipate

Bio-Inspired Design   325

before failure). How does nature achieve such a feat when industry tends to produce either strong brittle materials or weak tough ones? It appears that rather than developing entirely new synthetic routes or industrial processes, nature adapts the materials it already has to the current challenge. This is achieved by altering the types and proportions of relatively simple components (such as proteins, sugars and minerals), which taken alone have limited mechanical properties. This results in combinations of hard (ordered) and soft (disordered) phases of specific sizes (length scales), where the resulting *extrinsic* mechanical properties of the material is greater than the sum of the individual components *intrinsic* properties.

Hence, almost all natural materials are composites, i.e. combinations of multiple materials. As a result, examples of structural hierarchies which co-ordinate this multi-component integration are prolific in nature. Through the use of the following three examples of wood, bone and silk, the underlying common design motifs of controlling composition, size, shape, orientation and architecture will hopefully be illustrated and how these impart desirable mechanical properties.

Whilst wood will be discussed in more detail elsewhere in this book, the basic structure of this material at the nanoscale is a combination of stiff, ordered cellulose polymer chains embedded in a disordered amorphous lignin-hemicellulose matrix, akin to fibreglass, but on a size ~10,000 times thinner than a human hair (Figure 10.2) [18]. Within this matrix, the cellulose chains are further aligned alongside one another and combine to form microfibrils, providing orientational rigidity and optimal packing. Looking to larger length scales, on the micro-scale, this composite forms the outer coating of hollow cells, which are stacked upon one another in varying architectural patterns in response to the forces applied to the plant during its growth. These patterns can become remarkably complex at the millimetre scale, and for certain plants like bamboo or palm, the hollow fibres are more concentrated on the outside of the trunk, providing reinforcement where it is needed most, leading to a high resistance to flexure and a very lightweight structure [19].

Bone adopts a similar structure at the nanoscale, [20–22] but instead of a polysacharride (sugar), long chains of a protein, collagen, come together in threes and twist around one another forming fibrils which like wood combine into fibrillar assemblies (Figure 10.2). However, in this case the fibrils are interpenetrated with calcium mineral nanocrystals called hydroxyapatite, providing hard phases to the collagen's soft. Like wood, in response to loading during growth, these building blocks are developed into a range of complex architectures in the millimetre scale forming two different types of bone. In dense compact bone, these larger scale microfibrils are tightly packed together and arranged in layered ring structures, surrounding blood vessels called osteons. In spongy bone, found in the end of long bones (like the head of the femur), the structure is much less dense, but nature attempts to compensate for the relative weakness of this material by aligning the osteons like small rods (known as trabecula at this scale) along the lines of loading

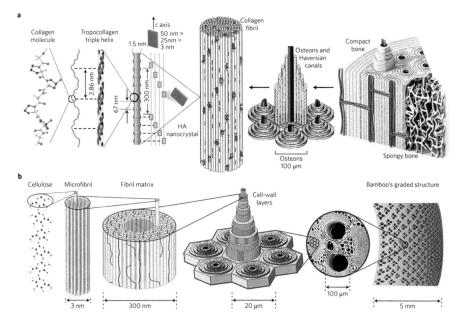

FIGURE 10.2
The structural hierarchies present in (a) animal bone and (b) plant wood/bamboo. Note how each length scale interacts with the one above and below it, building into an interlinked network of structures, which contribute together, from the atom to the organism to create properties greater than the sum of their parts. HA, hydroxyapatite. Reprinted by permission from Macmillan Publishers Ltd: Nature Materials, Wegst, U.G.K.; Bai, H.; Saiz, E.; Tomsia, A.P.; Ritchie, R.O., Bioinspired structural materials. *Nature Materials* **2015**, *14*(1), 23–36, copyright 2014.

in the bone. As an aside, it was this lightweight hierarchical structure that inspired Gustave Eiffel's design of the Eiffel tower in Paris in the late 1880's [23] (Figure 10.3).

Finally, like wood and bone, silk fibres also exhibit a structural hierarchy (Figure 10.4). At the nanoscale, silks are typically comprised of 1–2 long chain proteins called fibroins which self-assemble into nano and microfibrils aligned along the long axis of the fibre [24]. However, the subtle difference is that the silk proteins themselves have the ability to form both hard ordered and softer disordered phases depending on how they are processed [25–26]. This is advantageous not only as it reduced the number of components require to make the material, but hard and soft segments can occur within the molecule itself. This reduces the size of the ordered/disordered phases, increasing their degree of integration/interaction with one another and more effectively enabling the transfer of energy (i.e. stress), which increases the materials toughness (i.e. ability to dissipate energy). At the micrometre scale, silk is already in its final fibrous form. The fact that silk is so thin may also explain its strength, because at this scale it is unable to develop a

Bio-Inspired Design

FIGURE 10.3
Photograph of the Eiffel Tower, Paris, demonstrating the hierarchical nature of the metalwork's that facilitated its lightweight design and structural integrity.

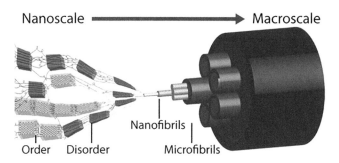

FIGURE 10.4
The hierarchical structures of silk extend from the nanoscale to macro-scale.

sufficient concentration of stress to allow a crack to propagate through it [27] (consider the difference between a glass fibre and a glass rod). In other natural materials such as those above which have larger scale architectures, they are known to employ a series of voids, which serve to blunt the tip of the crack and protect the material, such as the gaps for cells or vessels in wood and bone [28].

From these three examples, general themes emerge. It appears that most natural composite materials combine soft and hard phases at the nanoscale permitting effective dissipation of energy, then assembling larger nano- and micro-sized structures and orienting/ordering these in response to loading, thus facilitating strength, whilst at the same time inserting mechanisms to avoid cracks. This results in materials that exhibit both toughness and strength. Yet given such structural hierarchy is seemingly so elegant and simplistic, why haven't we been able to replicate it? The answer lies within two interrelated properties, the nanoscale size of the features and how they are manufactured.

Current industrial manufacturing excels at large scale structures yet finds it difficult to control features on the small scale. Take industrial plastics for example, whilst it is possible to align the long chain polymers along the axis of the fibre like silk, the hard crystallites formed during processing are much larger, leading to less efficient energy dissipation mechanisms [29]. The same is also known for metals, where efforts are focussed on producing smaller crystallite/grain size to impart improved mechanical properties [30–31].

Therefore, in order to develop new bio-inspired materials and enhance those we already have, we must look to not just the structures themselves but also *how they are formed*. I believe this alone is the greatest source of bio-inspiration for natural materials. By first understanding and then replicating the mechanisms employed during natural materials manufacture, it should be possible to manipulate and hopefully improve upon them, releasing said materials from the constraints imparted as a result of evolutionary trade-offs. In the following section, we will look at two examples, nacre (mother-of-pearl) and again silk, discussing how they represent different fabrication processes, how they can be manipulated and enhanced and how they have inspired artificial materials.

Of the two manufacturing routes employed by nature, growth is the most common, with structures being precisely laid down over time. Nacre is an organic–inorganic composite with components arranged in a brickwork fashion; elastic proteins serving as the mortar and calcium phosphate platelets the bricks (see [28] for an excellent review). Whilst nacre's growth process is not entirely known (highlighting the challenge for bio-inspiration perfectly), it appears that the mineral phase is first templated by the organic, with nanometre sized grains of minerals coming together to gradually form larger platelets (Figure 10.5).

This concept of templating is important in defining the nanoscale features of natural materials, and has already been copied by several researchers looking at replicating nacre since 2000 [32–34]. One recent example was achieved where researchers coated glass slides in a thin film of two polymers which they then made porous by dissolving one away [35]. This left tiny holes in the surface which was then used as a template to grow calcium phosphate minerals inside. The resulting thin coating had both the iridescent appearance and the mechanical properties of nacre.

Unfortunately, the use of directed growth via templating has one major limitation, time. These structures, regardless of how remarkable they are, still take a considerable time to produce, which is not surprising when you consider how long these materials can take to form in nature. Hence an exact replication does not lend itself readily to the human challenges of industrial scale-up and manufacture [36]. This has resulted in other approaches being used to manufacture nacre, even moving away from using

Bio-Inspired Design

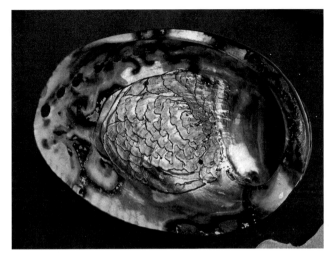

FIGURE 10.5
The Abalone shell's iridescent nacre is the result of light interacting with the millions of tiny platelet nanostructures. Photo credit: Rojer Flickr no changes made CC BY 2.0 (https://creativecommons.org/licenses/by/2.0/).

natural components altogether, instead replicating the structures found in nacre to create new bio-inspired composites that are material independent. There are numerous examples where non-naturally derived materials such as polymer–ceramics [37], metal–ceramics [38] and even fully ceramic-based hybrid materials [39] have resulted in materials appear to outperform the original nacre. But whilst these materials may indeed be bio-inspired and result in enhanced performance, the inspiration is not derived from the processing, but the structures deployed instead.

For inspiration on this level, we can look towards the other form of manufacturing in nature. Silks are quite unlike the other biological materials which are grown slowly over months and years (like hair, bone and nacre). Silk is spun. Spinning is the process of solidification whereby a liquid protein gel stored in specialised organs inside the animal is transformed within seconds into a solid fibre. This occurs as a result of the individual silk proteins being stretched as they flow down an ever-tapering tube (the spinning duct) prior to exiting the animal [24]. This molecular stretching results in several interconnected events at the nano and micro-scale that we can derive inspiration from with respect to our manufacturing. First, the stretching causes a build-up of energy in the proteins, which is dissipated by transferring it to the water molecules that surround them, causing them to 'boil off' as it were [40]. Once the silk proteins are dehydrated, they are able to change shape and begin associating with themselves and one another, aggregating together and forming hard phases known as ordered β-sheets and softer disordered amorphous phases [41]. These different phases exist on the nanoscale; but on the micro-scale, these proteins come together and form the nano/microfibrils, which align along the long axis of the fibre [42], imparting improved mechanical properties.

This summary, whilst seemingly simplistic, is the product of nearly 50 years of research from teams across the globe, and in so doing, we have uncovered not only how silk if formed but perhaps more importantly how it can be enhanced.

The first experiments that investigated the relationship between processing of silk and subsequent performance of the fibre, surrounded gently restraining the animals and pulling silk from them onto a spool at a known speed. This was known as 'forced reeling' and pioneered by Fritz Vollrath from Oxford (who's lab I studied for over a decade in). Using this approach, it is possible to control many aspects of the external environment in which silk is spun (i.e. temperature, humidity and reeling speed), and also given these animals are cold-blooded, the temperature of the internal processing as well.

These experiments revealed that by altering spinning conditions, the mechanical properties of the spun fibres can be changed in a very consistent and controlled manner. By pulling the silk out of the animal faster (>3 cm per second), a stiffer fibre was produced, but pulling it out slowly (<1 cm per second) resulted in a much weaker, more stretchy fibre [43] (Figure 10.6). Here, it is important to note that in both cases, the unspun liquid silk inside the animals were exactly the same, the only difference being how fast it was pulled through the spinning duct and thus to what degree the proteins were stretched. Later experiments investigating the molecular structure of these fibres revealed that faster spun silks not only showed greater alignment of

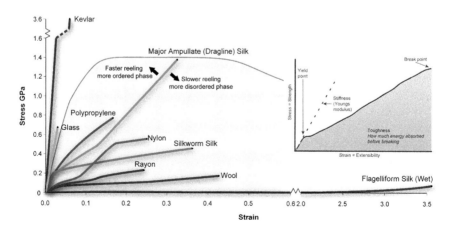

FIGURE 10.6
Engineering stress strain graph showing the properties of different natural and synthetic fibres under tension (glossary provided as an inset to the left). Note that by changing the processing conditions it is possible to change the mechanical properties of natural silks by altering the molecular order/disorder ratio. Spider Major Ampullate (dragline) silk is shown and note that by altering the reeling speed, it is possible to obtain fibre properties anywhere within orange-shaded region.

Bio-Inspired Design 331

fibrils along the long axis of the fibre; but on the nanoscale, the silks contained a higher proportion of the hard phase containing ordered β-sheets, which in turn made the fibre stiffer [44–45]. Thus, it was possible to generate bespoke fibres through forced reeling, and in some cases, enhancing the natural fibres by spinning at speeds not normally undertaken by the animal (which is seen in silkworms) [25,46]. It has also been possible to force reel one species in such a manner that it produces a fibre close to that of another! Taken together, this suggests that the processing is as important as the silk proteins themselves in determining final fibre properties (see [24] for a review).

This is strong evidence that the hierarchical structure of silk is formed in response to the forces encountered during processing, akin to bone and wood's response to their loading during growth but on a much shorter, now industrially more appropriate timescale. This is crucial to understand for bio-inspiration, as it outlines how nature is able to generate widely different materials using the same components, simply by processing them differently. Furthermore, it was found by looking at the flow properties of liquid silks, they shared a lot in common with our industrial polymer melts and even the way in which our own fibre spinning technologies have developed [47].

Yet, there is one key difference between natural and synthetic fibre spinning, energy management. When compared to silks, industrial polymers require over a thousand times more energy for solidification [48]. Hence, taking inspiration beyond just replicating the natural material and understanding *how* silks are processed, could unlock the door to improving our own materials in a sustainable, resource efficient manner.

In conclusion, there is a clear need for us to increase our use of natural materials in the future. This is in part because they represent a sustainable, often superior alternative to synthetic materials, provided they are deployed in a manner that appreciates their evolutionary context. However, natural materials are also the result of hundreds of millions of years of research and development under a life or death scenario (i.e. evolution through natural selection). Hence, they represent more than just their face value; bio-inspiration can be drawn from not just the structures in the materials themselves but also the processes by which they are made. Yet in order for us to harness the power of nature, we must understand it, which entails asking the right questions, working together at the interfaces and then adapting, extending and enhancing these solutions to our specific needs.

References

1. Valentinuzzi, M.E., Otto Herbert Arnold Schmitt (1913–1998), a pioneer (an overview of Schmitt's scientific production). *IEEE Engineering in Medicine and Biology Magazine* **2004**, *23*(6), 42–46. doi:10.1109/MEMB.2004.1378632.

2. Benyus, J., *Biomimicry: Innovation Inspired by Nature*. William Morrow and Company, Inc: New York, NY, 1997.
3. Romei, F.; Ricciardi, S.; Ricciardi, A., *Leonardo Da Vinci*. Minneapolis, MN: Oliver Press, 2008.
4. Gould, S.J.; Lewontin, R.C., The spandrels of San Marco and the panglossian paradigm: A critique of the adaptationist programme. *Proceedings of the Royal Society of London. Series B, Biological Sciences* **1979**, *205*(1161), 581–598.
5. Vincent, J.F.V., The trade-off: a central concept for biomimetics. *Bioinspired, Biomimetic and Nanobiomaterials* **2016**, *0*(0). doi:10.1680/jbibn.16.00005.
6. Mestral, G.D, Velvet type fabric and method of producing same, 1955.
7. Kobayashi, K., JFS biomimicry interview series: No.6 "Shinkansen Technology learned from an Owl?"—The story of Eiji Nakatsu. http://www.japanfs.org/en/news/archives/news_id027795.html (accessed 5/12/16).
8. Geim, A.K.; Dubonos, S.V.; Grigorieva, I.V.; Novoselov, K.S.; Zhukov, A.A.; Shapoval, S.Y., Microfabricated adhesive mimicking gecko foot-hair. *Nature Materials* **2003**, *2*(7), 461–463.
9. Vincent, J., Biomimetic patterns in architectural design. *Architectural Design* **2009**, *79*(6), 74–81. doi:10.1002/ad.982.
10. Vincent, J.F.V.; Bogatyreva, O.A.; Bogatyrev, N.R.; Bowyer, A.; Pahl, A.-K., Biomimetics: its practice and theory. *Journal of The Royal Society Interface* **2006**, *3*(9), 471–482. doi:10.1098/rsif.2006.0127.
11. Farji-Brener, A.G.; Chinchilla, F.A.; Rifkin, S.; SÁNchez Cuervo, A.M.; Triana, E.; Quiroga, V.; Giraldo, P., The 'truck-driver' effect in leaf-cutting ants: how individual load influences the walking speed of nest-mates. *Physiological Entomology* **2011**, *36*(2), 128–134. doi:10.1111/j.1365–3032.2010.00771.x.
12. Allan, E.; Weisser, W.; Weigelt, A.; Roscher, C.; Fischer, M.; Hillebrand, H., More diverse plant communities have higher functioning over time due to turnover in complementary dominant species. *Proceedings of the National Academy of Sciences* **2011**, *108*(41), 17034–17039. doi:10.1073/pnas.1104015108.
13. Conner, J.K., Artificial selection: A powerful tool for ecologists. *Ecology* **2003**, *84*(7), 1650–1660. doi:10.1890/0012–9658(2003)084[1650:ASAPTF]2.0.CO;2.
14. Pizzo, G.; Piscopo, M.R.; Pizzo, I.; Giuliana, G., Community water fluoridation and caries prevention: a critical review. *Clinical Oral Investigations* **2007**, *11*(3), 189–193. doi:10.1007/s00784-007-0111–6.
15. Offord, C.; Vollrath, F.; Holland, C., Environmental effects on the construction and physical properties of Bombyx mori cocoons. *Journal of Materials Science* **2016**, *51*(24), 10863–10872. doi:10.1007/s10853-016-0298-5.
16. Boulet-Audet, M.; Holland, C.; Gheysens, T.; Vollrath, F., Dry-spun silk produces native-like fibroin solutions. *Biomacromolecules* **2016**, *17*(10), 3198–3204. doi:10.1021/acs.biomac.6b00887.
17. Omenetto, F.; Kaplan, D., From silk cocoon to medical miracle. *Scientific American* **2010**, *303*, 76–77. doi:10.1038/scientificamerican1110–76.
18. Gibson, L.J., The hierarchical structure and mechanics of plant materials. *Journal of The Royal Society Interface* **2012**, *9*(76), 2749–2766. doi:10.1098/rsif.2012.0341.
19. Wegst, U.G.K., Bamboo and wood in musical instruments. *Annual Review of Materials Research* **2008**, *38*(1), 323–349. doi:10.1146/annurev.matsci.38.060407.132459.

Bio-Inspired Design 333

20. Reznikov, N.; Shahar, R.; Weiner, S., Bone hierarchical structure in three dimensions. *Acta Biomaterialia* **2014**, *10*(9), 3815–3826. doi:10.1016/j.actbio.2014.05.024.

21. Weiner, S.; Wagner, H.D., The material bone: Structure-mechanical function relations. *Annual Review of Materials Science* **1998**, *28*(1), 271–298. doi:10.1146/annurev.matsci.28.1.271.

22. Lakes, R., Materials with structural hierarchy. *Nature* **1993**, *361*(6412), 511–515.

23. Forbes, P., *The Gecko's Foot: Bio-Inspiration*. W. W. Norton & Company: New York, 2005.

24. Vollrath, F.; Porter, D.; Holland, C., The science of silks. *MRS Bulletin* **2013**, *38*(01), 73–80. doi:10.1557/mrs.2012.314.

25. Mortimer, B.; Guan, J.; Holland, C.; Porter, D.; Vollrath, F., Linking naturally and unnaturally spun silks through the forced reeling of Bombyx mori. *Acta Biomaterialia* **2015**, *11*, 247–255. doi:10.1016/j.actbio.2014.09.021.

26. Guan, J.; Wang, Y.; Mortimer, B.; Holland, C.; Shao, Z.; Porter, D.; Vollrath, F., Glass transitions in native silk fibres studied by dynamic mechanical thermal analysis. *Soft Matter* **2016**, *12*(27), 5926–5936. doi:10.1039/C6SM00019C.

27. Porter, D.; Guan, J.; Vollrath, F., Spider silk: Super material or thin fibre? *Advanced Materials* **2013**, *25*(9), 1275–1279. doi:10.1002/adma.201204158.

28. Wegst, U.G.K.; Bai, H.; Saiz, E.; Tomsia, A.P.; Ritchie, R.O., Bioinspired structural materials. *Nature Materials* **2015**, *14*(1), 23–36. doi:10.1038/nmat4089.

29. Mortimer, B.; Gordon, S.D.; Holland, C.; Siviour, C.R.; Vollrath, F.; Windmill, J.F.C., The speed of sound in silk: Linking material performance to biological function. *Advanced Materials* **2014**, *26*(30), 5179–5183. doi:10.1002/adma.201401027.

30. Hall, E.O., The deformation and ageing of mild steel: III discussion of results. *Proceedings of the Physical Society. Section B* **1951**, *64*(9), 747.

31. Petch, N.J., The cleavage strength of polycrystals. *Journal of the Iron and Steel Institute.* **1953**, *173*, 25–28.

32. Sellinger, A.; Weiss, P.M.; Nguyen, A.; Lu, Y.; Assink, R.A.; Gong, W.; Brinker, C.J., Continuous self-assembly of organic-inorganic nanocomposite coatings that mimic nacre. *Nature* **1998**, *394*(6690), 256–260.

33. Tang, Z.; Kotov, N.A.; Magonov, S.; Ozturk, B., Nanostructured artificial nacre. *Nature Materials* **2003**, *2*(6), 413–418. doi:10.1038/nmat906. http://www.nature.com/nmat/journal/v2/n6/suppinfo/nmat906_S1.html.

34. Kato, T., Polymer/calcium carbonate layered thin-film composites. *Advanced Materials* **2000**, *12*(20), 1543–1546. doi:10.1002/1521–4095(200010)12:20<1543::AID-ADMA1543>3.0.CO;2-P.

35. Finnemore, A.; Cunha, P.; Shean, T.; Vignolini, S.; Guldin, S.; Oyen, M.; Steiner, U., Biomimetic layer-by-layer assembly of artificial nacre. *Nature Communications* **2012**, *3*, 966. doi:10.1038/ncomms1970. http://www.nature.com/articles/ncomms1970#supplementary-information.

36. Corni, I.; Harvey, T.J.; Wharton, J.A.; Stokes, K.R.; Walsh, F.C.; Wood, R.J.K., A review of experimental techniques to produce a nacre-like structure. *Bioinspiration & Biomimetics* **2012**, *7*(3), 031001.

37. Munch, E.; Launey, M.E.; Alsem, D.H.; Saiz, E.; Tomsia, A.P.; Ritchie, R.O., Tough, bio-inspired hybrid materials. *Science* **2008**, *322*(5907), 1516–1520. doi:10.1126/science.1164865.

38. Liu, Q.; Ye, F.; Gao, Y.; Liu, S.; Yang, H.; Zhou, Z., Fabrication of a new SiC/2024Al co-continuous composite with lamellar microstructure and high mechanical properties. *Journal of Alloys and Compounds* **2014**, *585*, 146–153. doi:10.1016/j.jallcom.2013.09.140.

39. Bouville, F.; Maire, E.; Meille, S.; Van de Moortèle, B.; Stevenson, A.J.; Deville, S., Strong, tough and stiff bioinspired ceramics from brittle constituents. *Nature Materials* **2014**, *13*(5), 508–514. doi:10.1038/nmat3915. http://www.nature.com/nmat/journal/v13/n5/abs/nmat3915.html#supplementary-information.

40. Porter, D.; Vollrath, F., The role of kinetics of water and amide bonding in protein stability. *Soft Matter* **2008**, *4*, 328–336. doi: 10.1039/b713972a.

41. Porter, D.; Vollrath, F., Water mobility, denaturation and the glass transition in proteins. *Biochimica et Biophysica Acta (BBA)—Proteins & Proteomics* **2012**, *1824*(6):785–791.

42. Poza, P.; Perez-Rigueiro, J.; Elices, M.; Llorca, J., Fractographic analysis of silkworm and spider silk. *Engineering Fracture Mechanics* **2002**, *69*(9), 1035–1048.

43. Vollrath, F.; Madsen, B.; Shao, Z., The effect of spinning conditions on the mechanics of a spider's dragline silk. *Proceedings of the Royal Society—Biological Sciences (Series B)* **2001**, *268*(1483), 2339–2346.

44. Dicko, C.; Knight, D.; Kenney, J.M.; Vollrath, F., Structural conformation of spidroin in solution: A synchrotron radiation circular dichroism study. *Biomacromolecules* **2004**, *5*(3), 758–767.

45. Chen, X.; Shao, Z.; Vollrath, F., The spinning processes for spider silk. *Soft Matter* **2006**, *2*(6), 448–451.

46. Shao, Z.Z.; Vollrath, F., Surprising strength of silkworm silk. *Nature* **2002**, *418*(6899), 741–741.

47. Holland, C.; Terry, A.E.; Porter, D.; Vollrath, F., Natural and unnatural Silks. *Polymer* **2007**, *48*, 3388–3392.

48. Holland, C.; Vollrath, F.; Ryan, A.J.; Mykhaylyk, O.O., Silk and synthetic polymers: Reconciling 100 degrees of separation. *Advanced Materials* **2012**, *24*(1), 105–109. doi:10.1002/adma.201103664.

Index

A

Abductin, 52–54
Accoya® wood, 160
Acrylate resins, 80
Actuators, 292–298
Aesthetics, 169
  of design, 1–13
Algal type cellulose I_a, 35–36
Ambient vibration measurement,
  209–210
Animal-derived materials, 121–125
Antheraea pernyi moth silks, 50
Antler, 239–240
Araneus diadematus spider silks, 50
Automotive design, 165–177

B

Bacterial nanocellulose, 40
Bast fibre production process, 86
Beta-keratin, 49–50
Bio-based palette
  combining ingredients – recognising
    ultrastructure, 65–70
  inorganic components – biominerals
    calcium carbonate, 59–60
    calcium oxalate, 64–65
    dolomite, 62
    hydrated silica, 62–64
    hydroxyapatite, 60–61
    magnetite, 61–62
  organic components
    abductin, 52–54
    cellulose, 34–40
    chitin, 44
    collagen, 45–49
    elastin, 51
    fibroin, 50–51
    hemicelluloses and pectins, 41–43
    keratin, 49–50
    naturally occurring polymers,
      54–59
    polysaccharide, 34

proteins, 44–45
resilin, 51–52
Bio-based polyesters, 75
Biocomposites
  bioresins as matrices, 91–94
  man-made bio-derived organics
    acrylate resins, 80
    bio-based polyesters, 75
    bio-epoxy, 78–79
    cellulose, 72–73
    chitin, 74–75
    polyamides, 77
    polybutylene succinate, 76
    polyhydroxy alkanoates and
      polyhydroxy butyrate-valerate,
      76
    polylactic acid, 75–76
    proteins, 74
    thermoplastic polyurethanes,
      77–78
    thermoplastic starch, 76–77
    unsaturated polyester resins, 79
  materials, 70–72
    designing with, 94–96
  natural fibre composites
    fibre types, 83–91
Biodegradation opportunities,
  169–170
Bio-derived platform chemicals, 299
Bio-epoxy, 78–79
Biofibres; *see also* Biocomposites
Bio-inspiration, 30–32
  design, 323–333
Biomimetics, 220, 261
Biomimicry, 218–220
Biotemplating, 300–303
BIP, *see* British Industrial Plastics (BIP)
Bombyx mori silk, 50
British Industrial Plastics (BIP), 153

C

CaC_2O_4, *see* Calcium oxalate (CaC_2O_4)
$CaCO_3$, *see* Calcium carbonate ($CaCO_3$)

335

336 *Index*

Calcite, 60

Calcium carbonate ($CaCO_3$), 59–60

Calcium oxalate (CaC_2O_4), 64–65

Cambridge Engineering Selector (CES) EduPack
  database, natural organic materials in, 19–20
  density and cost, comparison of, 25–26
  durability, comparison of, 22–23
  environmental impact parameters, 23–25
  mechanical properties, 20–21
  thermal properties, 21–22

Carbon footprinting, 113

Cardboard, rediscovering natural materials in packaging, 185–186

Cardolite, 79

Cartilage, 241

Cashew nut shell liquid (CNSL)-based epoxy, 79

Cellular materials, 269–271

Cellulose
  bio-based palette organic components, 34–40
  I_a and I_b, 35
  man-made bio-derived organics, 72–73

CEMA, *see* Council for Encouragement of Music and the Arts (CEMA)

CES EduPack, *see* Cambridge Engineering Selector (CES) EduPack

Chemical bio-inspiration, 260–261

Chitin
  bio-based palette organic components, 44
  man-made bio-derived organics, 74–75

Chitosan, 74–75

CLT, *see* Cross-laminated timber (CLT)

CNSL, *see* Cashew nut shell liquid (CNSL)

COID, *see* Council for Industrial Design (COID)

Collagen, 45–49, 241–244

Combining bioresins; *see also* Biocomposites

Cork, 270

Cotton-ramie type cellulose I_b, 35–36

Council for Encouragement of Music and the Arts (CEMA), 149–150

Council for Industrial Design (COID), 148

Council on Tall Buildings and Urban Habitat (CTBUH), 200

Cross-laminated timber (CLT), 204–206

Crustacean exoskeletons, 44

Crystalline ferrihydrite, 61

CTBUH, *see* Council on Tall Buildings and Urban Habitat (CTBUH)

Cutin, 58–59

D

Denser woods, 22

DGEBA, *see* Diglycidyl ether of bisphenol-A (DGEBA)

Diglycidyl ether of bisphenol-A (DGEBA), 78

DMA, *see* Dynamic mechanical analysis (DMA)

Dolomite ($CaMg(CO_3)_2$), 62

$CaMg(CO_3)_2$, *see* Dolomite ($CaMg(CO_3)_2$)

'Double Diamond' process, 2

Dynamic mechanical analysis (DMA) of spider silks, 51

E

Eco-design standards, 112

Environmental Product Declaration (EPD) programme, 116

EPD programme, *see* Environmental Product Declaration (EPD) programme

Epoxy resins, 78–79

Eurocode, 208

Extrinsic shielding mechanisms, 238–239

F

Fiber Office chair, 12

Fibril associated collagens with triple helixes (FACIT collagens), 46

Fibrillar collagen, 241

Index 337

Fibroin, 50–51
Fibrous insulation products, 282
Fireproof construction, 206
Flax fibre reinforced cylinders, 252–253
Flexible cylinders, with helical fibre reinforcement, 278–279
'Flexible packaging,' 185
Fluid dynamics, 280–285
Foams, 271–280
Foldable origami cells, 279–280
Fracture toughness, 239

G

GHG emissions, *see* Greenhouse gas (GHG) emissions
Glass, rediscovering natural materials in packaging, 183–184
Glued laminated timber frame structures, 206–207
Glycerol, 74
Green chemistry; *see also* Biocomposites
Greenhouse gas (GHG) emissions, 117

H

Hand lay-up process, 86–87
Harnessing natural processes, 298–303
Hedgehog spines, 274
Helical fibre reinforcement, flexible cylinders with, 278–279
Hemicelluloses, 41–43
Honeycombs, 271–280
Hydrated silica ($SiO_2 \cdot nH_2O$), 62–64
Hydration, 48
Hydrostatic support, 285–292
Hydroxyapatite ($Ca_{10}(PO_4)_6(OH)_2$), 60–61
$Ca_{10}(PO_4)_6(OH)_2$, *see* Hydroxyapatite ($Ca_{10}(PO_4)_6(OH)_2$)
Hydroxyproline, 45–46

I

Inflatable soft cylinders, 278
Inorganic components - biominerals
 calcium carbonate, 59–60
 calcium oxalate, 64–65
 dolomite, 62
 hydrated silica, 62–64
 hydroxyapatite, 60–61
 magnetite, 61–62
Insect carapaces, 44
Insulation, 168–169
Internal triangular hinge ligament (ITHL), 52
Intrinsic toughening mechanisms, 236–238
ITHL, *see* Internal triangular hinge ligament (ITHL)

K

Keratin, 49–50

L

LCA, *see* Life cycle assessment (LCA)
Leather, 124–125
Life cycle assessment (LCA)
 and animal-derived materials, 121–125
 methodology, 113–116
 and plant fibre materials, 119–121
 wood and wood-based products, 116–119
Lightweighting, 168
Lotusan paint, 257
Lotus leaf effect, 255–258
Lowest density balsa, 22

M

Man-made bio-derived organics
 acrylate resins, 80
 bio-based polyesters, 75
 bio-epoxy, 78–79
 cellulose, 72–73
 chitin, 74–75
 polybutylene succinate, 76
 polyhydroxy alkanoates and polyhydroxy butyrate-valerate, 76
 polylactic acid, 75–76
 proteins, 74
 thermoplastic polyurethanes, 77–78
 thermoplastic starch, 76–77
 unsaturated polyester resins, 79
'Material Universe,' 15–16

338 *Index*

Metals, rediscovering natural materials
  in packaging, 184–185
Microfibrillated cellulose, 40
Microfibrils, 37
Mimic plant fibres, 252–253
Mineral composites
  antler, 240–241
  bone, 233–240
  mollusc shells, 226–229
  nacre, 229–231
  tensile and compressive strength,
    221–226
Mineralised tissues, 20
Moisture harvesting, 285
Molluscs, 44
Mollusc shells, 226–229

N

Nacre, 222–226, 229–231
Nanocellulose, 40
Nanocrystalline cellulose, 40
Natural fibre composites (NFCs), 21
  fibre types, 83–91
  to mimic plant fibres, 251–252
Natural fibre insulation (NFI), 120
Natural fibres, 19–20, 90
  rediscovering natural materials in
    packaging, 188–189
Naturally occurring polymers, 54–59
Natural organic materials, 16
NFCs, *see* Natural fibre composites
  (NFCs)
NFI, *see* Natural fibre insulation (NFI)
Non-lignified plant cells, 38–39

O

Organic components
  abductin, 52–54
  cellulose, 34–40
  chitin, 44
  collagen, 45–49
  fibroin, 50–51
  hemicelluloses and pectins, 41–43
  keratin, 49–50
  naturally occurring polymers, 54–59
  polysaccharide, 34
  proteins, 44–45

resilin, 51–52
Organic composites, microstructure of
  collagen, 241–244
  natural fibre composites to mimic
    plant fibres, 252–253
  plant cell walls as composites,
    245–247
  wood and plant stems, 247–250
Organic matrix
  layer, 231
  proteins, 229
Osteons, 327

P

Packaging, rediscovering natural
    materials in
  glass, 183–184
  metals, 184–185
  natural fibres, 188–189
  paper and cardboard, 185–186
  wood, 186–188
  wool, 189–190
    packaging for superior insulation,
      191–194
Paper-bag maker, 186
Paper, rediscovering natural materials
    in packaging, 185–186
PBS, *see* Polybutylene succinate (PBS)
Pectins, 41–43
PHAs, *see* Polyhydroxy alkanoates
    (PHAs)
PHB, *see* Polyhydroxy butyrate (PHB)
PHBV, *see* Polyhydroxy butyrate-
    valerate (PHBV)
PLA, *see* Polylactic acid (PLA)
Plant fibre, 244
  materials, 119–121
Plant stems, organic composites,
    247–251
Plastic age, 142–146
Polyamides, 77
Polybutylene succinate (PBS), 76
Polyhydroxy alkanoates (PHAs), 76
Polyhydroxy butyrate (PHB), 76
Polyhydroxy butyrate-valerate (PHBV),
    76
Polylactic acid (PLA), 75–76
Polysaccharide, 34

Index

Proline, 45–46
Proteins
    bio-based palette organic
        components, 44–45
    man-made bio-derived organics, 74
Pultrusion, 89

R

Radulae of chiton, 61
Regenerated cellulose fibres, 73
Resilin, 51–52
Restoring credibility
    design for future, 148–150
    emerging synthetics and natural
        precursors, 142
    global influences, 135
    industrial revolution, 139
    mass-market demand, 148
    plastic age, 142–146
    resistance to 'American way,' 150
    silk, 159
    wood, 160
    wool, 160–163
Rhombohedral polymorph of calcium
    carbonate, 60

S

Sclerotin, 44
Sea sponge skeletons, 63
Sea urchin spicules, 62
'Semi-flexible' packaging, 186
Silica, 62–64
$SiO_2 \cdot nH_2O$, *see* Hydrated silica
    ($SiO_2 \cdot nH_2O$)
SIPs, *see* Structural insulation panels
    (SIPs)
Sisal, 21
Slippage of microfibrils, 241
Small-scale fluid dynamics/diffusion,
    283–285
'Smart fibre,' 160–163
Softwood hemicelluloses, 43
Spider silk fibroin, 50–51
Spongy bone, 327–328
Stability system, 211–213
Structural insulation panels (SIPs), 120
Suberin, 58–59

Sugar-based bioresins, 94
Superhydrophobicity, 253–259
Surface texture, 253–259
Surf Clear EVO, 79
Synthesized polygon bilayers, 277

T

Tall timber buildings
    cross-laminated timber, 204–206
    dynamics and, 208–211
    fire and, 207–208
    glued laminated timber frame
        structures, 206–207
    stability system and, 211–213
    timber for, 201–202
    types of construction for, 202–204
Tendons, 47–48, 241
Texture, 253–260
Thermoplastic biopolymers, 92
Thermoplastic polyurethanes (TPS),
    77–78
Thermoplastic starch, 76–77, 92
Thermosetting matrices, 93–94
TPS, *see* Thermoplastic polyurethanes
    (TPS)
Tropocollagen, 46
Tunicate microfibrils, 40

U

UNIDO, *see* United Nations Industrial
    Development Organization
    (UNIDO)
United Nations Industrial Development
    Organization (UNIDO), 124
Unsaturated polyester resins (UPE), 79
UPE, *see* Unsaturated polyester resins
    (UPE)

V

Vaterite, 60
Venus flytrap, 288–289
Vickers hardness, 23
VOCs, *see* Volatile organic compounds
    (VOCs)
Volatile organic compounds (VOCs),
    124

340 *Index*

W

Wood
  composites, 19–20
  organic composites, 247–251
  rediscovering natural materials in
    packaging, 186–188

Wool, 122–124
  packaging for superior insulation,
    191–194
  rediscovering natural materials in
    packaging, 189–190
  restoring credibility, 160–163